Photosynthesis: Overviews on Recent Progress and Future Perspectives

Photosynthesis: Overviews on Recent Progress and Future Perspectives

Shigeru Itoh
Division of Material Science, Nagoya University, Nagoya, Japan

Prasanna Mohanty
Honorary Professor, Regional Plant Resource Center, Bhubaneswar, India

K.N. Guruprasad
School of Life Sciences, Devi Ahilya University, Indore, India

I.K. International Publishing House Pvt. Ltd.

NEW DELHI • BANGALORE

Published by

I.K. International Publishing House Pvt. Ltd.
S-25, Green Park Extension
Uphaar Cinema Market
New Delhi – 110 016 (India)
E-mail: info@ikinternational.com
Website: www.ikbooks.com

ISBN 978-93-81141-00-7

10 9 8 7 6 5 4 3 2 1

Published by Krishan Makhijani for I.K. International Publishing House Pvt Ltd., S-25, Green Park Extension, Uphaar Cinema Market, New Delhi – 110 016 and Printed by Rekha Printers Pvt. Ltd., Okhla Industrial Area, Phase II, New Delhi – 110 020

Dedication

A 2010 portrait of Govindjee in his home at Urbana, Illinois, USA.
Photo by Rajni Govindjee.

We dedicate this book to **Govindjee**, born on October 24, 1932, in Allahabad, Uttar Pradesh, India, and to his many distinguished students and associates for their pioneering discoveries in oxygenic photosynthesis, particularly on Photosystem II (PS II), the waterplastoquinone oxido-reductase, and on the understanding and the relationship of the light emission by chlorophyll *a* (Chl *a*) (prompt fluorescence, delayed fluorescence and thermoluminescence) with various specific reactions of photosynthesis by plants, algae and cyanobacteria. Govindjee is best known for discovering that a 670 nm–absorbing form of Chl *a* belongs to PSII, for the discovery of the *two-light effect* in Chl *a* fluorescence, and in NADP reduction, and for the discovery of the specific role of the bicarbonate anion in photosynthetic electron flow to (and protonation of) a plastoquinone molecule on the electron acceptor side of PS II. In particular, we, honour him for his love and passion in communicating the concepts of *photosynthesis*, and in identifying the scientists behind their discoveries, to both young and old people around the world. He achieved these goals through truly inspiring lectures and through various publications that include books, reviews, Scientific American articles, and many edited volumes. For further information, see the *Foreword* in this book by George C. Papageorgiou, his first PhD student.

Govindjee's honours include: Fellow of the American Association of Advancement of Science (AAAS); Distinguished Lecturer of the School of Life Sciences, University of Illinois at Urbana-Champaign (UIUC); Fellow and Life Member of the National Academy of Sciences (Allahabad, India); President of the American Society for Photobiology (1980–1981); Honorary President of the 2004 International Photosynthesis Congress (Montréal, Canada); first recipient of the Lifetime Achievement Award of the Rebeiz Foundation for Basic Biology, 2006; recipient of the Communication Award of the International Society of Photosynthesis Research, 2007; and the Alumni Lifetime Achievement Award of the College of the Liberal Arts and Sciences, UIUC, 2008.

Foreword

Photosynthesis: Overviews on Recent Progress and Future Perspectives, is dedicated to Professor Govindjee of the University of Illinois at Urbana-Champaign, Illinois, USA, by his students and colleagues in recognition of his great contributions to the advancement and the interpretation of the science of photosynthesis. Its inspiration was an International Conference on "Photosynthesis in the Global Perspective" held in honour of Govindjee at the University of Indore, India, in November 27-29, 2008 (see Jajoo et al. 2009). The book, however, is not a proceedings volume. It is a collection of research and review articles invited by its editors, Professors S. Itoh, Division of Materials Science, University of Nagoya, Japan, P. Mohanty, formerly of the Jawaharlal Nehru University of New Delhi and presently at the Department of Agricultural Biotechnology, Bhubaneswar, Orissa, India and K. N. Guruprasad at the School of Life Sciences, Devi Ahilya University, Indore, India from selected participants of the conference.

The chapters of the book pertain to various subjects, such as to the effects of environmental stresses on photosynthesis, to regulatory mechanisms triggered by ground state redox signals or by carotenoid absorption, to the application of optical probes to photosynthesis research, notably of chlorophyll fluorescence, and to the historical evolution of the concepts of modern photosynthesis, particularly with respect to photosynthetic oxygen evolution. Although the theme of the global perspective of photosynthesis – namely exploiting science for improved food production and energy storage – was behind this book, none of its chapters addresses such themes. Instead, they provide a glimpse of how far the science of photosynthesis has advanced to ensure the successful confrontation of these crucial and controversial global issues.

The honoured person here, Professor Govindjee, was born in Allahabad, Northern India, in 1932, the youngest child of (Mrs) Savitri Devi Asthana and (Mr) Vishveswar Prasad Asthana. He finished elementary school, high school, college, and university up to the level of MSc (subjects: chemistry, botany, zoology) in Allahabad, and in 1956 he went to the University of Illinois at Urbana-Champaign as a Fulbright Scholar, where he was accepted as graduate student in the laboratory of the famous Robert Emerson of the Botany Department.

From the early 1930s through late 1950s Robert Emerson was a dominant personality in photosynthesis research. A gifted experimentalist he made radical discoveries that seemingly violated fundamental and firmly established chemical laws and concepts (see reviews by Myers, 1974; Govindjee, 2000; Govindjee and Krogmann 2004; and Govindjee and Björn, this volume). Thus, his discovery with William Arnold of a stoichiometric ratio of ~2480 chlorophylls per 1 O_2 evolved (Emerson and Arnold, 1932a, b) not only was way out exopragmatic, but also against what one would expect from an organic-magnesium compound like chlorophyll, for which a Chl.CO_2 complex, analogous to those of the organic-magnesium Grignard reagents form ($RMgX + CO_2 \rightarrow RCOOMgX$ where X is a halogen atom) seemed more likely. Likewise, the discovery that the unicellular green alga *Chlorella* requires a minimum of 8 ~12 quanta in order to release 1 O_2 (Emerson and Lewis, 1941, 1943) appeared to be in direct conflict with the Stark-Einstein law of photochemical equivalence, which prescribes 1 chemical event per absorbed photon. And lastly, Emerson's idea of two separate, but closely cooperating, photochemical systems (Emerson and Rabinowitch, 1960) had no precedent in photobiology (although and quite remarkably, it was predicted by Eugene Rabinowitch; see Govindjee and Björn, this volume).

These three discoveries of Emerson and coworkers, together with van Neil's (1941) discovery that both the anoxygenic (bacteria) and oxygenic (algae, plants) photosyntheses do not photo-deoxygenate CO_2 but they do photo-reduce it by photo-dissociating hydrogen donors are today the mainstays of the modern theory of photosynthesis. At the time of Govindjee's arrival in Urbana, however, they were still debated. Govindjee started working on his doctoral thesis under Emerson, seeking further evidence to consolidate the concept of the two light reactions in photosynthesis. Emerson, however, was killed in 1959, in an airplane accident, so Govindjee continued and completed his thesis requirements in 1960 under the supervision of Emerson's associate, Professor Eugene Rabinowitch, a physical chemist of many talents and countless interests including photosynthesis (see Brody 1995, 2002). Its appropriate title was *"Action Spectra of the Emerson Enhancement Effect in Algae"*.

Like an impatient race car at the starting line, Govindjee set off in his scientific race at top speed. In the years that followed he devoted his energy, mainly, to four areas of intellectual activity: education, scientific research, book and journal editing and in the chronicling of emerging concepts, trends and of the individuals behind them that occurred in the Gnostic area of photosynthesis. He is recognized as a truly inspirational educator, with a talent of adding drama to his teaching, as the numerous students he proselytized to photosynthesis research do prove. The first such proselyte (in 1963) was the author of this Foreword, and shortly after, Prasanna Mohanty, one of the editors of this volume, did join in. Since that magic time, both of us, Mohanty and Papageorgiou, are in continuous contact and collaboration with Govindjee. From the very beginning, through his retirement and beyond, his research output was always near or in the moving frontline of advancing photosynthesis. Govindjee is best known for his research on the various functional manifestations of the Photosystem II (PS II) supercomplex (light-driven water-plastoquinone oxidoreductase), including excitation energy transfer, light emission, primary photochemistry and electron transfer. His major achievements further include an understanding of the basic relationships between Chl a fluorescence and photosynthetic reactions; an unique role of the bicarbonate ion in protonations in the QB–binding region; a theory of thermoluminescence in plants; the first picosecond measurements on the primary photochemistry of PS II; and the use of fluorescence lifetime imaging microscopy (FLIM) of Chl *a* fluorescence to understand photoprotection by plants against excess light. He was and is an innovative editor, having edited several important volumes, but most remarkable for the editorial guidance he provided to the journal *Photosynthesis Research* and to the series of volumes for Kluwer/Springer under the title *Advances in Photosynthesis and Respiration*. As a chronicler, he initiated and steered the *Historical Corner in Photosynthesis Research* (Govindjee, 2006). Although, a section like this may be more appropriate for a periodical or for a magazine, and not for a technical scientific journal like Photosynthesis Research, the plethora of the contributed historical articles over the years and the fact that succeeding editors (Robert Blankenship and David Knaff) kept it on, shows beyond doubt that the Historical Corner was indeed quite well received by the photosynthesis community.

All these, and much more, including rich anecdotal material contributed by past students and colleagues of Govindjee, are presented in detail in two excellent guest editorials published in the journal Photosynthesis Research in 2007 and 2008 by his former student, Julian Eaton-Rye, of the University of Otago, New Zealand. His early times in the family home at Allahabad, with his parents, two brothers and one sister, come alive in an informal family volume published by Govindjee in 2007.

Govindjee retired from the University of Illinois to become an Emeritus Professor in 1999, after 38 years of service. With his retirement, the famous Photosynthesis Laboratory of Emerson–Rabinowitch–Govindjee in the Botany Department was dissolved and its historical equipment was donated at best, or somehow disposed of. However, even without a laboratory, Govindjee continued unabatedly his research activity, mostly by orchestrating international collaborations, as well as his editorial leadership of the Advances in Photosynthesis and Respiration series, as well as in several volumes and special issues of journals. After retirement, he received various distinctions and awards for outstanding contributions to the advancement of the science of photosynthesis and for the interpretation of its mechanism and its global significance for the benefit of the general scientific and non-scientific public (Eaton-Rye 2007a; also *http://www.life.illinois.edu/govindjee/*).

The 2008 International Conference held at the University of Indore, and the present volume are two such distinctions.

George C. Papageorgiou
National Center of Scientific Research, Institute of Biology, Athens, Greece 153 10
E-mail: gcpap@bio.demokritos.gr & gcpap@ath.forthnet.gr

REFERENCES

Brody SS (1995) We remember Eugene: Rabinowitch and his laboratory during the fifties.. Photosynth Res 43: 67-74

Brody SS (2002) Fluorescence yields and energy transfer spectrum in photosynthesis, 1950-1960. Photosynth Res 73: 127-132

Eaton-Rye JJ (2007a) Celebrating Govindjee's 50 years in photosynthesis research and his 75th birthday. Photosynth Res 93:1-5

Eaton-Rye JJ (2007b) Snapshots of the Govindjee lab from the late 1960s to the late 1990s, and beyond ... Photosynth Res 94:153-178

Emerson R and Arnold W (1932a) A separation of the reactions in photosynthesis by means of intermittent light. J Gen Physiol 15: 391-420

Emerson R and Arnold W (1932b) The photochemical reaction in photosynthesis. J Gen Physiol 16: 191-205

Emerson R and Lewis CM (1941) Carbon dioxide exchange and measurement of the quantum yield of photosynthesis. Amer J Bot 28: 789-804

Emerson R and Lewis CM (1943) The dependence of the quantum yield of Chlorella photosynthesis on wavelength of light. Am J Bot 30: 165-178

Emerson R and Rabinowitch E (1960) Red drop and role of auxiliary pigments in photosynthesis. Plant Physiol 35: 477-485

Govindjee (2000) M Younis, U Pathre and P Mohanty (eds) Milestones in Photosynthesis Research. In: Probing Photosynthesis. Taylor & Francis, UK, pp 9-39 - An electronic file is available at: <http://www.life.illinois.edu/govindjee/papers/milestones.html>

Govindjee (2006) Editorial: Celebrating 20 years of historical papers in Photosynthesis Research. Photosynt Res 87: 151-158.

Govindjee (2007), (ed) Amma and Babuji. Our Life at Allahabad (PDQ Printing, Urbana, Ill.)

Govindjee and Krogmann D (2004) Discoveries in oxygenic photosynthesis (1727-20030: a perspective

Govindjee and Bjorn LO (2011) This volume

Jajoo A, Guruprasad KN, Mohanty P (2009) International conference "Photosynthesis in the Global Perspective" held in honour of Govindjee, November 27-29, 2008, Indore, India. Photosynth Res 100: 49-55

Myers J (1974) Conceptual developments in photosynthesis, 1924-1974. Plant Physiol 34: 420-426

Van Niel CB (1941) The bacterial photosynthesis and their importance for the general problem of photosynthesis. Adv Enzymol 1: 263-328

Preface

Photosynthesis has been a challenging scientific topic since it deals with the process by which a wide range of organisms from bacteria to higher plants convert solar energy to chemical energy for producing food, fibre and fuel out of small molecules of carbon dioxide and water. Deeper and greater understanding of this process holds a promise for a more sustainable Earth with enjoyable pollution-free environment with new technologies for food and cleaner energy for a peaceful and progressive society. Thus there have been global efforts to mimic the photosynthetic process to meet the needs of the future world.

Topics of photosynthesis are unusually complex and difficult for students and laymen to understand. Luckily there are quite a few inspired teachers of photosynthesis research spread all over the world that makes this process of photosynthesis fascinating and understandable to all. One such teacher is Govindjee, Professor Emeritus of Biophysics, Biochemistry and Plant Biology at the University of Illinois at Urbana – Champaign, USA. His colleagues and comrades, students and friends, from around the world gathered on Nov. 26, 2008 at Devi Ahilya University, Indore, India to honour him for his contributions to photosynthesis research and his teaching by holding a International Conference entitled "Photosynthesis in the Global Perspective".

This book entitled *Photosynthesis – Overviews on Recent Progress and Future Perspective* is an extension of the conference for honouring Govindjee for his numerous contributions to our understanding of the many critical processes in evolving the popular Z – scheme of Electron Transport in Photosynthesis. This book contains 20 selected overviews on several important aspects of Photosynthesis contributed by the participants at the Indore Conference. These 20 essays provide both tutorial content and critical insight in to the future of photosynthesis research.

We are immensely grateful to all our contributors for writing these chapters for this book. We also understand the limitations of some of the participants of the conference who were unable to provide us their valuable contributions for this volume.

We acknowledge with gratitude the Foreword by George C Papageorgiou that provides an excellent introduction for this book. We have also received encouragement and support from numerous friends and co-authors of Govindjee. We expect that this book would receive the attention of teachers and students of photosynthesis and by people who are interested in the process of photosynthesis by which living organisms are sustained on this planet.

The chapters of this volume have been reviewed by scientists active in the field. We thank them for their valuable comments. This is consistent also with the content of the present volume and all contributors deserve our sincerest thanks in this joint endeavour to disseminate 'information' and 'knowledge' about photosynthesis.

Shigeru Itoh
Prasanna Mohanty
Guruprasad Kadur

Acknowledgements

Reviewers

We thank all the reviewers listed here for their valuable comments on the articles included in this book.

1. A. Suleyman, Institute of Basic Biological Problems, Russian Academy of Sciences, Puschino, Moscow, Russia
2. A. Louis, University of Illinois, Chicago, USA
3. A. Jajoo, School of Life Sciences, Devi Ahilya University, Indore, India
4. B. Biswal, School of Life Sciences, Sambalpur University, Sambalpur, India
5. C. H. Lee, Department of Microbiology, Pusan National University, Republic of Korea
6. G. C. Papageorgiou, Institute of Biology, Athens, Greece
7. G. K. Joshi, Department of Biotechnology, HNB Garhwal University, Srinagar, Uttarakhand, India
8. G.R. Rout, Regional Plant Research Centre, Bhubaneswar, India
9. K. Nath, Department of Microbiology, Pusan National University, Republic of Korea
10. M. D. Gupta, Bose Institute, Kolkata, India
11. M. K. Rawal, School of Life Sciences, Sambalpur University, Sambalpur, India
12. Prakash J.S.S., School of Life Sciences, University of Hyderabad, India
13. R. K. Mishra, Biochemistry Department, Lucknow University, Lucknow, India
14. S. Das, Indian Statistical Institute, Kolkata, India
15. T. Hashimoto, Kobe Women's University, Kobe, Japan
16. U. Pathre, National Botanical Research Institute, Lucknow, India
17. V. D. Kreslovski, Institute of Basic Biological Problems, Russian Academy of Sciences, Puschino, Moscow, Russia
18. V. Sharma, Department of Biotechnology, Banasthali University, Banasthali, Rajasthan, India.

Editorial Assistance

We thank Ms Priyanka Singh and Ms Sonika Sharma from Photobiology Lab, School of Life Sciences, Devi Ahilya University, Indore, India for their editorial assistance.

Contents

Contributors

Anjali Agarwal (India): Chapter 16, p. 229
anjaliaggarwal25@yahoo.com

Rachna Agarwal (India): Chapter 2, p. 28
rachna@barc.gov.in

Samira A. Aliyeva (Azerbaijan): Chapter 7, p. 101
aliyevasamira27@hotmail.com

Suleyman I. Allakhverdiev (Russia and Japan): Chapter 14, p. 189
suleyman@issp.serpukhov.su

Mamta Baunthiyal (India): Chapter 16, p. 229
mamtabaunthiyal@yahoo.co.in

Bishnupriya Behera (India): Chapter 15, p. 217
bishnu_priyabt@yahoo.co.in

Sudhakar Bharti (India): Chapter 11, p. 146
sbharti2010@hotmail.com

Basanti Biswal (India): Chapter 18, p. 245
basanti_b@hotmail.com

Udaya C. Biswal (India): Chapter 18, p. 245
biswaluc@yahoo.co.in

Lars Olof Björn (Sweden; China) : Chapter 1, p. 1
Lars_Olof.Bjorn@cob.lu.se

Rainer Bode (Canada): Chapter 8, p. 108
rbode2@uwo.ca

Robert Carpentier (Canada): Chapter 14, p. 189
Robert.Carpentier@uqtr.ca

Anath Bandhu Das (India): Chapter 15, p. 217
a_b_das@hotmail.com

Challabathula Dinakar (India): Chapter 12, p. 158
challadina@rediffmail.com

Young-Jae Eu (Korea): Chapter 10, p. 130
eu2lee@googlemail.com

Irina R. Fomina (Russia and USA): Chapter 14, p. 189
irafomi@rambler.ru

Yoshimasa Fukushima (Japan): Chapter 6, p. 86
fukushima.yoshimasa@gmail.com

Ralphreed A. Gasanov (Azerbaijan): Chapter 7, p. 101
ra38hasan@yahoo.com

Govindjee (USA): Chapter1, p. 1
gov@illinois.edu

Kadur N. Guruprasad (India): Chapter 19, p. 261
knguruprasad@hotmail.com

Norman P.A. Hüner (Canada): Chapter 8, p. 108
nhuner@uwo.ca

Shigeru Itoh (Japan): Chapter 6, p. 86
itoh@bio.phys.nagoya-u.ac.jp

Anjana Jajoo (India): Chapter 11, p. 146
anjanajajoo@hotmail.com

Navasard V. Karapetyan (Russia): Chapter 4, p. 60
nkarap@inbi.ras.ru

Asako Kawamori (Japan): Chapter 5, p. 73
kawamori-a@tulip.sannet.ne.jp

Renu Khanna-Chopra (India): Chapter 13, p. 167
renu.chopra3@gmail.com

Vyacheslav V. Klimov (Russia): Chapter 14, p. 189
klimov@issp.serpukhov.su

Vladimir D. Kreslavski (Russia): Chapter 14, p. 189
vkreslav@rambler.ru

Marianna Krol (Canada): Chapter 8, p. 108
mkrol@uwo.ca

Choon-Hwan Lee (Korea): Chapter 10, p. 130
chlee@pusan.ac.kr

Wenze Li (Canada): Chapter 8, p. 108
WLi@danforthcentre.org

Dmitry A. Los (Russia) : Chapter 14, p. 189
losnet@mail.ru

Fikret Mamedov (Sweden): Chapter 7, p. 101
fikret.mamedov@fotomol.uu.se

Gururaj Maralihalli (India): Chapter 17, p. 237
gmarali@barc.gov.in

Denis P. Maxwell (Canada): Chapter 8, p. 108
dmaxwell@uwo.ca

Michael Melzer (Germany): Chapter 2, p. 28
melzer@ipk-gatersleben.de

Mamoru Mimuro (Deceased): Chapter 14, p. 189
mamo-mi@mm1.mbox.media.kyoto-u-ac.jp

Hiroyuki Mino (Japan): Chapter 5, p. 73
mino@bio.phys.nagoya-u.ac.jp

Sujata R. Mishra (India; Korea): Chapter 3, p. 38; Chapter 10, p. 130
sujatar12@hotmail.com

Prasanna Mohanty (India): Chapter 3, p. 38; Chapter 14, p. 189; Chapter 15, p. 217
photosis@rediffmail.com

Pranab K. Mohapatra (India): Chapter 18, p. 245
pkmohapatras@yahoo.co.in

Ashok K. Munjal (India): Chapter 16, p. 229
ashokmunjal70@yahoo.co.in

Krishna Nath (Korea): Chapter 10, p. 130
krishnanath125@hotmail.com

Kollipara Padmasree (India): Chapter 12, p. 158
kpssl@uohyd.ernet.in

George C. Papageorgiou (Greece): Foreword, p. vii
gcpap@bio.demokritos.gr;
gcpap@ath.forthnet.gr

Subramanyam Rajagopal (India): Chapter 3, p. 38
psrajagopal@yahoo.com

Marina G. Rakhimberdieva (Russia): Chapter 4, p. 60
rakhimberd@mail.ru

Mukesh K. Raval (India): Chapter 18, p. 245
mraval@yahoo.com

Steven R. Rodermel (USA): Chapter 8, p. 108
rodermel@iastate.edu

Dominic Rosso (Canada): Chapter 8, p. 108
drosso2011@dents.uwo.ca

Jayashree K. Sainis (India): Chapter 2, p. 28; Chapter 17, p. 237
jksainis@barc.gov.in

Iti Sharma (India): Chapter 16, p. 229
iti4in@yahoo.com

Vinay Sharma (India): Chapter 16, p. 229
vinaysharma30@yahoo.co.uk

Louis A. Sherman (USA) : Chapter 9, p. 120
lsherman@purdue.edu

Vladimir V. Shubin (Russia): Chapter 4, p. 60
shubin@inbi.ras.ru

Priyanka Singh (India): Chapter 19, p. 261
priyankaphotobiol@gmail.com

Nellaepalli Sreedhar (India): Chapter 4, p. 38
samsree@gmail.com

Singh **Srivalli** (India): Chapter 13, p. 167
valli_singh@hotmail.com

Tina C. Summerfield (USA): Chapter 9, p. 120
tina.summerfield@otago.ac.nz

Irina V. Terekhova (Russia): Chapter 4, p. 60
Terekhova@yahoo.com

Jorg Toepel (USA): Chapter 9, p. 120
joerg.toepel@uni-bielefeld.de

Altanzaya Tovu (Korea): Chapter 10, p. 130
zaya@pusan.ac.kr

Bhuminath N. Tripathi (India): Chapter 16, p. 229
bhuminathtripathi@hotmail.com

Ismayil S. Zulfugarov (Korea): Chapter 10, p. 130
iszulfugarov@pusan.ac.kr

CHAPTER 1

Dissecting Oxygenic Photosynthesis: The Evolution of the "Z"-Scheme for Thylakoid Reactions

Govindjee[1] **and Lars Olof Björn**[2, 3]

[1]*University of Illinois at Urbana-Champaign, Urbana, Illinois, USA*

E-mail: gov@illinois.edu

[2]*School of Life Science, The Key Laboratory of Ecology and Environmental Science in Guangdong Higher Education, South China Normal University, Guangzhou 510631, China*

[3]*Lund University, Department of Biology, Sölvegatan 35, SE-22362 Lund, Sweden*

ABSTRACT

From a primitive beginning we have come to our present understanding of what is going on in the thylakoid membranes. We start our overview with the determination of the size of the photosynthetic unit. This unit gave the concept of light absorbing antenna and the reaction centers, where the primary photochemistry takes place. We continue with the determination of the maximum quantum yield and the realization that two light reactions and two photosystems are involved in oxygenic photosynthesis. This is then followed by the discovery of the "period four oxygen clock" and the structure of the oxygen-evolving center. After a brief mention of the various protein components in the thylakoid membrane and a comparison of the two photosystem complexes, we provide here a summary and the evolution of the so-called "Z" scheme for the electron flow from water to the pyridine nucleotide $NADP^+$, nicotinamide adenine dinucleotide phosphate. We conclude with a picture of the thylakoid membrane domains in the chloroplast, and the intricate machinery where all the reactions from water to $NADP^+$ reduction takes place. Our chapter is not a review, but a viewpoint that we favour.

Keywords: Electron Transfer Pathway; Emerson Enhancement Effect; Oxygen Clock; Photosynthetic Unit; Red Drop in Photosynthesis; Two Light Reactions; Two Photosystems; Z-Scheme of Photosynthesis.

INTRODUCTION

Photosynthesis is a very complicated process, and in particular, this holds for the kind of photosynthesis carried out by plants, algae and cyanobacteria, i.e., oxygen-producing (oxygenic) photosynthesis (Rabinowitch and Govindjee 1969; Blankenship 2002). Initially, it was thought that oxygen was released by the photochemical splitting of carbon dioxide:

$$CO_2 + \text{light} \Rightarrow C^* + O_2 \quad \text{(the asterisk on C indicates an activated state)}$$

Then the carbon would combine with water to form carbohydrate:

$$C^* + 6H_2O \Rightarrow C_6H_{12}O_6$$

This is the origin of the term carbohydrate, still in use, despite the fact that this scheme has been proven to be wrong. For the description and historical evaluation of this early and erroneous hypothesis, see Nickelsen (2010). In the following section we shall discuss the 'photosynthetic unit (i.e., the existence of a large number of light-harvesting pigment molecules and a very few reaction center molecules where the primary photochemistry takes place); the minimum quantum requirement, i.e., the minimum number of photons of light needed for the evolution of one molecule of oxygen (its inverse being the maximum quantum yield, i.e., the number of oxygen molecules evolved per photon of light absorbed); and the two-light reaction and two-pigment scheme of photosynthesis. We conclude with a picture of the thylakoid membrane domains in the chloroplast, the intricate machinery where all the reactions from water to NADP reduction take place.

We refer the readers to Govindjee (2000) for an earlier overview of the concepts discussed here; Govindjee and Krogmann (2004) for a timeline of oxygenic photosynthesis; and Orr and Govindjee (2010) for an entry into all literature on photosynthesis at all levels.

PHOTOSYNTHETIC UNITS

Engelmann (1884) was one of the first to quantitate the relationship between absorption of light and photosynthesis.The story of how the complex photosynthetic machinery has been dissected and its components (at least many of them, now well characterized) started with a paper by Blackman (1905), in which he studied the rate of photosynthesis in different combinations of light intensity and carbon dioxide concentration. He drew the conclusion that the photosynthetic process could be divided into a "light reaction" and a "dark reaction", and that the "dark reaction" limits the rate when the light is strong enough. (The "dark reaction", in fact, as it turned out, a series of a great number of reaction steps, had often been referred to, in the past, by many as the "Blackman reaction".)

In two very important papers, Emerson and Arnold (1932a, 1932b) conclusively separated and characterized the photochemical step and the Blackman reaction. They built on earlier observations by Richard Willstätter (Nobel laureate in Chemistry, 1915) and Otto Warburg (Nobel laureate in Physiology or Medicine, 1931) in which they had found that if the light driving photosynthesis was divided into short pulses, the amount of photosynthesis per total time of the experiment was not very much decreased, while the amount of photosynthesis during the time in light was increased. Taking the technical possibilities of the time into account, we marvel at the ingenuity of the apparatus that Emerson and Arnold (1932a) describe (see the historical appreciation by Myers (1994)). With their apparatus they could deliver very short and intense repetitive light flashes, and study how photosynthesis varied when intensity, flash duration, and flash interval were varied. Using the unicellular green alga, *Chlorella*, they found that as the flash intensity increased they approached light saturation, which could not be overcome even if the flash interval was increased (so that the average amount of light per unit time was decreased). They drew the conclusion that with strong flashes they would activate all "photosynthetic units", which would then, during the dark interval, use the collected energy to drive the "Blackman reaction" and complete photosynthesis. By comparing the maximum yield of oxygen per flash with the content of chlorophyll, they could calculate that one photosynthetic unit contains about 2400 chlorophyll molecules (Fig. 1.1).

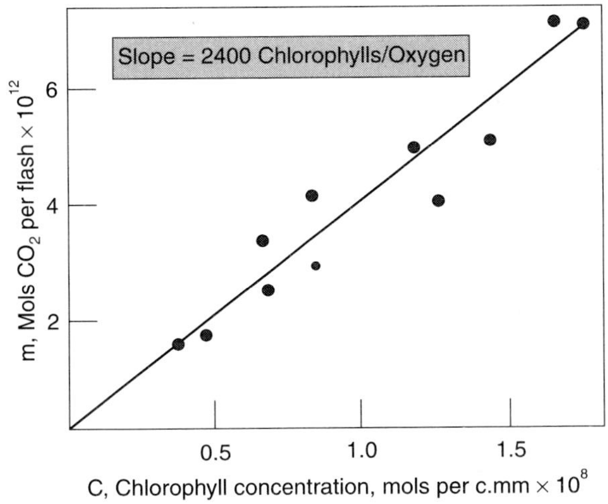

Fig. 1.1 The Photosynthetic Unit in the green alga *Chlorella* (1932). With sufficiently strong and short light flashes, and with appropriate optimal dark periods between repetitive flashes, the amount of oxygen given off (considered equivalent to carbon dioxide uptake) did not depend on the flash intensity, but only on the amount of chlorophyll in algal suspensions illuminated by the flashes. The slope of the line (C/m) in the diagram corresponded to 2400 molecules of chlorophyll per molecule of oxygen evolved. Thus the chlorophyll content of one photosynthetic unit (PSU) was considered to be 2400 per oxygen molecule evolved (or carbon dioxide taken up). This number should be divided by four, since four electrons must be moved to produce one oxygen molecule, so the photosynthetic unit in this experiment may be regarded as containing 600 chlorophyll molecules per electron transferred. However, since we know now that there are two pigment systems and two light reactions, this number would be 300 chlorophylls per electron transferred. Redrawn from Emerson and Arnold (1932b).

Emerson and Arnold (1932b) wrote:

> *"We need only suppose that for every 2480 molecules of chlorophyll there is present in the cell one unit capable of reducing one molecule of carbon dioxide each time it is suitably activated by light."*

It is obvious to us that although Emerson and Arnold had measured oxygen, they talked about CO_2 since they were still using, for description, the early ideas of oxygen coming from activated CO_2. However, it was in this experiment that the concept of what we now call the antenna and the reaction center was born. The unit could very well be what was called the 'photoenzyme', and in today's language, we may equate it to the 'reaction center' with the caveat that the ~2400 chlorophyll molecules be divided by at least 8, the minimum number of photons needed to evolve one oxygen molecule. Soon thereafter, Arnold and Kohn (1934) called this unit "chlorophyll unit". On the other hand, Kohn (1936) suggested that the absorbing unit was ~500 Chls, since he assumed that 4 photons were needed for the evolution of one molecule of oxygen. However, Gaffron and Wohl (1936) really understood these experiments and calculated that in a *Chlorella* suspension that Emerson and Arnold (1932a, 1932b) had used, where oxygen evolution begins immediately upon exposing the samples to low light, it would have taken an average of an hour or more to collect several photons (4-12) on the same chlorophyll molecule necessary for evolving one molecule of oxygen. Gaffron and Wohl (1936) explained this paradox by suggesting that there must be 'funneling' of quanta absorbed by different molecules on to one common 'center' where the chemistry would begin. We now know that it is not the photons that are being transferred but the "excitation energy", or the excitons (see Clegg *et al.* 2010). We also know that 2400 is not a magic number in all photosynthetic organisms. Schmid and Gaffron (1968) showed that in several algae, cyanobacteria and higher plants, the number of chlorophyll molecules evolved per oxygen, under different physiological conditions, was in multiples of approximately 300: ~300, ~600, ~1200, ~1800, ~2400 and ~4800.

The method of dividing light into flashes in order to characterize light-driven processes introduced by Emerson and Arnold (1932a, 1932 b) has been used by others. For example, Björn (1965) used it to characterize steps in the differentiation of plastids, and Sundqvist (1969) used it to study steps in the formation of chlorophyll. But more importantly, it was used in a more elaborate form, as single flashes, with appropriate dark times, by Joliot *et al.* (1969) and Kok *et al.* (1970) to further characterize the oxygen evolution steps of photosynthesis. We shall return to this later.

QUANTUM YIELDS AND ACTION SPECTRA OF OXYGENIC PHOTOSYNTHESIS

Transformation of two water molecules, plus one molecule of carbon dioxide, to one molecule of oxygen and a one-carbon equivalent of carbohydrate, involves the transfer of 4 electrons (2 electrons per oxygen atom). The Stark-Einstein equivalence law (Stark 1908, Einstein 1912) states that one photon reacts with one molecule, so splitting the carbon dioxide to C* and O_2, according to the old and naive theory mentioned earlier, would need only one photon, provided it had sufficient energy. However, Cornelis B. van Niel (see e.g., van Niel 1941) had, by comparison of non-oxygenic (referred to as 'anoxygenic') bacterial photosynthesis with the oxygenic plant, algal and cyanobacterial photosynthesis, come to the conclusion that both could be summarized with one formula:

$$2\,H_2A + CO_2 + \text{light} \Rightarrow 2A + CH_2O + H_2O$$

where, in the case of anoxygenic photosynthesis, A stands for, e.g., S, and in the case of oxygenic photosynthesis for O (or ½ O_2), implying that the oxygen released comes from water and not from carbon dioxide. This makes it likely that one photon moves one electron, so that 4 photons would be required for the release of one molecule of oxygen and assimilation of one molecule of carbon dioxide.

Otto Warburg and coworkers reported that this, so-called, minimum quantum requirement of photosynthesis (the ratio of minimum number of photons required per oxygen evolved) was 4, or in later publications, under certain conditions even a little less (see e.g., Warburg *et al.* 1950). Warburg wanted to harmonize the thermodynamic requirement of at least 3 red photons with the Stark-Einstein one-to-one relation between photons and molecules. He worked out a theory by which three photons produced three oxygen molecules with the help of extra energy; this extra energy was obtained by reoxidation of assimilate, using up two of the oxygen molecules produced (see e.g., Damaschke *et al.* 1953, Warburg 1958). In contrast to Warburg's minimum values of 3-4 photons per oxygen, many investigators obtained values that were in the range of 8-12 (see e.g., Arnold 1935, 1949; Magee *et al.* 1939). Emerson and Lewis (1941) discovered that one of the reasons for Warburg's low minimum quantum requirement (inversely high maximum quantum yield) was the CO_2 burst in the first minutes of illumination that Warburg had counted as oxygen. Emerson and Chalmers (1955) substantiated this error in Warburg's work in a long and a thorough paper. Emerson and Lewis (1943), using the green alga *Chlorella* as Warburg had used, determined the minimum number of photons to be ~11 per oxygen evolved, and the same result was obtained later for the diatom *Navicula* by Tanada (1951) and the red alga *Porphyridium* by Brody and Emerson (1959). (See Emerson, 1958, for a review.) Emerson died in a plane crash

Frenkel A (1954) Light-induced phosphorylation in cell-free preparations of photosynthetic bacteria. 76: 5568-5569

Gaffron H, Wohl K (1936) Zur Theorie der Assimilation. Naturwissenschaften 24: 81-90; 103-107

Golbeck JH (ed) (2006) Photosystem I: The Light-driven Plastocyanin: Ferredoxin Oxidoreductase. Springer, Dordrecht

Gorman DS, Levine RP (1966) Photosynthetic electron transport chain of *Chlamydomonas reinhardtii*.VI. Electron transport in mutant strains lacking either Cyt 559 or plastocyanin. Plant Physiol 41: 1648: 1656

Govindjee (1999a) On the requirement of minimum number of four versus eight quanta of light for the evolution of one molecule of oxygen in photosynthesis. Photosynth Res 59: 249-254

Govindjee (1999b) In: HA Frank, AJ Young, G Britton, RJ Cogdell (eds) Carotenoids in photosynthesis: An historical perspective. In: The Photochemistry of Carotenoids, Kluwer Academic Publishers (now Springer), Dordrecht, Netherlands pp 1-19

Govindjee (2000) Milestones in Photosynthesis Research. In: Probing Photosynthesis. Edited by M. Younis, U. Pathre and P. Mohanty. Taylor & Francis, UK, pp 9-39 [An electronic file is available at: <http://www.life.illinois.edu/govindjee/papers/milestones.html>]

Govindjee (2004a) Chlorophyll a Fluorescence: A Bit of Basics and History. In: G Papageorgiou and Govindjee (eds) Chlorophyll a Fluorescence: A Probe of Photosynthesis, Kluwer Academic (now Springer), Dordrecht, The Netherlands, pp 2-42

Govindjee (2004b) Robert Emerson, Eugene Rabinowitch: Understanding Photosynthesis. In: Lillian Hoddeson (editor). "No Boundaries: University of Illinois Vignettes", Chapter 12, University of Illinois Press, Urbana and Chicago, pp 181-194

Govindjee, Rabinowitch E (1960) Two forms of chlorophyll *a* in vivo with distinct photochemical functions. Science 132: 159-160

Govindjee R, Thomas JB, Rabinowitch E (1960) Second Emerson effect in the Hill reaction of *Chlorella* cells with quinone as oxidant. Science 132: 421

Govindjee, Ichimura S, Cederstrand C, Rabinowitch E (1960) Effect of combining far-red light with shorter wave light in the excitation of fluorescence in *Chlorella*. Arch Biochem Biophys 89: 322-323

Govindjee R, Govindjee, Hoch G (1962) The Emerson enhancement effect in TPN photoreduction by spinach chloroplasts. Biochem Biophys Res Comm 9: 222-225

Govindjee, Owens O v H, Hoch G (1963) A mass spectroscopic study of the Emerson enhancement effect. Biochim Biophys Acta 75: 281-284

Govindjee R, Govindjee, Hoch G (1964) Emerson enhancement effect in chloroplast reactions. Plant Physiol 39: 10-14

Govindjee, Govindjee R (1965) Two different manifestations of enhancement in the photosynthesis of *Porphyridium cruentum* in flashing monochromatic light. Photochem Photobiol 4: 401-415

Govindjee, Bazzaz M (1967) On the Emerson enhancement effect in the ferricyanide Hill reaction in chloroplast fragments. Photochem Photobiol 6: 885-894

Govindjee R, Rabinowitch E, Govindjee (1968) Maximum quantum yield and action spectrum of photosynthesis and fluorescence in *Chlorella*. Biochim Biophys Acta 162: 539-544

Govindjee and Govindjee R (1975) Introduction to Photosynthesis, In: Govindjee (ed) Bioenergetics of Photosynthesis, Academic Press, NY, pp 1-50

Govindjee, Coleman WJ (1990) How plants make oxygen. Scientific American 262: 50-58

Govindjee and Krogmann D (2004) Discoveries in oxygenic photosynthesis (1727- 2003): a perspective: Dedicated to the memories of Martin Kamen (1920-2002) and William A Arnold (1904-2001). Photosynth Res 80: 15–57

Govindjee, Seibert M (2010) Picosecond spectroscopy of the isolated reaction centers from the photosystems of oxygenic photosynthesis: Ten years (1987-1997) fun. A tribute to Michael R Waisielewski on his 60th birthday. Photosynth Res 103: 1-6

Govindjee, Kern JF, Messinger J, Whitmarsh J (2010) Photosystem II. In: Encyclopedia of Life Sciences (ELS). John Wiley & Sons Ltd: Chichester. DOI: 10.1002/9780470015902.a0000669.pub2 (See http://www.life.illinois.edu/govindjee/recent_papers.html)

Govindjee, Shevela D (2011) Adventures with cyanobacteria: a personal perspective. Frontiers in Plant Science. DOI: 3389/fpls.2011.00028, 17 pages, available free online

Grabolle M, Dau H (2005) Energetics of primary and secondary electron transfer in photosystem II membrane particles of spinach revisited on basis of recombination-fluorescence measurements. Biochim Biophys Acta 1708: 209-218

Guskov A, Kern J, Gabdulkhakov A, Broser M, Zouni A, Saenger W (2009) Cyanobacterial photosystem II at 2.9 Å resolution and the role of quinones, lipids, channels and chloride. Nature Structural and Molecular Biology 16: 334-342

Homann M, Liebisch P, Muller C, Barra M, Grabolle M, Dau H (2005) Photosynthetic O_2 formation tracked by time-resolved X-ray experiments. Science 310: 1019-1021

Haxo FT, Blinks LR (1946) Photosynthetic action spectra in red algae. Am J Bot 33: 836-837

Haxo FT, Blinks LR (1950) Photosynthetic action spectra of marine algae. J Gen Physiol 33: 389-422

Hill R, Bendall F (1960) Function of the cytochrome components in chloroplasts: a working hypothesis. Nature 186: 136-137

Homann PH (2005) Chloride and calcium in photosystem II: from effects to enigma. In: Govindjee, JT Beatty, H Gest and JF Allen (eds) Discoveries in Photosynthesis, Springer, Dordrecht, pp 383-389

Joliot P (1956) Dispositif ampérometrique de measure de la photosnthèse. CR Acad Sci Paris 243: 677-680

Joliot P (1960) Contribution à l'étude des phénomènes d'induction de la photosynthèse. Thèse de Doctorat es-Sciences, Faculté de Sciences de Paris

Joliot P (1965a) Cinetiques des reactions liees a l'emission d'oxygene photosynthetique. Biochim Biophys Acta 102: 116-134

Joliot P (1965b) Etudes simultanees des cinetiques de fluorescence et d'emission d'oxygene photosynthetique. Biochim Biophys Acta 102: 135-148

Joliot P (1968) Kinetic studies of photosystems in photosynthesis. Photochem Photobiology 8: 461-463

Joliot P, Barbieri G, Chabaud R (1969) Un nouveau modele des centres photochimiques du systeme II. Photochem Photobiol 10: 309-329

Joliot P, Kok B (1975) Oxygen evolution in photosynthesis. In: Govindjee (ed), Bioenergetics of Photosynthesis, Academic Press, New York, pp 387-412

Jordan P, Fromme P, Witt HT, Klukas O, Saenger W, Krauss N (2001) Three-dimensional structure of cyanobacterial photosystem I at 2.5 Å resolution. Nature 411: 909-917

Junge W, Sielaff H, Engelbrecht S (2009) Torque generation and elastic power transmission in the rotary FoF1-ATPase. Nature 459: 364-370

Ke B (2001) Photosynthesis: Photobiochemistry and Photobiophysics. In Govindjee (Series ed) Advances in Photosynthesis, volume 10, Springer, Dordrecht

Kohn HI (1936) Number of chlorophyll molecules acting as an absorbing unit in photosynthesis. Nature 37: 706-706

Kok B (1956) On the reversible absorption change at 705 mμ in photosynthetic organisms. Biochim Biophys Acta 22: 399-401

Kok B (1957) Absorption changes induced by the photochemical reaction of photosynthesis. Nature 179: 583-584

Kok B (1959) Light-induced absorption changes in photosynthetic organisms. II. A split-beam difference spectrophotometer. Plant Physiol 34: 184-192

Kok B, Hoch G (1961) Spectral changes in photosynthesis. In : WD McElroy and B Glass (eds.) Light and Life, The Johns Hopkins Press, Baltimore, Maryland, pp 397-423

Kok B (Chairman), Jagendorf AT (Organizer) (1963) Photosynthetic Mechanisms of Green Plants. Publication #1145. National Academy of Sciences—National Research Council, Washington, DC, see 'Foreword', p. ix; papers from many groups were presented and there was a general agreement on the basic scheme(s)

Kok B, Forbush M, McGloin M (1970) Cooperation of charges in photosynthetic O_2 evolution 1. Photochem Photobiol 11: 457-475

Levine RP (1969) The analysis of photosynthesis using mutant strains of algae and higher plants. Annu Rev Plant Physiol 20: 523-540

Levring T (1947) Submarine daylight and the photosynthesis of marine algae. Göteborgs K Vetenskaps-och Vitterhets-samhällets Handl 5(6): 1-90

Loll B, Kern J, Saenger W, Zouni A, Biesiadka J (2005) Towards complete cofactor arrangement in the 3.0 A resolution structure of photosystem II. Nature 438: 1040-1044

Losada M, Whatley FR, Arnon DI (1961) Separation of two light reactions in non-cyclic photophosphorylation of green plants. Nature 190: 606-610

Magee JL, de Witt TW, Smith EC, Daniels F (1939) A photocalorimeter. The quantum efficiency of photosynthesis in algae. J Amer Chem Soc 61: 3529-3533

Mar T and Govindjee (1972) Kinetic models of oxygen evolution in photosynthesis. J Theoret Biol 36: 427-446

McConnell IL, Eaton-Rye JJ and van Rensen JJS (2011) Regulation of photosystem II electron transport by bicarbonate. In: Eaton-Rye JJS, Tripathy BC, Sharkey TD (eds) Photosynthesis: Plastic Biology, Energy Conversion and Carbon Assimilation, Vol 34, Advances in Photosynthesis and Respiration, Springer, Dordrecht, in the press

Mitchell P (1961a) Coupling of photophosphorylation to electron and hydrogen transfer by a chemiosmotic type of mechanism. Nature 191: 144-148

Mitchell P (1961b) Chemiosmotic coupling in oxidative and photosynthetic phosphorylation. Biochem J 79: 23-24

Myers J (1971) Enhancement studies in photosynthesis. Annu Rev Plant Physiol 22: 289-312

Myers J (1994) The 1932 experiments. Photosynth Res 40: 303-310

Myers J, French CS (1960) Evidences from action spectra for a specific participation of chlorophyll *b* in photosynthesis. J Gen Physiol 43: 723-736

Najafpour MM, Govindjee (2011) Oxygen evolving complex in photosystem II: better than excellent. Dalton Transactions DOI: 10.1039/c1dt10746a, 9 pages, Online

Nickelsen K (2010) Of Light and Darkness—Modelling Photosynthesis 1840-1960.Habilitationschrift-eingereicht der Phil.-nat.Fakultat der Universutat Bern, Switzerland, Kluwer Academic Press, in press

Nield J (2010) See Web page accessed August 8, 2010: http://www.queenmaryphotosynthesis.org/nield/psIIimages/oxygenicphotosynthmodel.html

Ono T, Noguchi T, Inoue Y, Kusunoki M, Matsushita T, Oyanagi H (1992) X-ray detection of the period-four cycling of the manganese cluster in photosynthetic water oxidizing enzyme. Science 258: 1335-1337

Orr L, Govindjee (2010) Photosynthesis Online. Photosynth Res DOI 10.1007/s11120-010-9570-8, 34 pages (See http://www.life.illinois.edu/govindjee/photoweb/)

Pirson A (1937) Ernährungs- und stoffwechselphysiologische Untersuchungen an Fontinalis und Chlorella. Z Botan 31: 193-267

Rabinowitch E (1945) Photosynthesis and Related Processes. Volume I. Interscience Publishers, Inc., New York, pp 599 see p 162

Rabinowitch E (1956) Photosynthesis and Related Processes, Vol II (Part 2). Kinetics of Photosynthesis (continued); Addenda to Vol I and Vol II, Part 1 Interscience Publishers, New York, pp 1211-2088. see p 1862

Rabinowitch E (1961) Robert Emerson (1903-1959) Biographical Memoirs of the National Academy of Sciences, National Academy of Sciences, Washington, DC, USA 25: 112-131

Rabinowitch E, Govindjee (1961) Different forms of chlorophyll *a* in vivo and their photochemical function. In: McElroy WD, Glass B (eds) A Symposium on Light and Life, The Johns Hopkins Press, Baltimore, Maryland, pp 378-391.

Rabinowitch E, Govindjee (1969) Photosynthesis. John Wiley & Sons, New York. See Chapters 11; 13; 14; and 16. The book is available free on the internet at http://www.life.illinois.edu/govindjee/photosynBook.html

Renger, G. (ed) (2008) Primary Processes of Photosynthesis: Principles and Apparatus, Part I, Photophysical Principles, Pigments and Light Harvesting/Adaptation/Stress, Royal Society of Chemistry, Cambridge

Renger G (2010) The light reactions of photosynthesis. Current Science 98: 1305-1319

Ruben S (1943) Photosynthesis and phosphorylation. J Amer Chem Soc 65: 279-282

Satoh K, Wydrzynski T, Govindjee (2005) Introduction to Photosynthesis. In: Wydrzynski T and Satoh (eds) photosystem II: The Light-driven Water-plastoquinone Oxidoreductase, Springer, Dordecht, pp 11-22

Schmid G H, Gaffron, H (1968) Photosynthetic units. J Gen Physiol 52: 212-239

Shinkarev VP, Wraight CA (1993) Kinetic factors in the bicycle model of oxygen evolution by photosystem II. Photosynth Res 38: 315-321

Shinkarev V P, Xu C, Govindjee, Wraight CA(1997) Kinetics of the oxygen evolution step in plants determined from flash-induced chlorophyll *a* fluorescence. Photosynth Res 51: 43-49

Siegbahn PEM (2008) A structure-consistent mechanism for dioxygen formation in photosystem II. Chem Eur J 27: 8290-8302

Siegbahn PEM (2009) Structures and energetics for oxygen formation in photosystem II. Accnts Chem Res 42: 1871-1880

Skillman JB (2008) Quantum yield variation across the three pathways of photosynthesis: not yet out of the dark. J Exp Bot 59: 1647-1661

Spencer D, Possingham JV (1961) The effect of manganese deficiency on photophosphorylation and the oxygen-evolving sequence in spinach chloroplasts. Biochim Biophys Acta 52: 379-38

Sproviero EM, Gascon JA McEvoy JP, Brudvig GW, Batista VS (2008a) Quantum mechanics/molecular mechanics study of the catalytic cycle of water splitting in photosystem II. J Amer Chem Soc 130: 3428-3442

Sproviero EM, Gascon JA, McEvoy JP, Brudvig GW, Batista VS (2008b) A model of the oxygen-evolving center of photosystem II predicted by structural refinement based on EXAFS simulations. J Am Chem Soc 130: 6728-6730

Srivastava A, Strasser RJ and Govindjee (1999) Greening of peas: parallel measurements of 77K emission spectra, OJIP chlorophyll a fluorescence transient, period four oscillation of the initial fluorescence level, delayed light emission, and P700. Photosynthetica 37: 365-392

Stark J (1908) The atomising effect of light and optical sensitising. Physikal Z 9:894900

Strehler BL (1952) Photosynthesis-energetics and phosphate metabolism. In: McElroy WD, Glass HB (eds) Phosphorous Metabolism II. Johns Hopkins Press, Baltimore, Md, pp 491-502

Strehler BL (1953) Firefly luminescence in the study of energy transfer mechanisms II: Adenosine triphosphate and photosynthesis. Arch Biochem Biophys 43: 67-79

Stroebel D, Choquet Y, Popot J-L, Picot D (2003) An atypical haem in the cytochrome b_6f complex. Nature 426: 413-418

Sugiura M, Rappaport F, Hillier W, Dorlet P, Ohno Y, Hayashi H, Boussac A (2009) Evidence that D1-His332 in photosystem II from *Thermosynechococcus elongatus* interacts with the S3-state and not with the S2-state. Biochemistry 48: 7855-7866

Sundqvist C (1969) Transformation of protochlorophyllide, formed from exogenous δ aminolevulinic acid, in continuous light and in flashlight. Physiol Plantarum 22: 147-156

Suzuki H, Suggiura M, Noguchi T (2009) Monitoring proton release during photosynthetic water oxidation in photosystem II by means of isotope-edited infrared spectroscopy. J Am Chem Soc 131: 7849-7857

INTRODUCTION

Molecular architecture of any living organism is unique for its form and function. This architecture plays an important role in dictating the precision of the biochemical reactions in living systems which occur in a defined space and time in this universe. The conventional biochemical research hardly takes note of such organization and most of our knowledge about the chemistry of living organisms remains grounded in the information obtained from isolated systems. The field of photosynthesis, which is discussed here, is no exception to this.

Sun is an eternal source of energy on our earth, fuelling all the life processes. This energy is converted into usable chemical energy by the process of **PHOTOSYNTHESIS.** During photosynthesis, light energy is used to drive electrons and protons from water to NADP through electron transport chain. Oxygen is evolved in the process with the concomitant reduction of carbon dioxide. This process involves two distinct types of reactions: **light reactions,** the pigment protein complexes of which reside in the membranous structures called thylakoids; and **dark reactions,** the machinery of which is believed to be randomly distributed in the cytosol of the prokaryotic cell or stroma of the eukaryotic chloroplast. The process of photosynthesis is performed by all green plants, algae, diatoms, certain bacteria and cyanobacteria, which have evolved unique molecular architectures of their photosynthetic machinery to adapt to light and other environmental conditions so as to the make their best for sustaining their life styles.

Due to the insolubility of proteins in biological membranes, the associations of membrane proteins have always been thought to be obvious. The molecular architectures of photosystem I, photosystem II, cytochrome b_6f and ATP synthase are well studied and better understood (Nelson and Ben-Shem, 2004; Dekker and Boekema, 2005). The role of these complexes is production of ATP and NADPH through sequential transfer of electrons between a series of acceptors and donors. Biochemically PSII is water plastiquinone oxidoreductase consisting of a heterodimer of D1 and D2 proteins along with proteins of oxygen evolving complex and many other small proteins. It also has antennae complexes in higher plants called light harvesting complexes comprising of chlorophylls and carotenoids. Cytb_6f complex is a plastoquinol-plastocyanin reductase minimally composed of cytochrome b_6, carrying a low- and a high-potential heme group (b_L and b_H), cytochrome *f* with one covalently bound heme *c*, Rieske iron-sulfur protein (ISP) containing a single [Fe_2S_2] cluster, and subunit IV (17 kDa protein). It transfers electrons between the two reaction center complexes, photosystem II and photosystem I, and also creates an electrochemical gradient for ATP generation. Photosystem I is a pigment protein supercomplex accepting electrons from cytb_6f complex to generate NADPH. CF_0-CF_1 ATP synthase complex is a multi subunit super complex, which utilizes the electrochemical gradient for the generation of ATP.

The associations of these sequential and non-sequential components of the electron transport chain and their structure-function relations are also well investigated (Staehelin and DeWit, 1984; Vermeglio and Joliot, 2002). Characterization of these stable, intramembranous complexes is done by biochemical methods, fluorescence studies, and EPR-spectroscopy (Lagoutte and Mathis, 1989; Nilsson *et al.*, 1992; Tang and Diner, 1994; Eshaghi *et al.*, 1999). The advancements in mass spectrometry, X-ray crystallography and cryo-electron microscopy have made the studies on molecular architecture of these complexes more precise (Boekema *et al.*, 1994; Krauss *et al.*, 1993; Zolla and Timperio, 2000).

Apart from the structural and functional studies on the isolated complexes, the distribution and diffusion of these complexes in intact thylakoid membranes has also been explored and thought to be functionally significant. Lateral heterogeneity in distribution of photosystems in thylakoids membranes of spinach chloroplast, and radial asymmetry in distribution of PSII, PSI and ATP synthase in photosynthetic membranes of cyanobacteria is well established (Andersson and Andersson, 1980; Sherman *et al.*, 1994). Cytochrome b_6f complex is uniformly distributed throughout the thylakoids. Separation of the two pigment-protein systems is thought to be important in preventing unregulated excitation energy flow between them. Without this, PS I, which is much faster than PS II, would disturb the balance of energy distribution between the two photosystems. Fluorescence recovery after photobleaching (FRAP) and confocal laser scanning microscopy (Mullineaux, 2004) have shown that phycobilisomes, the accessory pigment protein complexes of cyanobacteria, are also mobile.

ENVIRONMENT WITHIN CHLOROPLASTS

The environment inside the chloroplasts is unusual. The soluble protein concentration in the stroma is around 400 mg/ml, out of which around 250 mg/ml alone is due to RuBisCO (Robinson and Walker, 1981). A dense packing of large protein clusters was observed in the stroma portion of the chloroplast using cryo-scanning electron microscopy, suggesting that there should be a tight packing of stromal proteins, minimizing the random mobility (Süss and Sainis, 1997). However, in contrast to membrane proteins, research on associations of soluble proteins did not find much appreciation. The stroma harbours enzymes of several metabolic pathways such as oxidative pentose-phosphate cycle, nitrate and ammonia assimilation, amino acid and fatty acid biosynthesis, ribosomes and the entire protein synthetic machinery along with Calvin cycle enzymes. In addition, multiple copies of DNA, mRNA, starch grains etc. also reside in the chloroplast. Furthermore, Dilley and Vernon (1965) had shown that chloroplasts show reversible changes in volume in response to light conditions, resulting in changes in ion composition and water content of stroma. The ultrastructure of thylakoid membranes was also shown to change in response to light affecting the rates of electron transport and photophosphorylation (Miller and Nobel, 1972). Thus exposure to light was found to result in an increase in macromolecular crowding in chloroplast and a reduction in free water. These light dependent changes in the ultrastructure of chloroplast would in turn regulate photosynthetic activities probably through changing the dynamic associations of metabolic proteins. The efficiency of metabolism will be difficult to maintain in this scenario unless there is an organization among sequential enzymes, which will facilitate the sequestering of desired metabolites (Wolosiuk *et al.*, 1993).

Therefore, the concept that the components of light and dark reaction are mutually exclusive in function due to their exclusive spatial location needs to be revisited since the real situation *in vivo* is far more complex than expected from *in vitro* experiments. Though according to the third law of thermodynamics, entropy of the physical world always increases, the situation is reversed in living systems, where organization or decrease in entropy are the key features of their survival.

SOLUBLE SUPERCOMPLEXES IN CHLOROPLASTS

Chloroplasts are capable of utilizing light energy for the synthesis of several organic compounds using oxides of carbon, sulphur, phosphorus etc. Many of the enzymes involved in these processes have been isolated from stroma of chloroplasts and are considered as functioning independently of each other. Among these metabolic pathways in chloroplasts, enzymes of Calvin cycle have been extensively investigated, albeit mainly in isolation. The existence of a CO_2 fixing complex was predicted in the sixties (Van Noort and Wildman, 1964). Muller (1972) showed that some of the CO_2 fixing enzymes might be associated in the form of a labile complex. Several years later, the associations among Calvin cycle enzymes were discovered by a variety of procedures. Sainis and Harris (1986) and Sainis *et al.* (1989) observed the association of phosphoriboisomerase (RPI) and phosphoribulokinase (RPK) with RuBisCO on a sucrose gradient. Gontero *et al.* (1988) purified a functional five enzyme complex of the consecutive enzymes of Calvin cycle, viz., RPI, RPK, RuBisCO, PGK and GAPDH by using DEAE Tris-acryl, Sephadex G-200 and hydroxyapetite. Persson and Johansson (1989) had reported the partition behaviour of six Calvin cycle enzymes viz. RuBisCO, PGK, GAPDH, TPI, aldolase and FBPase using countercurrent distribution in the aqueous two phase system, which suggested a trend to exist as a protein-protein complex among these enzymes. Nicholson *et al.* (1991) reported a stable complex between GAPDH and RPK from chloroplasts. Süss *et al.* (1993) were able to isolate a multienzyme complex containing RPI, RPK, RuBisCO, GAPDH, sedoheptulose-1,7-bisphosphatase and also Ferredoxin NADP reductase (FNR) using molecular sieve and anion exchange chromatography. Mouche *et al.* (2002) used a multi-technique approach to study multienzyme complex of GAPDH and RPK. This bi-enzyme complex uses ATP and NADPH produced by the primary reactions in photosynthesis.

Recently nearest-neighbour analysis was used to study co-localization of several Calvin cycle enzymes in the stroma of pea chloroplasts. These studies indicated that carbonic anhydrase, RPK and PGK occur in the neighbourhood of RuBisCO (Anderson and Carol, 2004); GAPDH, triose-P isomerase and aldolase are also located close to one another; aldolase is located close to SBPase (Anderson *et al.*, 2005); and transketolase, xylulose-5-P 3-epimerase and RPI exist near each other (Anderson *et al.*, 2006).

Thus, there is adequate data from these *in vitro* studies to demonstrate that many of the sequential Calvin cycle enzymes can be isolated as supramolecular complexes. Not much information, however, is available about the supercomplexes of enzymes functioning in chloroplasts in other metabolic pathways (Hrazdina and Jensen, 1992).

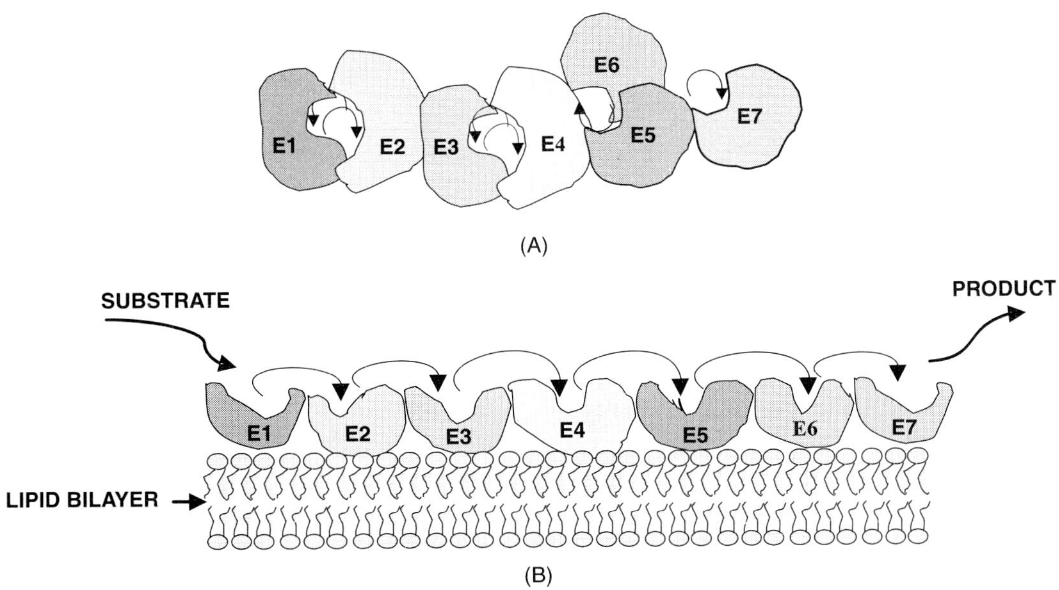

Fig. 2.1 Molecular architecture of multienzyme organization; **A:** Tight coupling of active sites of sequential enzymes of a metabolic pathway. The tight coupling of active sites of sequential enzymes will not allow the intermediates to diffuse to sequential enzymes of a pathway. **B:** Thin layer of sequential enzymes of a metabolic pathway on the surface of membranes. The thin layer of sequential enzymes on the surface of membranes would facilitate channelling of intermediates among sequential enzymes and avoid unwanted diffusion of substrates.

MEMBRANE ASSOCIATION OF SOLUBLE ENZYMES

Such multienzyme supercomplexes would have a diffusion limitation within the intracellular soluble phases owing to their large sizes. The traditional concept of multienzyme organization revolves around the notion of interlocking the active sites of sequential enzyme (Fig. 2.1 A). This situation may not be very useful *in vivo* as the intermediates have to diffuse to the consecutive sequential enzymes and yet not disseminate all over in the soluble phases in the cell. This makes it plausible to hypothesize that sequential enzymes would be structured to optimize the diffusion of intermediates. If organized as a thin layer on membranes, the membrane will provide partially non-aqueous environment, thereby reducing the superfluous diffusion of intermediates and also helping in channelling among sequential enzymes (Fig. 2.1 B). In case of Calvin cycle, besides CO_2; ATP and NADPH, which are produced by the components of light reactions residing in thylakoid membranes are also required. Hence, if the Calvin cycle enzyme supercomplexes are positioned next to the machinery of light reactions, ATP and NADPH would be easily and efficiently channelized for photosynthetic carbon assimilation pathway. Additionally, thylakoid membranes may act as anchoring surfaces or a guide for organization of sequential enzymes of Calvin cycle. Figure 2.2 shows immunolocalization of RuBisCO in *Synechococcus* (Fig. 2.2 A) cells and in the dimorphic chloroplasts of maize (Fig. 2.2 B and C). About 70-80% of gold particles were observed to be located near thylakoid membranes. This was true for several other soluble enzymes of Calvin cycle and other metabolic pathways. However, very few investigations were done on the association of Calvin cycle enzymes with isolated thylakoid membranes. In 1967, Howell and Moudrianakis reported the structure and properties of membrane-bound particle that are active in the dark reactions of photophosphorylation. The dual localization of many of these enzymes was shown by Mori *et al.* (1984) and Hermoso *et al.* (1992). The intricate and specific nature of these unusual interactions was subsequently evaluated by Raghavendra *et al.* (1981) and Wolosiuk *et al.* (1993). Süss *et al.* (1993) and Sainis *et al.* (2003) predicted the involvement of thylakoid membranes in the organization of Calvin cycle enzymes.

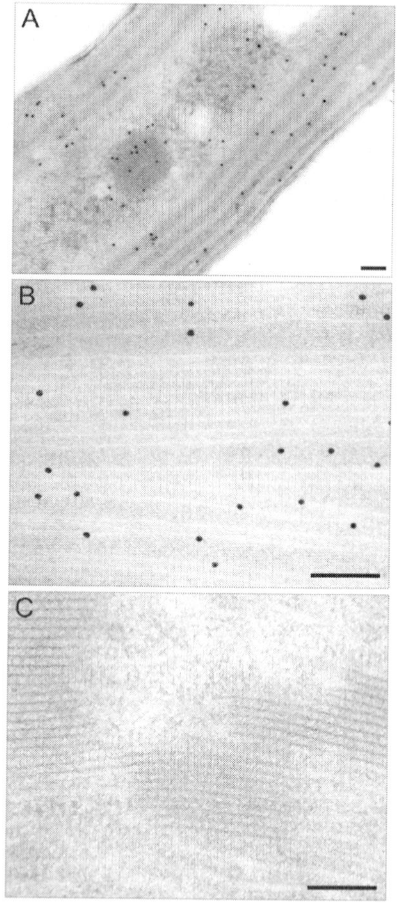

Fig. 2.2 Immunogold localization of RuBisCO in *Synechococcus* cells and maize chloroplasts. Predominant immunogold localization of RuBisCO is seen adjacent to the thylakoid membranes (A) in *Synechococcus* cells (5 nm protein-A gold) and (B) in maize Bundle Sheath chloroplasts (10 nm protein-A gold). (C) Mesophyll chloroplasts of maize did not show any labelling. Bar = 100 nm.

However, in all these investigations, the involvement of thylakoids was inferred indirectly. It was necessary to show that indeed thylakoid membranes can be isolated with Calvin cycle enzymes and the cycle can function using ATP and NADPH produced by electron transport in light. To address this hypothesis, we chose cyanobacteria as the model system. The thylakoids from higher plants are known to show heterogeneity, seen as stroma and grana lamellae. These spatially separated thylakoids have differential distribution of the two photosystems, with PSI present preferentially in stroma lamellae and PSII located in appressed regions of the grana (Trissl and Wilheim, 1993). Cyanobacteria are photosynthetic prokaryotes having similarity in pigment protein complexes to higher plants. However, there are several dissimilarities. The prominent among them is that there is no stacking of thylakoid membranes in these organisms and parallel unstacked thylakoid membranes lie in the cytosol. There are a few reports on non-uniform distribution of photosystems in cyanobacterial thylakoids. It was observed that the native thylakoids from cell-free extracts of cyanobacteria could be fractionated by successive ultracentrifugation at 40,000×g, 90,000×g and 150,000×g (Dani and Sainis, 2005). Among these, the thylakoid membrane fraction sedimenting at 150,000×g (150k fraction) showed the presence of five sequential Calvin cycle enzymes viz. phosphoriboisomerase, phosphoribulokinase, ribulose-1,5-bisphosphate carboxylase/oxygenase, 3-phospho glycerate kinase and glyceraldehyde-3-phosphate dehydrogenase in addition to chlorophyll *a* and phycocyanins. This thylakoid fraction showed linked activities of all the five enzymes measured as R-5-P+ATP dependent GAPDH activity (Dani and Sainis, 2005; Agarwal *et al.*, 2009). It also showed another novel linked activity in which CO_2 fixation could be measured in light if provided with R-5-P + ADP + Pi. This was termed photophosphorylation dependent carboxylation in which the ATP produced by light reactions was utilized by the associated phosphoribulokinase to generate RuBP, the substrate for CO_2 fixation (Dani and Sainis, 2005). The 150k fraction showed the highest rates of these linked activities compared to other thylakoid fractions sedimenting at 40,000×g, 90,000×g (40k and 90k fractions). Thus, the 150k fraction appeared as a unique thylakoid fraction having all the components of electron transport chain and ATP synthase as well as the five sequential enzymes of Calvin cycle, viz., phosphoriboisomerase, phosphoribulokinase, RuBP carboxylase, 3-phosphoglycerate kinase and glyceraldehyde-3-phosphate dehydrogenase along with soluble proteins such as phycobilisomes, ferredoxin-NADP reductase and CP12.

Recently the existence of Calvin cycle enzymes in the neighbourhood of thylakoids was corroborated by Agarwal *et al.* (2009) using immunoelectron microscopy. Immunogold labelling of this fraction showed the presence of the above five enzymes as well as the catalytic portion of the chloroplast H^+-ATP synthase in this membrane fraction. Notably the gold labelling of this thylakoid membrane fraction was observed even without the use of cross linkers and in spite of the normal washing steps used during standard immunolabelling (Agarwal *et al.*, 2009). The recent proteomic analysis has also substantiated these findings (Agarwal *et al.*, 2010).

EVOLVING THE CONCEPT OF THE PHOTOSYNTHETIC MODULE

The results of thylakoid fractionation showed that the thylakoids in cyanobacteria are heterogeneous in composition and the enzymes of Calvin cycle of photosynthesis may be organized as supercomplexes on or near to the particular thylakoid membranes *in vivo.* Using simple biochemical cell fractionation techniques, we could isolate a thylakoid fraction, which showed photophosphorylation dependent CO_2 fixation. This fraction, therefore, can be viewed as a functional photosynthetic module or photosynthesome (Dani and Sainis, 2004). The photosynthesome represents a functional module of photosynthesis where the components of the photosynthetic apparatus would come together in a specific manner to facilitate the coupling between photosynthetic electron transport and CO_2 fixation for an enhanced efficiency and more sensitive regulation of their function. Such an organization will also minimize unwanted water diffusion into the enzyme layer, thus creating a non-polar microenvironment to assist in enzyme catalysis.

These observations propose a paradigm shift in biology, leading to exploration of supramolecular organizations of functionally related molecules, regarded as modules. A module is an organization of macromolecules performing a synchronous function in a given metabolic pathway. The complex metabolic reactions occurring inside the chloroplasts can be best described by the perception of the molecular architecture in modular fashion. Figure 2.3 A and B show the modular organization of different metabolic pathways in chloroplasts. Large metabolic networks can be imagined as a web of several interconnected modules, which may or may not influence each other. Elucidating the physical associations and the dynamics of the interactions in these predicted networks of modules would be a challenge to the forthcoming technologies.

Networking of biochemical reactions by specific spatio-temporal associations of participating components will answer several fundamental questions in biology. It will explain why several variants of the same proteins, generated by multi-gene families, isozymes, isoforms, alternate splicing of genes or variety of post-translational modifications are required. Single nucleotide polymorphisms may also result in minor structural variations in proteins. The necessity for interaction with other proteins may drive the need for such modifications without affecting the functionality. Several sites on the surface of enzyme would be required for generation and maintenance of 3-D superstructures. Variations in protein conformation will be tolerated as long as they do not affect the function of a given module, putting a limitation on toleration of mutations. Networking of proteins in modules may also explain the gene silencing that occurs in transgenics due to overexpression of a gene, resulting in suppression of a given phenotype. Overexpression of one of the proteins in the network may result in differential titration of interacting partners, and hence in inhibition of the phenotype. Modular biology will help to understand the phenomenon of homeostasis where a perfect balance is maintained between anabolism and catabolism, synthesis and destruction of proteins, and other macro- and micromolecules. Modules would result in caging and isolation of intermediates to avoid unwanted reactions in living cells. Such a microenvironment will explain how several competing and contrasting metabolic reactions occur simultaneously in a crowded atmosphere *in vivo.* The differential degree of recruitment of modules will explain the phenomenon of up and down regulation of metabolic processes as also metabolic compensation. Thus metabolic regulation may have to be looked from a different perspective in the background of modularity.

The finer details of the molecular architecture of modules, their interaction and organization may decide the final form of an organism. This will explain the fact that though evolutionarily, there are similarities in the gene sequences of several proteins in a variety of living organisms, the final forms of these organisms are quite different.

Modular biology offers an intellectually demanding perception of metabolism and will amalgamate the conventional biochemical studies and the recent high-throughput technologies to understand the hierarchy of networks and molecular architectures and their dynamism in biological systems. This will test the limits of old and new technologies in biology. A leap in faith in imagination and technology would be required to conceptualize the modular molecular architectures *in vivo* for precise functions and regulation of metabolic pathways.

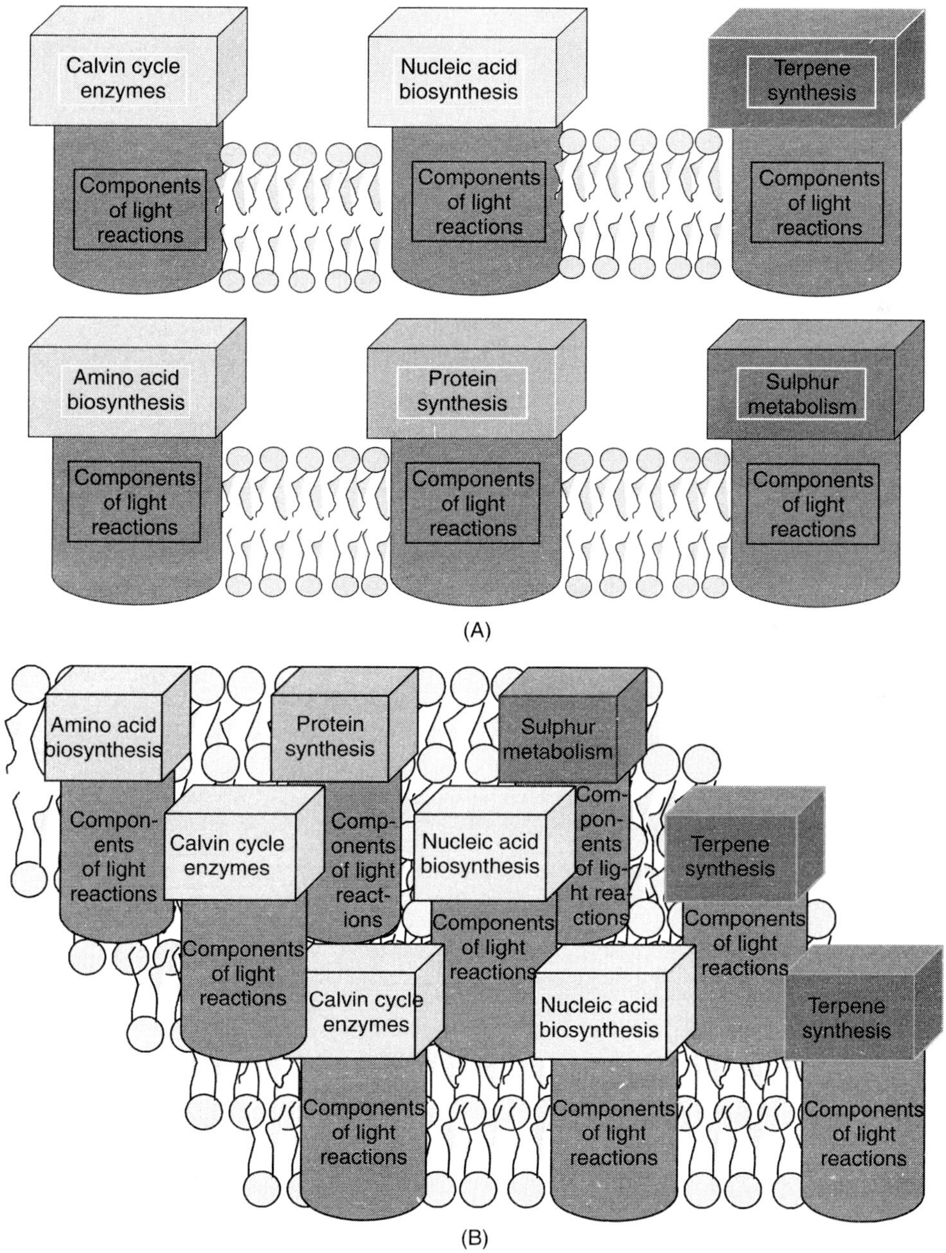

Fig. 2.3 Modular organization. **A** shows modules of different metabolic pathways requiring energy inputs generated from electron transport. A module is a discrete structural and functional entity, an ensemble of physically interacting proteins, which deal with a specific metabolic process. The composition, location and function of a module would vary depending on cellular requirements. Modules thus represent a dynamic entity of a group of molecules that occur and function together. The modules may exist as isolated structure or may be interconnected *in vivo*, forming a network (Hartwell *et al.*, 1999) **(Figure 2.3 B)**. Functional properties of modules would arise from the properties of the underlying components and their spatial interactions. The modules would involve only a small fraction of the total molecules. Different molecules of same protein may belong to different modules, at different time and cellular locations. Small structural differences in the enzyme molecules will be significant for their involvement in different modules. Alternately, same protein may exhibit difference in function depending on the environment in the module. Change in orientation or activity of individual components would disturb the harmony in function of a module.

REFERENCES

Agarwal R, Ortleb S, Sainis J, Melzer M (2009) Immunoelectron Microscopy for Locating Calvin Cycle Enzymes in the Thylakoids of Synechocystis 6803. Molecular Plant 2(1): 32-42

Agarwal R, Matros A, Melzer M, Mock H, Sainis J (2010) Heterogeneity in thylakoid membrane proteome of Synechocystis 6803. J Proteomics 73: 976-991

Andersson B, Anderson J. (1980) Lateral heterogeneity in the distribution of chlorophyll-protein complexes of the thylakoid membranes of spinach chloroplasts. Biochim Biophys Acta 593: 427-440

Anderson L, Carol A (2004) Enzyme co-localization with rubisco in pea leaf chloroplasts. Photosynthesis Res 82: 49-58

Anderson L, Gatla N, Carol A (2005) Enzyme co-localization in pea leaf chloroplasts: glycerladehyde-3-P dehydrogenase, triose-P-isomerase, aldolase and sedoheptulose bis phosphatase. Photosynthesis Res 83: 317-328

Anderson L, Chrostowski J, Carol A (2006) Enzyme co-localization with transketolase, xylulose-5-P 3-epimearse and phosphoriboisomerase in pea leaf chloroplasts. Plant Science 171: 686-698

Boekema E, Boonstra A, Dekker J, Rögner M (1994) Electron microscopic structural analysis of photosystem I photosystem II and the cytochrome b_6f complex from green plants and cyanobacteria. J Bioenerg Biomembr 26: 17-29

Dani D, Sainis J (2004) Putative photosynthesome from *Anacystis nidulans*: A supermolecular organization of the photosynthetic apparatus. In: A. van der Est and D. Bruce ed. Photosynthesis: Fundamental Aspects to Global Perspectives 779-780.

Dani D, Sainis J (2005) Isolation and characterization of a thylakoid membrane module showing partial light and dark reactions. Biochim Biophys Acta 1669(1): 43-52

Dekker J, Boekema E (2005) Supramolecular organization of thylakoid membrane proteins in green plants. Biochim Biophys Acta 1706: 12-39

Dilley R, Vernon L (1965) Ion water transport processes related to light dependent shrinkage of spinach chloroplasts Arch Biochem Biophys 111: 365-375

Eshaghi S, Andersson B, Barber J (1999) Isolation of a highly active PSII-LHCII supercomplex from thylakoid membranes by a direct method. FEBS Lett 446: 23-26

Gontero B, Cardenas M, Ricard J (1988) A functional five-enzyme complex of chloroplasts involved in the Calvin cycle. Eur J Biochem 173: 437-443

Hartwell L, Hopfield J, Leibler S, Murray A (1999) From molecular to modular cell biology. Nature 402: C47-C52

Hermoso R, Fonolla J, de Felipe M, Vivo A, Chueca A, Lazara J, Lopez Gorge J (1992) Double immunogold localization of thioredoxin f and photosynthetic fructose-1, 6-bisphosphatase in spinach leaves. Plant Physiol Biochem 30: 39-46

Howell S, Moudrianakis E (1967) Function of the 'quantasome' in photosynthesis: Structure and properties of membrane-bound particle active in the dark reactions of photophosphorylation. Proc Natl Acad Sci USA 58: 1261-1268

Hrazdina G, Jensen R (1992) Spatial organization of enzymes in plant metabolic pathways. A Rev Plant Physiol Plant Molec Biol 43: 241-267

Krauss N, Hinrichs W, Witt I, Fromme P, Pritzkow W, Dauter Z, Betzel C, Wilson K, Witt H, Saenger W (1993) Three-dimensional structure of system I of photosynthesis at 6 Å resolution. Nature 361: 326-331

Lagoutte B, Mathis P (1989) The photosystem I reaction center: structure and photochemistry. Photochem Photobiol 49: 833-844

Miller M, Nobel P (1972) Light induced changes in the ultrastructure of pea chloroplasts *in vivo*. Plant Physiol 49: 535-541

Mori H, Takabe T, Akazawa T (1984) Loose association of ribulose-15-bisphosphate carboxylase/oxygenase with chloroplast thylakoid membranes. Photosynthesis Res 5: 17-28

Mouche F, Gontero B, Callebaut I, Mornon J, Boisset N (2002) Striking conformational change suspected within the phosphoribulokinase dimer induced by interaction with GAPDH. J Biol Chem 277: 6743-6749

Muller B (1972) A labile CO_2 fixing enzyme complex from spinach leaves. Z Naturforsch 27B: 925-932

Mullineaux C (2004) FRAP analysis of photosynthetic membranes. J Expt Bot 55: 1207-1211

Nelson N, Ben-Shem A (2004) The complex architecture of oxygenic photosynthesis. Nat Rev Mol Cell Biol 5: 971-982

Nicholson S, Easterby J, Powls R (1991) Properties of multimeric protein complex from chloroplasts possessing potential activities of NADPH dependent glyceraldehyde-3-phosphate dehydrogenase and phosphoribulokinase. Eur J Biochem 162: 423-431

Nilsson F, Kleoniki G, Styring S, Andersson B (1992) Isolation and characterization of oxygen-evolving. Photosystem II membranes from the cyanobacterium *Synechocystis* 6803. Biochim Biophys Acta 1100: 251-258

Persson L-O, Johansson G (1989) Studies of protein-protein interaction using countercurrent distribution in aqueous two-phase systems. Partition behavior of six Calvin cycle enzymes from a crude spinach (*Spinacia oleracea*) chloroplast extract. Biochem J 259: 863-870

Raghavendra A, Carrillo N, Vallejos R (1981) The pattern and characteristics of membrane bound Ribulose bisphosphate carboxylase in spinach chloroplasts. Proceedings 5th Int. Congr. Photosynthesis. IV G. Akoyunoglou eds. (Balban International Science Services) pp 31-38

Robinson S, Walker D (1981) Photosynthetic carbon. reduction cycle. In: Hatch MD Boardman NK eds. The Biochemistry of Plants in A Comprehensive Treatise. New York Academic Press, Vol 8. 193-236

Sainis J, Harris G (1986) The association of Ribulose-1, 5-bisphosphate carboxylase with phosphoriboisomerase and phosphoribulokinase. Biochem Biophys Res Comm 139: 947-954

Sainis J, Merriam K, Harris G (1989) The association of Ribulose-1, 5-bisphosphate carboxylase/oxygenase with phosphoribulokinase. Plant Physiol 89: 368-374

Sainis J, Dani D, Dey G (2003) Involvement of thylakoid membranes in supramolecular organization of Calvin cycle enzymes in *Anacystis nidulans.* J Plant Physiol 160: 23-32

Sherman D, Troyan T, Sherman L (1994) Localization of membrane proteins in the cyanobacterium *Synechococcus* sp. PCC 7942. Plant Physiol 106: 251-262

Staehlin L, DeWit M (1984) Correlation of structure and function of chloroplast membranes at the supramolecular level. J Cell Biochem 24: 261-269

Süss K-H, Arkona C, Manteuffel R, Adler K (1993) Calvin cycle multienzyme complexes are bound to chloroplast thylakoid membranes of higher plants *in situ*. Proc Natl Acad Sci USA 90: 5514-5518

Süss K-H, Sainis J (1997) Supramolecular organization of water-soluble photosynthetic enzymes in chloroplasts in Handbook on photosynthesis Pessarakli M. eds. (Marcel & Dekker Inc. New York Basel Hong Kong) pp 305-314

Tang X-S, Diner B (1994) Biochemical and spectroscopic characterization of a new oxygen-evolving photosystem II core complex from the cyanobacterium *Synechocystis* 6803. Biochemistry 33: 4594-4603

Trissl H, Wilheim C (1993) Why do thylakoid membranes from higher plants form grana stacks? Trends Biochem Sci 18: 415-419

Van Noort G, Wildman S (1964) Proteins of green leaves IX. Enzymatic properties of Fraction-1 protein isolated by specific antibody. Biochim Biophys Acta 90: 309-317

Vermeglio A, Joliot P (2002) Supramolecular organization of the photosynthetic chain in anoxygenic bacteria Biochim Biophys Acta 1555: 60-64

Wolosiuk R, Ballicora M, Hagelin K (1993). The reductive pentose phosphate cycle for photosynthetic CO_2 assimilation: enzyme modulation. The FASEB J 634: 622-637

Zolla L, Timperio A-M (2000) High performance liquid chromatography-electrospray mass spectrometry for the simultaneous resolution and identification of intrinsic thylakoid membrane proteins. Proteins: Structure Function and Genetics 41: 398-406

CHAPTER 3

State Shifts in Photosynthetic Organisms: Tracking Traits and Techniques

Prasanna Mohanty[1, 3, 4], N Sreedhar[2], Sujata Mishra[1] and Subramanyam Rajagopal[2, 4]

[1]Regional Plant Resource Center, Bhubaneswar, India

[2]Department of Biochemistry, University of Hyderabad, India

[3]Jawaharlal Nehru University, New Delhi, India

[4]Department of Plant Sciences, University of Hyderabad, Hyderabad, India

ABSTRACT

State transitions, in general, refer to the short term physiological adaptation by the photosynthetic organisms to adjust to the changing light conditions in the environment and efficiently sharing the absorbed light energy by photosystem I (PS I) and photosystem II (PS II) under limiting light conditions. State transition or state shift occurs by well defined mechanism(s) in plants, algae, and cyanobacteria by one or other mechanisms, defined as spillover and/or change in absorption cross section in the proximity of PSI and PSII called α change. Besides balancing of absorbed excitation between the two photosystems, the state transition also serves as a regulatory valve between light harvest and its optimization in the assimilatory processes through redox controlled poising of cyclic and non-cyclic electron transfer. Here we discuss some of the landmarks on the discovery and the mechanisms of state transitions. The mechanism of state transition has been elucidated for green plants and cyanobacteria. Attention has been paid to the differences in light harvesting mechanism between the prokaryotic oxygenic photosynthetic organism (cyanobacteria), green plants and red algae. Some of their mutants widely used in state transition or state shift studies have been highlighted. Mutants generated in *Arabidopsis* and *Chlamydomonas* related to state transition strengthened the concepts of state transition. Many techniques played crucial role in the discovery of state transitions. Apart from chl *a* fluorescence spectroscopy, some other techniques like kinetics of oxygen evolution, photoacoustic techniques, atomic force microscopy (AFM), fluorescence recovery after photobleaching (FRAP), single particle electron microscopy, and fluorescence video imaging have contributed richly to our understanding of state transition. These techniques have been briefly discussed in this review. Recent studies on structural changes in thylakoid membranes and the change in absorption cross-section of PSI and PSII supercomplexes during state shift have also been discussed. In this review we also detailed the significance of cations and anions in determining the structure of thylakoid membranes (granal stacking) during state shift. The main focus of this review is to highlight the landmarks of state shift, methods and techniques that unfolded the mechanism of this vital energy regulating as well as signaling process.

Keywords: State shift, PQ pool, LHC II antenna, Chl *a* fluorescence, Cyt b_6f.

Abbreviations

PA: Photoacoustic, *NPQ:* Non-photochemical quenching, *FRAP:* Fluorescence recovery after photobleaching, *FTIR: Fourier transform infrared spectroscopy.*

INTRODUCTION

The quality and quantity of electromagnetic radiation varies depending on the dynamic behavior of atmosphere and climate. Daylight, the combination of sunlight and skylight (blue sky), is the natural source for sustenance of living organisms especially oxygenic photosynthetic organisms, which convert light energy into chemical energy by a process called photosynthesis (Blankenship, 2002). In general, the daylight spectral distribution peaks at different wavelengths depending on the time of the day. The spectral distribution of blue-sky peaks at 400 nm (Rechtsteiner and Ganske, 1998). When the sun is low in the sky (sunrise and sunset), spectral distribution has less intensity in the blue end of the visible region and more in the red region. In the sunset spectrum, the red "dips" are prominent at any given time. The quantity of this red light reaching the atmosphere is not always the same (Rechtsteiner and Ganske, 1998) and atmospheric gases play a major role in determining the quality and quantity of the electromagnetic spectrum reaching the earth.

In oxygenic photosynthetic organisms, the conversion of light energy into chemical energy, involving photooxidation of water and reduction of CO_2 to carbohydrates is mediated by two interactive photochemical reactions named photosystem I (PSI) and photosystem II (PSII). These two photosystems are continuously being exposed to different regimes of sunlight. The availability of light for crop plants, woody trees and epiphytes is significantly higher when compared to the shade loving plants. The availability of solar radiation is poor for shade loving plants, and the light reaching these plants is enriched with far red light and less of blue light (Green and Parson, 2003). The reaction centers (RC) of PSII and PSI, designated as P680 and P700, respectively, are different. The two photosystems absorb solar radiation differentially resulting in unbalanced energy absorption by two photosystems which require a mechanism by which the absorbed energy can be balanced for optimal photosynthesis.

The light harvesting complex II (LHCII), which acts as a major light harvesting antenna of PSII, is enriched in Chl *b* and less Chl *a* pigment, most of which is absorbed in 670–676 nm region (Van and Novoderezhkin, 2006). The PSI antenna, on the other hand, has less number of Chl *b* and it possesses a number of red Chl *a* pigments with absorption expanding to 690–710 nm region. This difference dictates a different spectral design of the respective antenna systems resulting in the PSII antenna to have more blue absorption relative to the PSI. Earlier studies have shown that PSII exhibits maximal absorption in blue, red region and less absorption in far red region, whereas PSI shows maximal absorption at red and far red region and less at blue region of electromagnetic spectrum. This differential absorption of PSI and PSII makes differential excitation leading to the imbalance in energy distribution. So, all types of plants, green algae, red algae and cyanobacteria have evolved a mechanism by which equal distribution of absorbed energy occurs between the two photosystems to ensure the optimal photosynthetic electron flow (Fig. 3.1) and thus, optimization of the quantum yield of photosynthesis (Green and Parson, 2003).

These oxygenic photosynthetic organisms have different types of antennae, but similar electron transport carriers and mechanisms for electron transport, phosphorylation and CO_2 fixation (Fig. 1). The

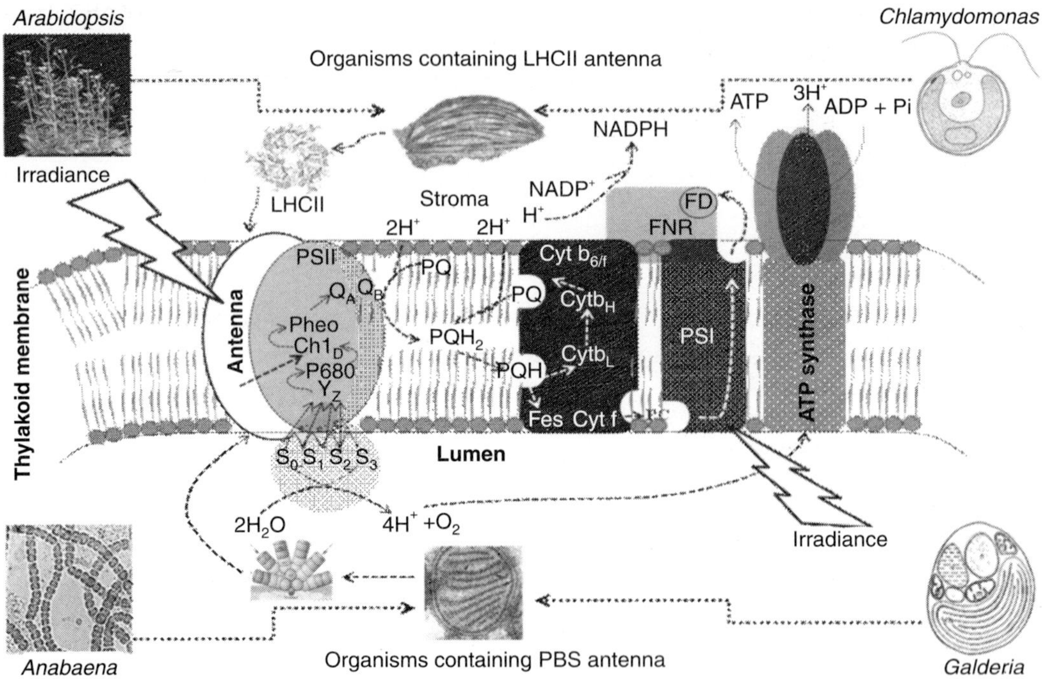

Fig. 3.1 Sketch of the common electron transport chain shared in oxygenic photosynthetic organisms having different types of light harvesting antennae (PBS or LHCII).

redox state of plastoquinone pool (PQ) represents signaling element for state transition, although redox states of other carriers have been suggested for the signal transduction mechanisms of state shift in green plants, green algae, red algae and cyanobacteria (Allen, 2002; Mao *et al.*, 2002). Furthermore it has also been shown that the redox state of PQ pool, which regulates state shift, also correlated the differential gene expressions of plastidal and nuclear genes (Slesak *et al.*, 2002; Pfannschmidt 2003; Steinbrenner and Linden, 2003; Slesak *et al.*, 2003; Chang *et al.*, 2004). Thus state shift is coupled to regulated gene expression and development (Pfannschmidt, 2003; Zer and Ohad, 2003; Chang *et al.*, 2004).

The term "state transition" which has been coined by the discoverer is in vogue in photosynthetic literature. This term is also used in plant respiration and physics. We would like to introduce the term "state shift", which is the substitute for state transition in this review.

THE REDOX STATE OF PQ POOL AND PHOSPHORYLATION/ DEPHOSPHORYLATION OF LHCI! ANTENNA

The preferential excitation of PSII makes the PQ pool in the electron transport chain more reduced (Allen *et al.*, 1981; Horton *et al.*, 1981). These conditions favor the docking of PQH_2 to quinol oxidation (Qo) site (Fig. 3.1) of cytochrome b_6f complex, (Zito *et al.*, 1999; Mao *et al.*, 2002), which in turn activates an enzyme, thylakoid protein kinase Stn7 in higher plants and green algae (Stt7 is a ortholog of Stn7 in *Chlamydomonas*) (Pesaresi *et al.*, 2009). Redox activated kinase Stn7/Stt7 phosphorylates LHCII and gets associate with PSI, thereby increasing the absorption cross section (α) (Fork and Satoh, 1986) of

PSI complex (Allen, 1992c; Kargul *et al.*, 2003; Takahashi *et al.*, 2006; Lemeille *et al.*, 2009; Takahashi *et al.* (2006) have shown that Stt7 kinase gets associated with photosynthetic membrane complexes including LHCII, PSI and cytochrome b_6f complex. Reversible phosphorylation of LHCII and state shift are crucial for plant fitness under natural light conditions. The preferential excitation of PSI makes the PQ pool oxidized, which is the signal for the activity of phosphatase to dephosphorylate LHCII (Allen *et al.*,1981). Recently, it has been shown that thylakoid-associated phosphatase TAP38 is required for LHCII dephosphorylation and shift from state II to state I in *A. thaliana* (Pribil *et al.*, 2010).

The Mechanism of State Shift in Cyanobacteria and Red Algae

The mechanism of state shift has been well documented in higher plants (Fig. 3.2d); however, in other organisms like cyanoes and red algae it remains to be established. Figure 3.2 gives the mechanisms of various models. Similar to LHCII of green plants, the antenna phycobilisomes (PBS) play a crucial role in PBS containing organisms like cyanobacteria and red algae. Four models have been proposed to explain state shifts in these organisms:

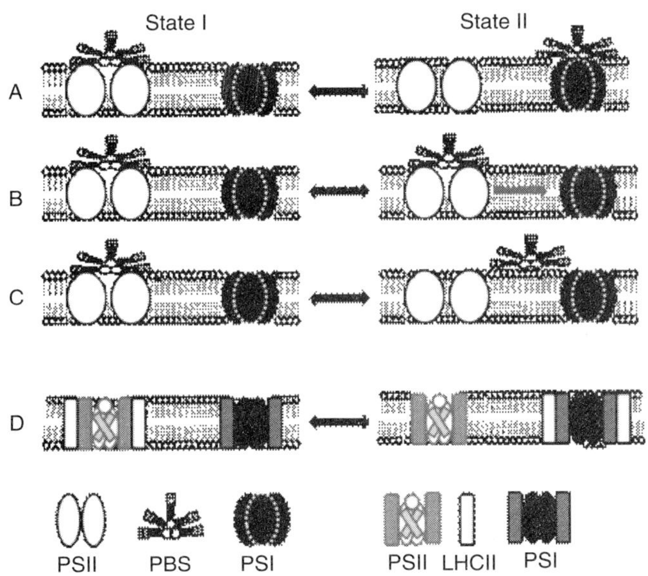

Fig. 3.2 Schematic representation of state shift models both in cyanobacteria and higher plants and green algae; (A-C models of cyanobacterial state shifts; A. In the mobile model PBS could diffuse on the surface of the membrane and reversibly migrate from PSII in state I to PSI in state II, B. Spillover model, excitation energy transferred to PSI from PSII; C. Dissociation model explains about partial dissociation of PBS and reassociation without docking to PSI; D. Model of higher plant and *Chlamydomonas* state shift, in which LHCII in the antennae shuttles between PSII and PSI.

1. The mobile model PBS could diffuse on the surface of the membrane and reversibly migrate from PSII in state I to PSI in state II (Allen *et al.*, 1985; Allen, 1992a; Mullineaux *et al.*, 1997) (Fig. 3.2a-c).
2. Spillover model, excitation energy transferred to PSI from PSII (McConnell *et al.*, 2002; Tsimilli *et al.*, 2009);

3. Dissociation model explains the partial dissociation of PBS and reassociation without docking to PSI.

Non-Photochemical Quenching (NPQ) and State Shift

As early as 1968, Papageorgiou and Govindjee, with their studies on Chl *a* fluorescence yield changes or quenching during fluorescence induction using cyanobacterium *Anacystis* (now *Synechosystis*) and green alga *Chlorella*, recognized the contribution of state shift to the yield changes or quenching process. This became evident when fluorescence yield changes occurred even in the presence of DCMU that prevented reduction and subsequent oxidation of PQ pool (Papageorgiou and Govindjee, 1968a, b). The fluorescence yield changes occur due to redox state of $Q_{A,}$ the primary electron acceptor of PSII and its oxidation by PQ pool via Q_B, the photochemical quenching or also by mechanisms not related to the oxidation of Q_A, the non-photochemical quenching (NPQ). NPQ is a protective mechanism against overexcitation and dissipates excess energy as heat. It occurs in almost all eukaryotic photosynthetic organisms and also in red algae and cyanobacteria (Papageorgiou, 2004). NPQ occurs *in vivo* under physiological conditions by three separate mechanisms, namely:

1. Energy dependent quenching caused by transthylakoid proton gradient acidification, also known as ΔpH dependent NPQ, relaxes very fast within seconds to minutes (Murata and Sugahara, 1969).
2. Photoinhibitory quenching related to photoinhibition (qI) and shows very slow relaxation kinetics (Krause and Behrend, 1986).
3. Quenching related to state shifts regulated by phosphorylation of LHCII, relaxes within tens of minutes (Bennett *et al.,* 1980; Horton and Black, 1980). NPQ related to state shifts (qT) seems to be very small in magnitude. Adjustment of energy distribution is important in low limiting light when qT may become the major qN component.

Recent studies on PSII reaction center quenching in high light, which is a zeaxanthin-independent pathway for dissipation of excess light energy and it may also play a significant role in photoprotection (Weis and Berry, 1987; Walters and Horton, 1993; Ivanov *et al.,* 2008). Reaction center quenching complements photoprotection through antenna quenching that relates to state shift.

LANDMARKS IN "STATE SHIFT" STUDIES

While tracking the path of progress in state shift in plants, algae and cyanos, one finds several landmarks, and we in this review discuss only a few of them. The readers may like to see some of the reviews, both past and present, on the history of state shift research (Fork and Satoh, 1986; Allen, 2002; Murata, 2009). The fundamental aspects of energy transfer mechanism in photosynthetic organisms were developed by Duysens (1952) based on the quantum mechanical theory of energy transfer mechanism proposed by Forster, which has laid the foundation for many concepts in photosynthesis (Forster, 1948); one of them being state transition. It has been demonstrated that differences in absorption crosssections of PSI and PSII lead to the imbalance in energy distribution, and this imbalance could be overcome by the state shift process.

Chl *a* Fluorescence and State Shift Signature

Chl *a* fluorescence emission is an important parameter that reveals the signature of the photosynthetic apparatus, its structure and function (Krause and Weis, 1991; Papageorgiou and Govindjee, 2004). Consequently, Chl *a* fluorescence induction (transients) and emission characteristics, both at room and low temperatures, have been used in monitoring state shifts. Particularly the low temperature spectral analyses are useful because at room temperature the fluorescence emission from PSI is not appreciable and most fluorescence emanates from PSII. Upon cooling to liquid nitrogen temperature (77K) both PSII and PSI emit appreciably and thus, the ratio of PSII to PSI emission is a quantitative parameter for monitoring the extent of state shift (Fig. 3.3).

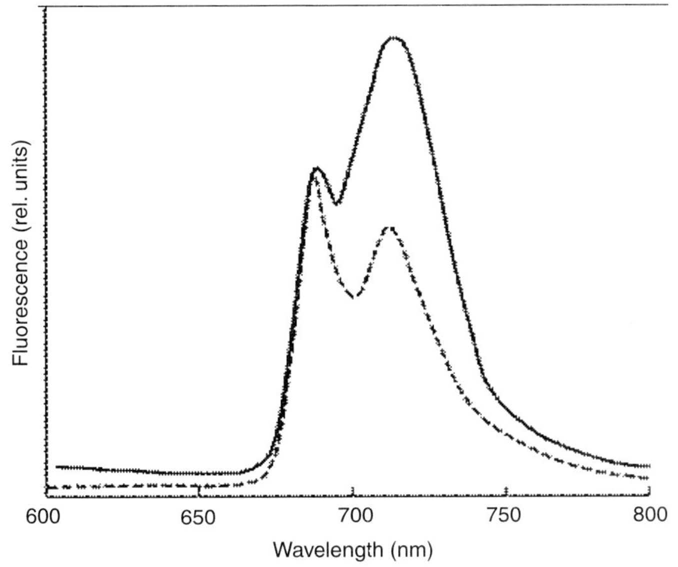

Fig. 3.3 Sketch of 77K emission spectra of thylakoid membranes under state I (dotted line) and state II (dark line) light conditions. In this sketch the spectra were normalized at F685.

Murata (1969a), using the red alga *Porphyridium cruentum,* found that pre-illumination of phycoerythrin pigment in the PBS excited at 433 nm decreased the yield of fluorescence emissions at 685 nm and 695 nm originating from PSII and increased at 715 nm emission originating from PSI (Fig. 3.4b). Pre-illumination of phycoerythrin excited at 533 nm yielded fluorescence emissions at 648 nm and 664 nm originating from PBS remained unchanged (Fig. 3.4a). Also on pre-illumination with 694 nm, preferentially excited Chl *a*, the fluorescence yields showed little change. He described this observation as "a control mechanism of excitation transfer between the two pigment systems" (Murata *et al.,* 1966; Murata, 1969a; Murata, 1969b). These series of observations suggested that light induced the control of excitation energy transfer, which enables the organisms to utilize absorbed light energy efficiently in oxygenic photosynthesis.

A similar conclusion was arrived at independently by Bonaventura and Myers (1969) using the green algae *Chlorella pyrenoidosa* and monitored both oxygen and fluorescence transients (Fig. 3.5). These authors used both light induced Chl *a* fluorescence and oxygen evolution. They observed that state I light condition was characterized by a low O_2 yield and high fluorescence yield, while in state II light condition, the oxygen yield was high and fluorescence yield was low. Upon switching on far red light to drive state II to state I and then back to state II upon switch off, there was a transient parallel increase in both oxygen and fluorescence yield. When the time dependent changes in fluorescence and oxygen yield in *Chlorella* were followed simultaneously, the kinetics was not always antagonistic or anti-parallel as one would expect from the notion that the two processes were always competitive. However, Bonaventura

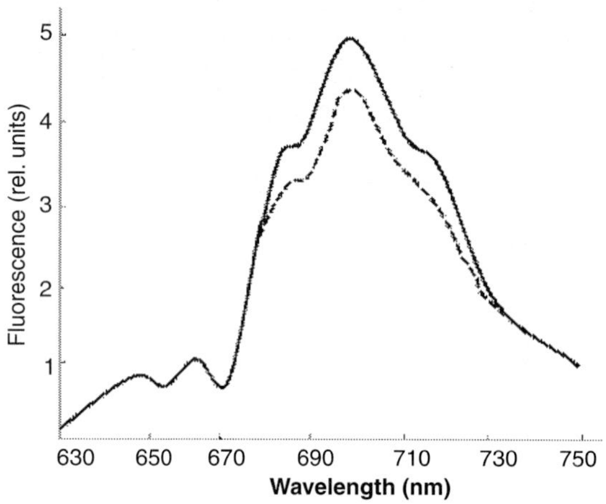

Fig. 3.4a Effect of pre-illumination at room temperature on the 77K emission spectrum of *Porphyridium*. Excitation light, 553 nm. —No pre-illumination; --- pre-illumination with 567 nm light (10 nE.cm^{-2}. sec^{-1}, 60 sec) (redrawn after Murata 1969a).

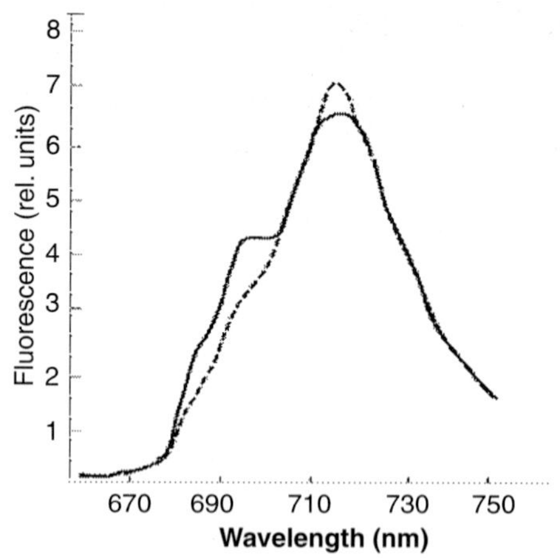

Fig. 3.4b Effect of pre-illumination at room temperature on the 77K emission spectrum of *Porphyridium*. Excitation light, 433 nm. —No pre-illumination; --- pre-illumination with 567 nm light (13 nE.cm^{-2}. sec^{-1}, 60 sec) (redrawn after Murata 1969a).

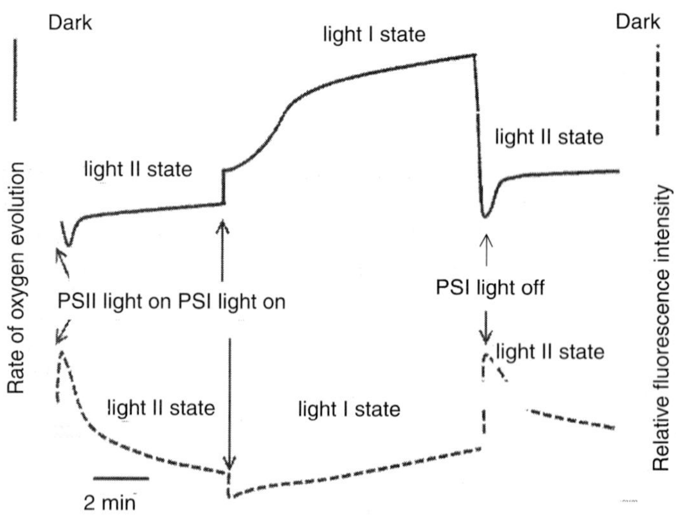

Fig. 3.5 Changes in the rates of oxygen evolution and chl *a* fluorescence during shifting from state II to state I and state I to state II on *Chlorella* (redrawn after Bonaventura and Myers 1969; Allen, 1992a).

and Myers (1969) observed both anti-parallel and parallel changes in both O_2 and fluorescence yields, suggesting the shifting of states (Fig. 3.4). Thus, from these two classic experiments of Murata and that of Bonaventura and Myers, the concept of energy sharing between the two photosystems was pitched and the state shift refers to change of the two systems from one adaptive state low energy sharing, state I to the high energy sharing, state II and each state has its characteristic features (Table 3.1).

Table 3.1 General characteristic features of state I and state II in higher plants and green algae.

Parameters	*State I*	*State II*
Oxygen yield	Low	High
Fluorescence yield	High	Low
Emerson enhancement effect	Low	High
Emission peak ratio of F730/F696 at 77K	Low	High
Fluorescence quenching by far red light	Low	High
NPQ (state shift dependent quenching)	High	Low
pH	Neutral or alkaline	Acidic
Thylakoid protein kinase activity	No	Yes
Phosphorylation of LHCII (thylakoid kinase)	Starts	Ends
LHCII trimerization	Trimerized	Monomerized
LHCII mobilization	No	Yes
Thylakoid phosphatase activity	High	Less
Dephosphorylation of LHCI	Ends	Starts
Granal stacking	Stacking of thylakoid membranes enhanced	Un-stacking of thylakoid membranes enhanced
Absorption cross section of PSI	Low	High
Absorption cross section of PSII	High	Low

Earlier Chl *a* fluorescence induction rise in DCMU poisoned *Chlorella* and *Synechocystis* also lead to the conclusion that fluorescence yield changes could occur independent of the redox state of stable electron acceptor Q_A of PSII, and this non-Q linked fluorescence yield changes were stated to be associated with intersystem energy transfer process similar to state shift (Fig. 3.6) (Papageorgiou and Govindjee, 1968a, b). It was observed that light induced fluorescence transient excited by PSII light showed different extent of quenching by low intensity far red PSI light during the slow phase of fluorescence induction curve in *Chlorella* (Fig. 3.7) (Mohanty *et al.,* 1970). The on and off phase of PSI light application during the slow phase of the fluorescence transient exhibited state II to state I type fluorescence fluctuations (Fig. 3.7).

Antenna Phosphorylation and State Changes

As stated earlier, the PSII absorbs more of photosynthetic active radiation (PAR) from sunlight than PSI because of relatively more abundance of accessory pigments in their antennas resulting in a downhill energy transfer from PSII to PSI (Haldrup *et al.,* 2001). The actual mechanism of energy transfer from PSII to PSI was proposed to be by two processes called α change (sharing of antennae) and spillover. Bennett demonstrated a light induced activation of phosphorylation of light harvesting chlorophyll protein complex by thylakoid protein kinase in algal thylakoids (Bennett, 1977; 1979a,b). Both redox state of PQ (Allen *et al.,* 1981) and cytochrome b_6f were involved in phosphorylating chl *a/b* complex (Bennett

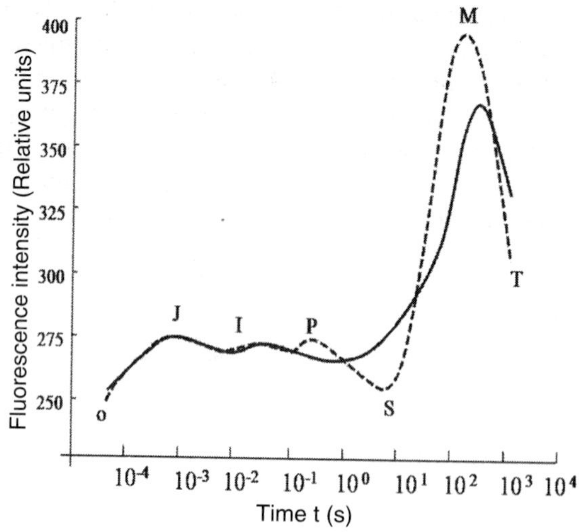

Fig. 3.6 The fluorescence transient of DCMU-treated *Synechococcus* sp. PCC 7942 cells (dark line) and control cells (dotted line) plotted on logarithmic time scale (adapted from Tsimilli *et al.*, 2009).

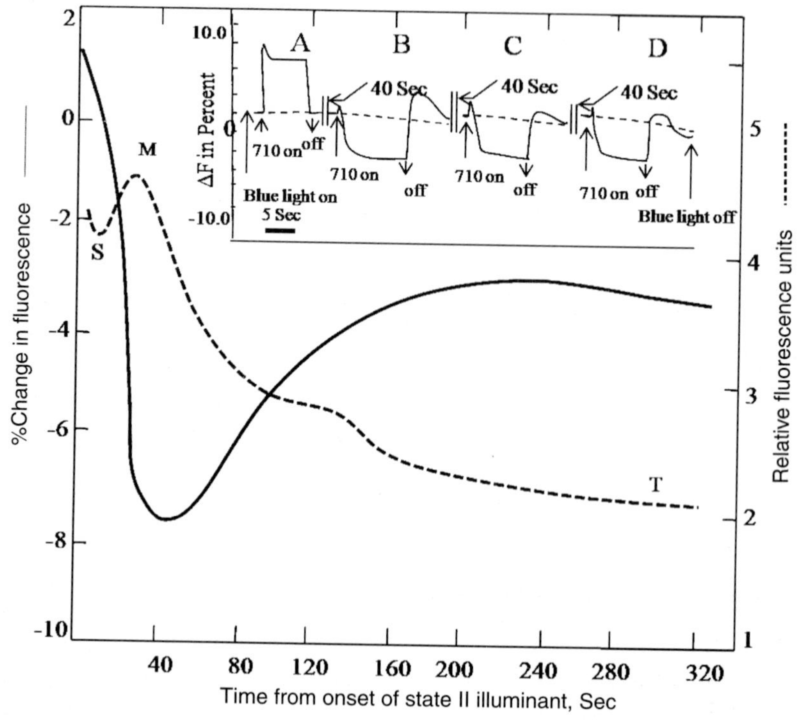

Fig. 3.7 The differential extent of quenching caused by the quenching far red (710 nm) PSI light during OPSMT fluorescence transient in *Chlorella* (adapted from Mohanty *et al.*, 1970).

et al., 1980, 1987, 1988). Either the presence of PSII inhibitor DCMU in light or the electron donation by reduced DCPIP to PSI did not cause phosphorylation of LHCII, suggesting the involvement of cytochrome b_6f in light induced phosphorylation of LHCII antennae. These experiments brought forward the idea that the regulation of excitation energy transfer involves protein phosphorylation and migration (Bennett, 1977; Bennett *et al.*, 1980). These conclusions were confirmed by the use of single turnover flashes that reduced PQ and activated the thylakoid kinase that phosphorylated threonine residue of LHCII (Bennett, 1977, 1979b; Bennett *et al.*, 1980). From all their studies, these authors established that phosphorylation/dephosphorylation of LHCII was due to thylakoid bound protein kinase/phosphatase dependent upon illumination and/or by Mg^{+2} ions (Bennett, 1979b; Bennett *et al.*, 1980). In this model, the α change involved migration of LHCII antennae protein from PSII located in the grana region to stroma lamellae located PSI by phosphorylation and released from PSI upon dephosphorylation by the enzyme phosphatase (Pribil *et al.,* 2010) (Fig. 3.2d). From these landmark discoveries on state shift mechanisms, attempts were made to characterize in detail the shifts from state I to state II and vice versa. It is to be noted that the uphill transfer of excitation energy would involve energy transfer from PSI to PSII, which is energetically unfavorable.

Role of Cytochrome b_6f in State Shift: The Role of Q_O Site

Cytochrome b_6f is a membrane bound complex that transfers electrons from lipophilic quinones to hydrophilic plastocyanin, and in the process, transfers proton across the thylakoid membrane and develops proton gradient, thereby controlling not only electron transfer, but also energy transduction (Leegood *et al.,* 2000). It is now known that this membrane protein also holds a key role in state shifts (Zito *et al.,* 1999).

It has been shown that cytochrome b_6f less mutant of *Lemna* lacks LHCII phosphorylation (Gal *et al.,* 1987). Further, it has also been shown in *Acetabularia* that the inhibition of quinol oxidase site of the cytochrome b_6f complex inactivated the LHCII kinase in the dark, and also in the light, or in the presence of duroquinol when the plastoquinone pool was reduced. Inhibition of the quinone reductase site of the cytochrome b_6f complex has practically no effect in the dark and stimulates the kinase activity in the light. Gal *et al.* (1988) from their studies with the mutant *Acetabularia* lacking cyt b_6f complex proposed that the activity of LHCII kinase was regulated by the redox state of a cytochrome b_6f complex component(s) that balances electron flow from PSII to PSI via plastoquinone pool. Similar results were obtained by Bennet *et al.* (1987, 1988) using plastoquinone antagonist type inhibitors in pea thylakoids. Further Vener *et al.* (1995) reported that LHCII phosphorylation could be activated in spinach thylakoids without light by low pH or by the addition of reductant in spinach thylakoids. The induction of LHCII phosphorylation by low pH or light treatment was completely inhibited by the oxidizing agent ferricyanide, suggesting that the redox state of cytochrome b_6f is the key regulator for the state shift. Following this line of research, this group proposed the activation/deactivation cycle of redox-controlled thylakoid protein phosphorylation to be the regulating feature in state shift. The occupancy of the quinol-oxidation site of the cytochrome b_6f complex by plastoquinol, and the redox state of protein thiol groups act as elements of the signal transducing chains (Vener *et al.,* 1998) (Fig. 3.1). In this line of research, Zito *et al.* (1999) demonstrated that Qo site of cytochrome b_6f complex possessed the affinity for plastoquinol, which activated LHCII kinase. These authors used a *Chlamydomonas* mutant designated *PWYE* (amino acid sequence at the Qo site) in which the Qo site of cytochrome b_6f was unable to bind plastoquinol. Photoacoustic and fluorescence studies confirmed that Qo site of cytochrome b_6f complex indeed controls the activation of LHCII kinase.

The PBS Antennae and State Shift

As stated earlier for LHC Chl *a/b*-containing organisms, the mechanism of state shifts has been well characterized (Allen 2000, 2002; Takahashi *et al.,* 2006), whereas the mechanism for cyanobacteria and red algae remains controversial (Biggins, 1983; Bennett *et al.*, 1984; Mullineaux and Allen, 1990).

Cyanobacteria and red algae are exposed to different light regimes in the water bodies that are relatively different from atmospheric light conditions. Most of the red and blue regions of PAR light get absorbed by green plants and green algae particularly in deeper water column, where light with longer wavelengths is less transmitted and therefore less available to chlorophylls, the remaining part of the spectral radiation (500-650 nm) which passes to the water is the only radiation available to these organisms. Under this ecological condition, these organisms develop a unique type of light harvesting antennae called "phycobilisomes" (PBs). The localization of PBs in the thylakoid membrane in red algae is different from that in cyanobacteria. PBs are multimeric extrinsic supramolecular complexes located on the outer surface of the thylakoid membranes (Fig. 3.2), while in the rhodophycian (red algae) algae, the PBs exist in the grana of chloroplasts (Gantt *et al.,* 1979).

Earlier studies with cyanobacterium *Synechocystis* revealed that upon the DCMU poisoning, the fluorescence was excited by the light that was predominantly absorbed by PBS. The fluorescence emission yield of both chl *a* as well as phycocyanin continue to increase for a prolonged period. Such type of non -Q_A dependent increase in fluorescence yield appears similar to fluorescence yield increase in DCMU poisoned thylakoids in the presence of cations (see later). Thus, the slow, long term fluorescence increase in cyanobacteria could be attributed to state change (Papageorgiou and Govindjee, 1968b; Mohanty and Govindjee, 1973a,). In recent years Tsimilli et al (2009) elaborated a similar mechanism in *Synechococcus* by ultra fast fluorescence technique of OJIPSMT kinetics (Fig. 3.6). In cyanobacteria, in dark, the PQ pool of the electron transport chain remains reduced due to the respiratory activity of photosynthetic and respiratory membranes (Fork and Satoh 1986; Sane *et al.,* 1982). Thus these organisms remain in state II in dark and not in state I unlike in state I in green plants. The state shift mechanism in cyanobacteria in terms of antennae protein phosphorylation has remained unsolved. The works of several authors indicated that unlike in higher plants, the protein phosphorylation and migration of phosphorylated pigment protein complexes, particularly involving the antennae PBS, does not occur, even though the general pattern of state shift is similar to green plants and green algae. No reversible protein phosphorylation event corresponding to state shift either in green plant chloroplasts or green algae has been detected in cyanobacteria, although many other proteins get phosphorylated (Piven *et al.,* 2005).

Gene Manipulation for Revealing State Shift

Genetic engineering techniques have helped to dissect the mechanism of state shift and the molecular interactions between state I and state II stoichiometry of thylakoid membranes. In late 1990 s and early 2000's, Scheller group generated transgenic *A. thaliana* plants lacking PSI peripheral antenna and other core complex subunits namely PSI-E, -F, -G, -H, -L, -N,-O or Lhca 4 (see for subunits of PSI, Blankenship, 2002). This work provided an evidence for the docking domain of phosphorylated LHCII on binding to PSI-I,-H,-O region at the "tip" of PSI supercomplex. Plants lacking PSI-H were unable to perform state shifts and remained arrested in State I, indicating PSI-H as the target binding site of LHCII under state II conditions (Lunde *et al.,* 2000, 2003), although they underwent LHCII phosphorylation (Delosme *et al.,* 1996). Mutant plants lacking PSI-O subunit showed, compared to the wild type (Wt), nearly 50% reduction in state I to state II transition, suggesting that PSI-O is also required for state shift (Jensen *et al.*, 2004). *A. thaliana* mutants lacking PSI-H or PSI-L subunits contained fewer but more extended granal stacks, implying that state shifts induced structural organization of thylakoid membranes (Lunde *et al.*, 2000). Further, in *A. thaliana,* the double mutant *stn7/stn8*, lacking kinases for phosphorylation of LHCII and D1/D2 of PSII RC, respectively showed no shift from state I to state II (Depege *et al.*, 2003; Bellafiore *et al.*, 2005). These mutational studies confirmed the α change, namely the change in absorption cross section of PSI-LHCI and PSII-LHCII of thylakoid membranes by phosphorylation triggered antennae during state shift in green plants and green algae. Below we list some of the mutants that have been used in elucidating the mechanisms of state shifts (Table 3.2).

Table 3.2 Some of the mutants widely used in demonstration of state shift mechanisms.

S.N	*Mutant*	*State shift*	*Description*	*Reference*
1	*stt7 (Chlamydomonas reinhardtii)*	*stt7* is unable to undergo state shift and remains blocked in state I.	Kinase responsible for LHCII phosphorylation is inhibited, thereby arresting the cells to state I.	J Biol Chem. 1999, 22; 274:30987-94.
2	*stn7 (Arabidopsis thaliana)*	Locked in state I.	*stn7* is an analogue of *stt7.* Thylakoid kinase responsible for LHCII phosphorylation was arrested.	Plant Molecular Biology, 2006, 62: 779-793.
3	*psi-h psi-l (Arabidopsis thaliana)*	Affected in their ability to perform state shift.	PSI-H, L were the docking site of LHCII of PSII. Lacking these subunits in plants greatly affects state shift.	Plant Cell Physiol. 2003, 44:44-54.
4	*stn7/stn8 double mutant (Arabidopsis thaliana)*	No shift from state I to state II.	Both LHCII and D1/D2 phosphorylation were arrested.	Nature, 2005. 433: 892-895
5	*PWYE (amino acid sequence at the Qo site). (Chlamydomonas reinhardtii)*	State shift arrested since Q_O site of cytochrome b_6f was unable to bind plastoquinol.	Q_O site of cytochrome b_6f complex controls the activation of LHCII kinase.	EMBO J, 1999. 18:2961-2969

The mechanism of state shift/state transition has been fairly elucidated for higher plants, green algae and cyanobacteria; the readers may refer to the following reviews for details (Fork and Satoh, 1986; Allen, 2000; Haldrup *et al.*, 2001; Allen, 2002; McConnell *et al.*, 2002; Lars *et al.*, 2008).

Mutants like *stn7, stn8, psaH, psaL, lhcII, PWYE, TAP38* (Table 3.2) helped to strengthen the concept of state shifts and also played a crucial role in determining the functional and structural organization of PSI-LHCI and PSII-LHCII supercomplexes during state transitions (Zito *et al.*, 1999; Lunde *et al.*, 2000; Ruban *et al.*, 2003; Jensen *et al.*, 2004; Pesaresi *et al.*, 2009; Tikkanen *et al.*, 2010).

Cation and Anion Regulating State Shifts *in vivo*

Alongside the discussion on state shift in intact cells and plants, it needs to be mentioned that cations, especially divalent cations, regulated the energy transfer and quantal distribution between the two photosystems (Rubin *et al.*, 1981; Barber, 1986). Addition of divalent cations in isolated chloroplasts

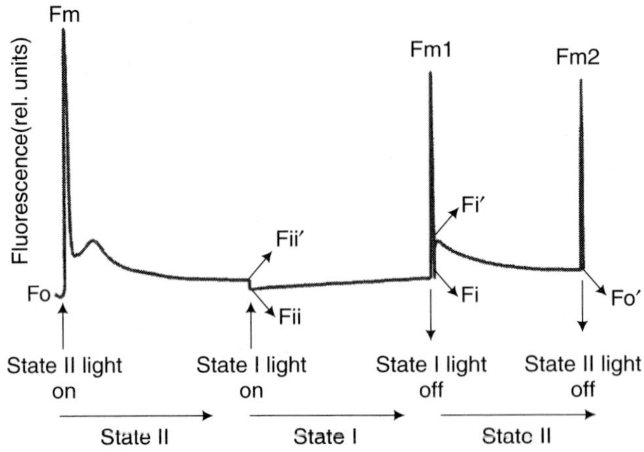

Fig. 3.8 Schematic representation of fluorescence signals during state shift using PAM. (Redrawn after Lunde *et al.*, 2000)

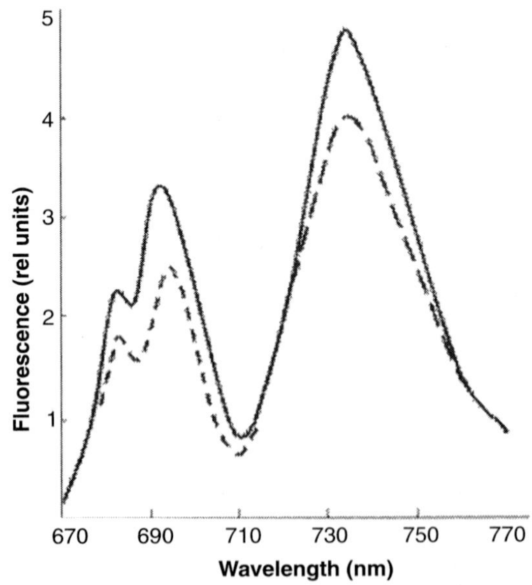

Fig. 3.9 77K emission spectra of spinach chloroplasts in the presence (--- 4 mM $MgCl_2$ was added before cooling) and absence of $MgCl_2$ (No $MgCl_2$ added). Excitation light 567 nm (Adopted from Murata, 1969).

blocked the energy distribution from PSII to PSI(Gross and Hess, 1973; Gross and Prasher, 1974). Besides cation induced Chl *a* fluorescence yield changes, the slow 90° scattering changes were observed for grana stacking and energy transfer from PSII to PSI. Notably the effect of Mg^{2+} in arresting the energy transfer from PSII to PSI has been shown to play a crucial role. Murata (1969) showed that the addition of Mg^{2+} increased the fluorescence emission at 684 nm and 694 nm of PSII and decreased at 735 nm of PSI (Fig. 3.9). Thus cation induced fluorescence yield changes and structural changes in granal stacking were assumed to be correlating events. Further the divalent and monovalent cations showed antagonistic effect (Wong and Govindjee, 1979). Alfonzo *et al.* (1980) demonstrated the absolute requirement of Mg^{2+} for protein kinase activity in chloroplast. Other cations like Mn^{2+} and Ca^{2+} could substitute for Mg^{2+} with diminished kinase activity. Not only cations but anions also were shown to induce state shift in isolated thylakoids (Jajoo *et al.*, 1998). While fluorescence yield changes linked to state shift were demonstrated with anions like fluoride and nitrate (Singh *et al.*, 2009), phosphorylation of thylakoid antennae proteins has not been shown unlike cations. It must be noted that grana stacking, which causes lateral separation of two photosystems and reduces its excitation transfer from PSII to PSI, enhances energy transfer among the PSII units in granal stacks (Briantais *et al.*, 1982). In view of these results, the cation induced scattering changes due to grana stacking and Chl *a* fluorescence yield and emission changes, though often measured together, are not essentially linked processes or interdependent processes but are independent of each other. However, phosphorylation of LHCII and 77K fluorescence changes are linked processes and constitute the index of state shift.

TECHNIQUES USED IN STATE SHIFT STUDIES

Fluorescence and Oxygen Evolution

The state I and state II, atleast in green plants and green algae, have some identifiable diagnostic parameters (Table 3.1) that have been characterized by oxygen evolution and fluorescence kinetics. Conventionally, and conveniently, the state shift is monitored by both oxygen evolution and fluorescence kinetics and by a few spectroscopic methods. Chl *a* fluorescence yield changes, both spectral features and transients at room temperature, have been used in studies in intact algal cells, protoplasts, leaves, tissues and chloroplast (Murata, 1969b; Bonaventura and Myers, 1969; Mohanty *et al.*, 1972, Mohanty and Govindjee, 1973b; Mohanty *et al.*, 1973). The pulse amplitude modulator (PAM) fluorescence yield changes have been a very useful method for monitoring state transition (Scheriber *et al.*, 1986). Figure 3.8 demonstrates the monitoring of fluorescence yield changes by single or multiple turnover flashes. Since fluorescence emission features of PSII and PSI are well resolved at low temperature, 77K emission spectral, analysis, as shown in Fig. 3.3 have also been used for monitoring the magnitude of state transitions (Murata, 1969; Mohanty *et al.*, 1973). From this 77K emission spectra the magnitude of state shift from state I (F730/F696) to state II by evaluating F730 emanating from PSI to F696 emanating from PSII can be monitored. David Fork's laboratory developed a procedure to measure simultaneous rise of F696 and F730 emission at 77K from samples adapted to state I or state II and the linear plot of F696 vs. F730 to obtain the extent of α change and spillover (Fork *et al.*, 1985). The slope of this plot yielded the value of changes in the antennae cross section (α change). The extrapolation of this plot to X-axis yielded a value indicating the extent of spillover. From such measurements it was pointed that in the green plants and green algae, the energy transfer from PSII to PSI by state shift was mostly due to change in absorption cross section linked to protein phosphorylation and migration, while in cyanobacteria the mechanism of energy transfer was linked to spillover and or α change (Fork and Satoh, 1986). In intact cells and isolated thylakoids, the fluorescence yield changes as well as the emission-spectral changes were used for monitoring state shifts. The slow fluorescence yield changes using lights predominantly absorbed by PSII and PSI separately and jointly exhibit characteristic state shift from state II to state I as shown in Fig. 3.6 with *Synechococcus*. As stated in the introduction, the oxygen evolution as well as fluorescence induction kinetics measured simultaneously with light absorbed predominantly by either PSII or PSI is an elegant technique for monitoring state shift, as was shown by Bonaventura and Myers (1969) in their classic experiment (Fig 3.5). It is to be noted that in state II, the PSII functions more efficiently than in state I. In recent years, besides the slow fluorescent induction representing P to S decline in fluorescence yield were shown to be particularly due to state shift. The fast OJIP fluorescence induction transients have been used in state shift studies by using Handy PEA. Tsimille group in 2009 have elegantly showed that the ultrafast OJIP transient would also detect state shift in *Synechococcus* (Fig. 3.6).

The rate of oxygen evolution, as stated in introduction for *Chlorella* cells, provides information on state shift by using light predominantly absorbed by PSII or PSI. The oxygen evolution rate in intact cells would be monitored by using a Joliot type oxygen rate electrode, as elegantly shown in Bonaventura and Myers (1969) (Fig. 3.5). Also, the oxygen yield or photochemical dye reduction catalyzed by PSII or PSI as a function of light intensity that excites either state I or state II (Murata, 1969; Sun and Sauer, 1971, 1972) is another technique.

For higher plants and green algae adapting to a specific state by pre-illumination or darkness, either state I or state II can be established. In this method, for example dark adaptation or by pre-illumination of PS I with far red light the state I can be adapted. In this condition, in higher plants and green algae, all Q_A remain in oxidized state. Upon illumination with moderately high light, the system can drive the system to state II. During this process, the kinase gets activated and phosphorylation of LHCII can be monitored by a variety of biochemical methods such as radiolabelling and immunoblotting (Elich *et al.*, 1997, Tikkanen *et al.*, 2010). The migration of phosphorylated component from grana to stroma region can also be visualized by preparative biochemical methods.

Other Spectroscopic Techniques

Photoacuostic (PA) spectroscopy

PA measures modulated heat (photothermal) and O_2 evolution (photobaric) from photosynthetically active samples when illuminated with modulated light (Cannani *et al.*, 1984; Delosme *et al.*, 1996; Bukhov *et al.*, 1999). This technique has been used to measure energy storage in state I and state II in cyanobacteria (Salehian and Bruce, 1992; Delosme *et al.*, 1996), red alga (Cannani *et al.*, 1984), and intact leaves (Canaani *et al.*, 1984; Veeranjaneyulu and Leblanc, 1994). Using PA technique, the action spectral analysis revealed that photochemical activity remained same in both state I and state II between 580 nm and 700 nm, but, it was lower in state II than in state I, in the wavelength range between 400 nm and 580 nm (Veeranjaneyulu and Leblanc, 1994). Because of the complexities involved in monitoring photothermal and photobaric signals by PA technique, this method of monitoring state I to state II has not been popular. However, using this technique, Cannani *et al.* (1984) demonstrated that in intact tobacco leaves the protein phosphorylation and migration of phosphorylated proteins controls light distribution between the two photosystems *in vivo* and it was similar to *in vitro* insolated thylakoids.

Fourier transform infrared spectroscopy (FTIR)

FTIR technique has also been used to study state shift mechanisms (Delosme *et al.*, 1971). Similarly, illumination with light predominantly absorbed by PSII, LHCII mobilization and its path of migration has been documented by using this technique. FTIR technique was also used to visualize that phosphorylated LHCII contained more of α helix and β turn structures and a lower content of β sheet structures than in the non-phosphorylated LHCII (Nilsson *et al.*, 1997). It is possible that other spectroscopic techniques, mainly Electron Spin Resonance (ESR) that monitors P700 of PSI (signal I) and the associated signal II of PSII could be used to monitor state I to state II shift in intact algal cells and leaf segments.

Fluorescence recovery after photobleaching (FRAP)

FRAP technique measures two-dimensional lateral diffusion of protein, lipid or carbohydrate by covalently attaching a fluorophore to the target molecule (Kirchhoff *et al.*, 2008). LHCII mobilization and its pathway of migration could be well defined by using this technique. Mullineaux *et al.* (1997) employed FRAP to show that PBS were capable of lateral diffusion along the thylakoid surface of the cyanobacterium *Dactylococcopsis salina*, but PSII reaction centers remained immobile. By FRAP method the mobility of PBS during state shift was assayed in *Synechococcus* (Mullineaux and Allen, 1990), thereby originating the concept of mobile PBS in cyanobacteria.

Microscopic Studies

Atomic Force Microscopy (AFM) and Electron Microscopy

AFM is designed to measure the topography of a nonconductive sample. This technique allows to measure the local resistivity, temperature, elasticity, tribology (the branch of engineering that deals with the interaction of surfaces in relative motion). The AFM works much the same way as phonograph works at a much smaller scale. A very sharp tip is dragged across a sample surface and the change in the vertical position (denoted the "z" axis) reflects the topography of the surface. By collecting the height data for a succession of lines, it is possible to form a three-dimensional map of the surface features. Studies with AFM revealed thylakoid membrane reorganization during state transitions. The shift from state I to state II led to significant changes in the network architecture of thylakoid membranes. AFM micrographs showed that the transition was accompanied by unstacking and disassembly of granal lamellae (Chuartzman *et al.,* 2008). In recent years, electron microscopy and single particle analysis of projection map of the PSI-LHCII supercomplex at 16 Å has been used to reveal that these particles were associated with LHCII (Kouril *et al.,* 2005). Further this group showed that PSI-LHCII supercomplex could be averaged and seen with a resolution of 16 Å resolution, and LHCII had a specific binding site at the PSI-A, -H, -L, and –K subunits (Kouril *et al.,* 2005). A transition from megacomplex to supercomplex and corecomplex during state 1 to state 2 has been recently well established using this technique in *Chlamydomonas* (Iwai *et al.,* 2008). Cryo-electron micrographs of PSI-LHCI supercomplexes showed 14 subunits around PSI core in state II and 11 subunits in state I from *Chlamydomonas*. The extra three subunits in state II represented the migration of protein subunit from PSII supercomplex. Kargul *et al.* 2005(1) suggested that phospho-CP29, possibly in a multiphosphorylated form, strongly associates with PSI in state II, adjacent to the PsaH protein. Thus, single particle electron microscopy is an effective method to determine the structure of macromolecular complexes with little or no symmetry (Thuman-Commike, 2001). This technique could also be used in determining state shifts. Fluorescence imaging is another technique used for screening the mutants deficient in state shift. Since the size of the mobile LHCII in *Chlamydomonas reinhardtii* is large, the fluorescence yield during state shift significantly changes. This feature has been exploited for screening the mutants (Fleischmann *et al.,* 1999).

Biochemical Analysis

Sucrose density gradient ultracentrifugation can separate complexes associated differentially in state I and state II. By poising the redox status of PQ pool, the state I and state II condition could be achieved and these two states of PSI could be separated by these techniques (Takahashi *et al.,* 2006; Rajagopal *et al.,* 2006). Using such technique, Takahashi *et al.,* (2006) have shown that sucrose density gradient ultracentrifugation could be used to quantify and identify the mobilized LHCII to PSI in state II condition in *Chlamydomonas reinhardtii.*

CONCLUSION

Recent studies on state shifts have revealed many new dimensions on the nature and its relevance of this dynamic shifts or changes in the regulations and functioning of photosynthetic machinery as a whole both in higher plants, algae and blue bacteria. Light induced Chl *a* fluorescence transients have been dissected to reveal the short term regulatory shifts of state I to state II. Also the redox driven state shift appear to control the extent of stoichiometry of photosystems, the rate of linear and cyclic electron flow and associated phosphorylation, differential gene expression that controls development and acclimation.

In view of these progresses, it is possible that the state transition or state shift may report the early events of abiotic stresses in plants.

ACKNOWLEDGEMENTS

The authors thank DST Indo-Hungary (DST/INT/HUN/P-8/06), DST (SR/SO/BB/34/2006) and UOH for support. They also acknowledge the help of Dr RJ Strasser and the encouragement of Dr AS Raghavendra, Dr GC Papageorgiou. The helpful comments of Dr PV Sane, Dr H Satoshi, Dr S Hoshina, Dr DG Maitrayee, Dr Krishna Nath, Dr M Tsimilli, and K Sireesha are acknowledged. PM acknowledges INSA for financial assistance. NS thank CSIR-UGC for fellowship.

REFERENCES

Alfonzo R, Nelson N, Racker E (1980) A light-dependent protein kinase activity of chloroplasts. Plant Physiol 65: 730-734

Allen JF (1992a) Molecular structural effects of protein-phosphorylation in regulation of photosynthesis. Photosynth Res 34: 118

Allen JF (1992b) Protein phosphorylation in regulation of photosynthesis. Biochim Biophys Acta Bioenergetics 1098: 275-335

Allen JF (2000) State transitions--a question of balance. Science 2003 299: 1530-1532

Allen JF (2002) Plastoquinone redox control of chloroplast thylakoid protein phosphorylation and distribution of excitation energy between photosystems: discovery, background, implications. Photosynth Res 73: 139-148

Allen JF, Bennett J, Steinback KE, Arntzen CJ (1981) Chloroplast protein phosphorylation couples plastoquinone redox state to distribution of excitation-energy between photosystems. Nature 291: 25-29

Allen JF, Sanders CE, Holmes NG (1985) Correlation of membrane protein phosphorylation with excitation-energy distribution in the cyanobacterium *Synechococcus*-6301. FEBS Lett 193: 271-275

Allen JF and Holmes NG (1986) A general model for regulation of photosynthetic unit function by protein phosphorylation. FEBS Lett 202: 175-181

Allen JF, Pfannschmidt P (2000) Balancing the two photosystems: Photosynthetic electron transfer governs transcription of reaction center genes. Philosophical Transaction of the Royal Society series B 355: 12351-1359

Barber J (1986) Regulation of energy transfer by cations and protein phosphorylation in relation to thylakoid membrane organisation. Photosynth Res 10: 243-253

Bellafiore S, Bameche F, Peltier G, Rochaix JD (2005) State transitions and light adaptation require chloroplast thylakoid protein kinase STN7. Nature 433: 892-895

Bennett J (1977) Phosphorylation of chloroplast membrane polypeptides. Nature 269: 344-346

Bennett J (1979a) Chloroplast phosphoproteins phosphorylation of polypeptides of the light-harvesting chlorophyll protein complex. Eur J Biochem 99: 133-137

Bennett J (1979b) Chloroplast phosphoproteins – protein kinase of thylakoid membranes is light-dependent. FEBS Lett 103: 342-344

Bennett J, Steinback KE, Arntzen CJ (1980) Chloroplast phosphoproteins - regulation of excitation-energy transfer by phosphorylation of thylakoid membrane polypeptides. Proc Natl Acad Sci USA 77: 5253-5257

Bennett J, Shaw EK, Bakr S (1987) Phosphorylation of thylakoid proteins and synthetic peptide analogs: Differential sensitivity to inhibition by a plastoquinone antagonist. FEBS Lett 210: 22-26

Bennett J, Elizabeth K, Michel H (1988) Cytochrome b_6f complex is required for phosphorylation of light-harvesting chlorophyll *a/b* complex II in chloroplast photosynthetic membranes. Eur J Biochem 171: 95-100

Biggins J (1983) Mechanism of the light state transition in photosynthesis. 1. Analysis of the kinetics of cytochrome-f oxidation in State-1 and State-2 in the red alga, *Porphyridium cruentum*. Biochim Biophys Acta 724: 111-117

Biggins J, Campbell CL, Bruce D (1984) Mechanism of the light state transition in photosynthesis. 2. Analysis of phosphorylated polypeptides in the red alga, *Porphyridium cruentum*. Biochim Biophys Acta 767: 138-144

Blankenship RE. Molecular mechanisms of photosynthesis, Blackwell Science (2002).

Bonaventura C, Myers J (1969) Fluorescence and oxygen evolution from *Chlorella pyrenoidosa*. Biochim Biophys Acta 189: 366-383

Briantais JM, Vernotte C, Maison B (1982) Influence of stacking on the distribution of light energy in the photosynthetic apparatus. Physiol Veg 20: 111-122

Bruce D, Salehian O (1992) Laser-induced optoacoustic calorimetry of cyanobacteria. The efficiency of primary photosynthetic processes in State 1 and State 2. Biochim Biophys Acta 1100: 242-250

Bukhov NG, Wiese C, Neimanis S (1999) Heber U. Heat sensitivity of chloroplasts and leaves: Leakage of protons from thylakoids and reversible activation of cyclic electron transport. Photosynth Res 59: 81-93

Canaani O, Barber J, Malkin S (1984) Evidence That phosphorylation and dephosphorylation regulate the distribution of excitation energy between the two photosystems of photosynthesis *in vivo* - photoacoustic and fluorimetric study of an intact leaf. Proc Natl Acad Sci USA 81: 1614-1618

Chang CCC, Ball L, Fryer MJ, Baker NR, Karpinski S, Mullineaux PM (2004) Induction of ascorbate peroxidase 2 expression in wounded *Arabidopsis* leaves does not involve known wound-signalling pathways but is associated with changes in photosynthesis. Plant J 38: 499-511

Chuartzman SG, Nevo R, Shimoni E, Charuvi D, Kiss V, Ohad I, Brumfeld V, Reich Z (2008) Thylakoid membrane remodeling during state transitions in *Arabidopsis*. Plant Cell 20: 1029-1039

Delosme R, Olive J, Wollman FA (1996) Changes in light energy distribution upon state transitions: An *in vivo* photoacoustic study of the wild type and photosynthesis mutants from *Chlamydomonas reinhardtii*. Biochim Biophys Acta Bioenerg 1273: 150-158

Depege N, Bellafiore S, Rochaix JD (2003) Rote of chloroplast protein kinase Stt7 in LHCII phosphorylation and state transition in *Chlamydomonas*. Science 299: 1572-1575

Duysens LNM . Transfer of excitation energy in photosynthesis. Doctoral thesis, Utrecht, 1952

Elich TD, Edelman M, Mattoo AK (1997) Evidence for light-dependent and light-independent protein dephosphorylation in chloroplasts. FEBS Lett 411: 236-238

Fleischmann MM, Ravanel S, Delosme R, Olive J, Zito F, Wollman FA, Rochaix JD (1999) Isolation and characterization of photoautotrophic mutants of *Chlamydomonas reinhardtii* deficient in state transition. J Biol Chem 274: 30987-30994

Fork DC, Prasanna M, Hoshina S (1985) The detection of early events in heat disruption of thylakoid membranes by delayed light emissions. Physiol Veg 23: 511-521

Fork DC, Satoh K (1986) The control by state transitions of the distribution of excitation energy in photosynthesis. Annu Rev Plant Phys 37: 335-361

Forster T (1948) Intermolecular energy transfer and fluorescence. Ann Physik 2: 55-75

Gal A, Shahak Y, Schuster G, Ohad I (1987) Specific loss of LHCII phosphorylation in the *Lemna* mutant 1073 lacking the cytochrome b_6f complex. FEBS Lett 221: 205-210

Gantt E, Lipschultz CA, Grabowski J, Zimmerman BK (1979) Phycobilisomes from blue-green and red algae: isolation criteria and dissociation characteristics. Plant Physiol 63: 615 620

Green BR, Parson WW. Light-harvesting Antennas in Photosynthesis. Kluwer Academic Publishers, 2003.

Gross EL, Hess SC (1973) Monovalent cation-induced inhibition of chlorophyll *a* fluorescence: Antagonism by divalent cations. Arch Biochem Biophys 159: 832-836

Gross EL, Prasher SH (1974) Correlation between monovalent cation-induced decreases in chlorophyll *a* fluorescence and chloroplast structural changes. Arch Biochem Biophys 164: 460

Haldrup A, Jensen PE, Lunde C, Scheller HV (2001) Balance of power: a view of the mechanism of photosynthetic state transitions. Trends Plant Sci 6: 301-305

Horton P, Black MT (1980) Activation of adenosine 5'-triphosphate-induced quenching of chlorophyll fluorescence by reduced plastoquinone. FEBS Lett 119: 141-144

Horton P, Allen JF, Black MT, Bennett J. (1981) Regulation of phosphorylation of chloroplast membrane polypeptides by the redox state of plastoquinone. FEBS Lett 125: 193-196

Ivanov AG, Sane PV, Vaugan H, Oquist G, Huner NPA (2008) Photosystem II reaction centre quenching: mechanisms and physiological role. Photosynth Res 98: 565-574

Iwai M, Takahashi Y, Minagawa J (2008) Molecular remodeling of photosystem II during state transitions in *Chlamydomonas reinhardtii*. Plant Cell 20: 2177-2189

Jajoo A, Bharti S, Govindjee (1998) Inorganic anions induce state changes in spinach thylakoid membranes. FEBS Lett 434: 193-196

Jensen P E, Haldrup A, Zhang S, Scheller HV (2004) The PSI-O subunit of plant photosystem I is involved in balancing the excitation pressure between the two photosystems. J Biol Chem 279: 24212-24217

Kana R, Prasil O, Komarek O, Papageorgiou GC, Govindjee (2009) Spectral characteristic of fluorescence induction in a model cyanobacterium *Synechococcus* sp. (PCC 7942) Biochim Biophys Acta 1787: 1170-1178

Kargul J, Nield J, Barber J (2003) Three-dimensional reconstruction of a light-harvesting complex I-photosystem I (LHCI-PSI) supercomplex from the green alga *Chlamydomonas reinhardtii* Insights into light harvesting for PSI. J Biol Chem 278: 16135-16141

Kargul J, Turkina MV, Nield J, Benson S, Vener AV, Barber J (2005) Light-harvesting complex II protein CP29 binds to photosystem I of *Chlamydomonas reinhardtii* under State 2 conditions. FEBS J 272: 4797

Kirchhoff H, Haferkamp S, Allen JF, Epstein D, Mullineaux CW (2008) Protein diffusion and macromolecular crowding in thylakoid membranes. Plant Physiol 146: 1571-1578

Kouril R, Zygadlo A, Arteni AA, Wit CD, Dekker JP, Jensen PE, Scheller HV, Boekema EJ (2005) Structural characterization of a complex of photosystem I and light-harvesting complex II of *Arabidopsis thaliana*. Biochemistry 44: 10935-10940

Krause GH, Behrend U (1986) pH-dependent chlorophyll fluorescence quenching indicating a mechanism of protection against photoinhibition of chloroplasts. FEBS Lett 200: 298-302

Krause GH, Weis E (1991) Chlorophyll fluorescence and photosynthesis: the basics. Annu Rev Plant Biol 42: 313-349

Lars D, Katharina B, Thomas P (2008) Photosynthetic acclimation: State transitions and adjustment of photosystem stoichiometry – functional relationships between short-term and long-term light quality acclimation in plants. FEBS J 275: 1080-1088

Lemeille S, Willig A, Fargeix DN, Delessert C, Bassi R, Rochaix JD (2009) Analysis of the Chloroplast Protein Kinase Stt7 during State Transitions. PLoS Biol 7: e1000045

Lunde C, Jensen PE, Haldrup A, Knoetzel J, Scheller HV (2000) The PSI-H subunit of photosystem I is essential for state transitions in plant photosynthesis. Nature 408: 613-615

Lunde C, Jensen PE, Rosgaard L, Haldrup A, Gilpin MJ, Scheller HV (2003) Plants impaired in state transitions can to a large degree compensate for their defect. Plant Cell Physiol 44: 44-54

Mao HB, Li GF, Ruan X, Wu QY, Gong YD, Zhang XF, Zhao NM (2002) The redox state of plastoquinone pool regulates state transitions via cytochrome b_6f complex in *Synechocystis* sp. *PCC 6803*. FEBS Lett 519: 82-86

McConnell MD, Koop R, Vasil'ev S, Bruce D (2002) Regulation of the distribution of chlorophyll and phycobilin-absorbed excitation energy in cyanobacteria. A structure-based model for the light state transition. Plant Physiol 130: 1201-1212

Mohanty P, Munday JC, Govindjee (1970) Time-dependent quenching of chlorophyll *a* fluorescence from (pigment) system II by (pigment) system I of photosynthesis in *Chlorella* Biochim Biophys Acta 223: 198-200

Mohanty P, Thornber JP, Braun BZ, Govindjee (1972) Chlorophyll fluorescence characteristics of System-I chlorophyll alpha-protein complex and System-II particles at room and liquid-nitrogen temperatures. Plant Cell Physiol 13: 81

Mohanty P, Braun BZ, Govindjee (1973) Light-induced slow changes in chlorophyll *a* fluorescence in isolated chloroplasts - Effects of magnesium and phenazine methosulfate. Biochim Biophys Acta 292: 459-476

Mohanty P, Govindjee (1973a) Light-induced changes in fluorescence yield of chlorophyll *a* in *Anacysti Nidulans* .1. Relationship of slow fluorescence changes with structural changes Biochim Biophys Acta 305: 95-104

Mohanty P, Govindjee (1973b) Light-induced changes in fluorescence yield of chlorophyll Alpha in *Anacystis-Nidulans*. 2. Fast changes and effect of photosynthetic inhibitors on both fast and slow fluorescence induction. Plant Cell Physiol 14: 611-629

Mullineaux CW, Allen JF (1990) State 1-State 2 transitions in the cyanobacterium *Synechococcus* sp. 6301 are controlled by the redox state of electron carriers between photosystems I and II. Photosynth Res 23: 297-311

Mullineaux CW, Tobin MJ, Jones GR (1997) Mobility of photosynthetic complexes in thylakoid membranes. Nature 390: 421-424

Murata N (1969a) Control of excitation transfer in photosynthesis. I. Light-induced change of chlorophyll *a* fluorescence in *Porphyridium cruentum*. Biochim Biophys Acta 172: 242

Murata N (1969b) Control of excitation transfer in photosynthesis. II. Magnesium ion-dependent distribution of excitation energy between two pigment systems in spinach chloroplasts. Biochim Biophys Acta 189: 171

Murata N (2009) The discovery of state transitions in photosynthesis 40 years ago. Photosynth Res 99: 155-160

Murata N, Nishimura M, Takamiya A (1966) Fluorescence of chlorophyll in photosynthetic systems III. Emission bands of chlorophyll *a* and the energy transfer between two pigment systems. Biochim Biophys Acta 126: 234-243

Murata N, Sugahara K (1969) Control of excitation transfer in photosynthesis. III. Light-induced decrease of chlorophyll Alpha fluorescence related to photophosphorylation system in spinach chloroplasts. Biochim Biophys Acta 189: 182-192

Nilsson A, Stys D, Drakenberg T, Spangfort MD, Forsqn S, Allen JF (1997) Phosphorylation controls the three-dimensional structure of plant light harvesting complex II. J Biol Chem 272: 18350-18357

Papageorgiou G C. (2004) Chlorophyll *a* Fluorescence: A Signature of Photosynthesis. Kluwer Academic Publishers

Papageorgiou G, Govindjee (1968a) Light-induced changes in fluorescence yield of chlorophyll Alpha *in vivo*. II. *Chlorella pyrenoidosa*. Biophys J 8: 1316-1328

Papageorgiou G, Govindjee (1968b) Light-induced changes in fluorescence yield of chlorophyll Alpha *in vivo*. I. *Anacystis nidulans*. Biophys J 8: 1299-1315

Pesaresi P, Hertle A, Pribil M, Kleine T, Wagner R, Strissel H, Ihnatowicz A, Bonardi V, Scharfenberg M, Schneider A, Pfannschmidt T, Leister D (2009) *Arabidopsis* STN7 Kinase provides a link between short-and long-term photosynthetic acclimation. Plant Cell 21: 2402-23

Pfannschmidt T (2003) Chloroplast redox signals: how photosynthesis controls its own genes. Trends Plant Sci 8: 33-41

Piven I, Ajlani G, Sokolenko A (2005) Phycobilisome linker proteins are phosphorylated in *Synechocystis* sp. PCC 6803. J Biol Chem 280: 21667-21672

Pribil M, Pesaresi P, Hertle A, Barbato R, Leister D (2010) Role of plastid protein phosphatase TAP38 in LHCII dephosphorylation and thylakoid electron flow. PLoS Biol 8: e1000288

Rechtsteiner GA, Ganske JA (1998) Using natural and artificial light sources to illustrate quantum mechanical concepts. The Chemical Educator 3: 1-12

Ruban AV, Wentworth M, Yakushevska AE, Andersson J, Lee PJ, Keegstra W, Dekker JP, Boekema EJ, Jansson S, Horton P (2003) Plants lacking the main light-harvesting complex retain photosystem II macro-organization. Nature 421: 648-652

Rubin BT, Chow WS, Barber J (1981) Experimental and theoretical considerations of mechanisms controlling cation effects on thylakoid membrane stacking and chlorophyll fluoresence. Biochim Biophys Acta Bioenerg 634: 174-190

Salehian O, Bruce D (1992) Distribution of excitation energy in photosynthesis: Quantification of fluorescence yields from intact cyanobacteria. J Luminescence 51: 91-98

Sane PV, Furtado D , Desai TS, Tatake VG (1982) A study of state changes in *Chlorella*: The effect of uncoupler and energy transfer inhibitors. Z Naturforsch Teil C 37: 458-63

Schreiber U, Schliwa U, Bilger W (1986) Continuous recording of photochemical and non-photochemical chlorophyll fluorescence quenching with a new type of modulation fluorometer. Photosynth Res 10: 51-62

Shyam R, Raghavendra AS, Sane PV (1993) Role of dark respiration in photoinhibition of photosynthesis and its reactivation in the cyanobacterium *Anacystis nidulans.* Physiol Plant 88: 446-452

Slesak I, Miszalski Z, Karpinska B, Niewiadomska E, Ratajczak R, Karpinski S (2002) Redox control of oxidative stress responses in the C_3-CAM intermediate plant *Mesembryanthemum crystallinum.* Plant Physiol Biochem 40: 669-677

Slesak I, Karpinska B, Surowka E, Miszalski Z, Karpinski S (2003) Redox changes in the chloroplast and hydrogen peroxide are essential for regulation of C3-CAM transition and photooxidative stress responses in the facultative CAM Plant *Mesembryanthemum crystallinum* L Plant Cell Physiol 44: 573-581

Steinbrenner J, Linden H (2003) Light induction of carotenoid biosynthesis genes in the green alga *Haematococcus pluvialis*: regulation by photosynthetic redox control. Plant Mol Biol 52: 343-356

Sun AS, K Sauer (1971) Pigment systems and electron transport in chloroplasts. I. Quantum requirements for the two light reactions in spinach chloroplasts. Biochim Biophys Acta 234:399-414

Sun AS, Sauer K (1972) Pigment systems and electron transport in chloroplasts. II. Emerson enhancement in broken spinach chloroplasts. Biochim Biophys Acta 256: 400-427

Takahashi H, Iwai M, Takahashi Y, Minagawa J (2006) Identification of the mobile light-harvesting complex II polypeptides for state transitions in *Chlamydomonas reinhardtii.* Proc Natl Acad Sci USA 103: 477-482

Thuman-Commike PA (2001) Single particle macromolecular structure determination via electron microscopy. FEBS Lett 505: 199-205

Tikkanen M, Grieco M, Kangasjärvi S, Aro EM (2010) Thylakoid protein phosphorylation in higher plant chloroplasts optimizes electron transfer under fluctuating light. Plant Physiol 152: 723–735

Tsimilli-Michael M, Stamatakis K, Papageorgiou GC (2009) Dark-to-light transition in *Synechococcus* sp. PCC 7942 cells studied by fluorescence kinetics assesses plastoquinone redox poise in the dark and photosystem II fluorescence component and dynamics during state 2 to state 1 transition. Photosynth Res 99: 243-255

Van GR, Novoderezhkin VI (2006) Energy transfer in photosynthesis: experimental insights and quantitative models. Phys Chem Chem Phys 8: 793-807

Veeranjaneyulu K, Leblanc RM (1994) Action spectra of photosystems I and II in State 1 and State 2 in intact sugar maple leaves. Plant Physiol 104: 1209-1214

Vener AV, Van Kan PM, Gal A, Andersson B, Ohad I (1995) Activation/deactivation cycle of redox-controlled thylakoid protein phosphorylation. J Biol Chem 270: 25225

Vener AV, Ohad I Andersson B (1998). Protein phosphorylation and redox sensing in chloroplast thylakoids. Curr Opin Plant Biol 1: 217-223

Walters RG, Horton P (1993) Theoretical assessment of alternative mechanisms for non-photochemical quenching of PSII fluorescence in barley leaves. Photosynth Res 36: 119-139

Weis E, Berry JA (1987) Quantum efficiency of photosystem II in relation to 'energy'-dependent quenching of chlorophyll fluorescence. Biochim Biophys Acta 894: 198–207

Wong D, Govindjee (1979) Antagonistic effects of mono-and divalent cations on polarization of chlorophyll fluorescence in thylakoids and changes in excitation energy transfer. FEBS Lett 97: 373-379

Zer H, Ohad I (2003) Light, redox state, thylakoid-protein phosphorylation and signaling gene expression. Trends Biochem Sci 28: 467-470

Zito F, Finazzi G, Delosme R, Nitschke W, Picot D, Wollman FA (1999) The Qo site of cytochrome b_6f complexes controls the activation of the LHCII kinase. EMBO J 18: 2961-2969

CHAPTER 4

Dissipation of Excess Absorbed Energy by the Photosynthetic Apparatus of Cyanobacteria

N.V. Karapetyan, V.V. Shubin, I.V. Terekhova and M.G. Rakhimberdieva

A.N. Bakh Institute of Biochemistry, Russian Academy of Sciences, 119071 Moscow, Leninsky pr. 33, Russia

ABSTRACT

The main pathway of dissipation of light excitation energy is non-photochemical quenching, which occurs in plants with participation of membrane bound light-harvesting complex LHCII and the violaxanthin cycle, thus protecting the photosynthetic apparatus against destruction. The pathways of energy dissipation in cyanobacteria differ significantly from that in higher plants because of the differences in pigment apparatus. The excessively absorbed energy in cyanobacteria is diverted away from the reaction centers by two different mechanisms. One of them relates to the decrease in the fluorescence yield of the longest-wavelength antenna chlorophylls by P700 cation radical in the PSI trimeric complexes. Another one deals with quenching of membrane-bound phycobilisome fluorescence. The conformation changes in phycobilisomes under illumination of cyanobacterial cells by intense blue-green light absorbed by the orange carotenoid-binding protein transform the non-quenched state of antenna into quenched state. Thus, energy dissipation, monitored as non-photochemical fluorescence quenching, occurs in different steps of energy transfer within the phycobilisomes or core antenna of PSI in cyanobacteria.

Keywords: Photoinhibition, Photosynthetic pigment, Cyanobacteria, Fluorescence quenching.

Abbreviations

APC: Allophycocyanin, *Chl:*Chlorophyll, *Chl710 (Chl740 etc):* Chlorophyll with absorption band peaked at10 nm (740 etc), *OCP:* Orange carotenoid-binding protein, *PBS:* Phycobilisome, *PC:* Phycocyanin, *700 (P700+):* Primary electron donor of the PSI reaction center in reduced (oxidized) state, *PSI (PSII):* Photosystem I (Photosystem II).

fluorescence quenching was ascribed preliminarily to PBS emission (El-Bissati *et al.*, 2000). In a PSII-less *Synechocystis* mutant, without variable fluorescence, illumination with intense 450-500 nm light was found to quench PBS fluorescence at 660 nm (Rakhimberdieva *et al.*, 2004). Similar fluorescence quenching was observed in wild-type *Synechocystis* cells with 3-(3,4-dichlorophenyl)-1,1-dimethylurea (DCMU) to block PSII activity (Rakhimberdieva *et al.*, 2007a; Scott *et al.*, 2006). This quenching of PBS fluorescence in cyanobacteria induced by blue-green light is independent of the pH of the lumen and treatment with uncouplers, observed only at physiological temperatures; the reversibility of quenching is more temperature dependent than light-on process (Rakhimberdieva *et al.*, 2004, 2007a; Wilson *et al.*, 2006; Scott *et al.*, 2006).

The action spectrum of blue light induced quenching for *Synechocystis* PSII-less mutant (Fig. 4.1) was characterized in spectral region 450-500 nm by three bands that allowed to determine photosensor as carotenoid (Rakhimberdieva *et al.*, 2004). Investigation of the origin and location of carotenoid that triggers blue light-induced fluorescence quenching of PBS in cyanobacteria, as well as an investigation of the interaction of carotenoids with partner in PBS revealed the new mechanisms of energy dissipation in cyanobacteria. Blue-green light induced fluorescence quenching was not observed in *Synechocystis* mutants lacking APC (Wilson *et al.*, 2006). Fluorescence quenching was also absent in mutant deficient in linker polypeptide ApcE (Rakhimberdieva *et al.*, 2006, 2007b) where the biosynthesis of PBS core is blocked (Shen *et al.*, 1993). Since action spectrum of PBS fluorescence quenching (Rakhimberdieva *et al.*, 2004; Scott *et al.*, 2006) was identical to absorption spectrum of 3'-hydroxyechinenone in OCP (Fig. 4.1) (Polivka *et al.*, 2005; Wilson *et al.*, 2008), it was proposed that this water-soluble OCP, localized out of membranes, was responsible for triggering of fluorescence quenching of PBS under blue-green light illumination (Wilson *et al.*, 2006, 2008).

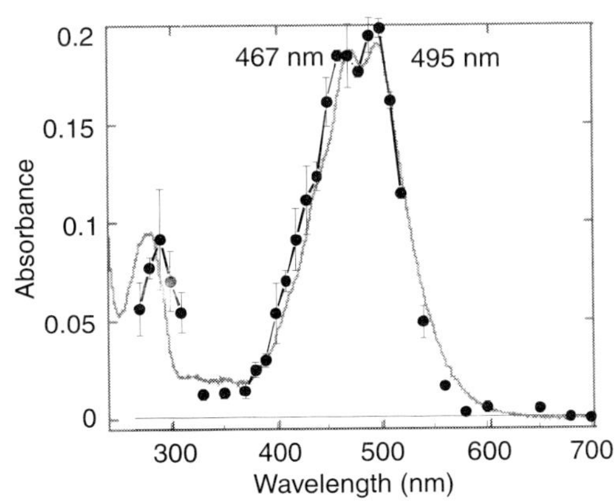

Fig. 4.1 Action spectrum (black circles) of OCP-induced quenching of PBS fluorescence in *Synechocystis* sp PCC 6803 (Rakhimberdieva *et al.*, 2004) and the absorption spectrum (solid line) of isolated OCP (Wilson *et al.*, 2008).

OCP, a 35 kDa two-domain protein, containing one molecule of 3′-hydroxyechinenone, was isolated as a dimer (Kerfeld *et al.*, 2003). This carotenoid found in cyanobacteria *A. maxima* and *Synechocystis* sp. PCC 6803, is present in *Synechocystis* as glycoside derivative (Kerfeld, 2004; Bouley *et al.*, 2008). Since blue-light induced quenching was more pronounced in cyanobacteria grown in iron depletion, at a certain period it was suggested that IsiA could be responsible for quenching; "free" PBS could interact with Chl-containing IsaA complex that may replace LHCII lacking in cyanobacteria (Cadoret *et al.*, 2004; Bailey *et al.*, 2005). However, quenching of PBS fluorescence by blue-green light was found for IsiA-less mutant of *Synechocystis* (Rakhimberdieva *et al.*, 2006, 2007b; Wilson *et al.*, 2006, 2007).

Thus, although IsiA complex is able to dissipate energy in cyanobacteria (Ihalainen *et al.*, 2005; van der Weij-de Wit *et al.*, 2007; Wilson *et al.*, 2007), it is not necessary for quenching of PBS fluorescence.

77K quenching spectrum (fluorescence spectrum of the untreated samples minus spectrum of samples treated at 288 K with blue-light and later cooled) of PSII-less *Synechocystis* mutant shows quenching of 683 nm band ascribed to terminal emitter (Rakhimberdieva *et al.*, 2007b), whereas at room temperature the fluorescence of APC at 660 and 676 nm is quenched (Rakhimberdieva *et al.*, 2004). Quenching of fluorescence at 679 nm ascribed to antenna Chls of PSII was also found in wild strain of *Synechocystis* at room temperature. It was suggested that these Chls are in energetic equilibration with terminal emitter of PBS (Rakhimberdieva *et al.*, 2007b). The effect of high glycerol concentrations on the time-course of quenching and the fluorescence spectra of PSII-less mutant strain of *Synechocystis* indicates that fluorescence of PBS energetically coupled with membrane is quenched (Rakhimberdieva *et al.*, 2007b). The increase in protein hydration as a result of high glycerol concentration causes the uncoupling of PBS from the membrane and damages their structure, suppressing blue-green light induced fluorescence quenching.

Time-resolved studies show that in PBS coupled to thylakoid membrane, energy from phycocyanin migrates to APC for 20-40 ps and then to Chls of PSII for 90-170 ps (Mullimeaux and Holzwarth, 1991; Vavilin *et al.*, 2003). When the PSII reaction centers are open, the life-time of the PSII excited state is less than the time of energy migration from APC to PSII, and energy migrates to Chls of PSII. But if the PSII activity is inhibited by DCMU, the life-time of excited Chls and the reverse energy migration from Chls to APC is of about 1500-1700 ps (Mullineaux *et al.*, 1993; Bittersmann and Vermaas, 1991), and excitation energy could migrate from Chls back to APC. Data on picosecond fluorescence decay obtained for dark adapted or treated by blue-green light cells of the cyanobacterium *Synechocystis* mutant confirm the quenching of APC fluorescence at room temperature. Under excitation at 650 nm, absorbed predominantly by APC, the decay component 25 ps that peaked at 660 nm was revealed in quenched cells (Scott *et al.*, 2006). The decay component 35 ps that peaked at 660 nm was also observed under excitation of dark-adapted cells by 407 nm light, which is equally absorbed by phycocyanin, APC, and Chl *a*. This component is ascribed to energy transfer from phycocyanin to APC; significant decrease in the amplitude of this component in quenched cells indicates fast quenching of APC excitation. Two peaks at 680 and 660 nm found for the second fast component (110 ps) may reflect energy transfer to terminal emitter or Chl *a*. Non-linear dependence of the rate constant of fluorescence quenching in PSII-less *Synechocystis* mutant versus temperature was observed, and the value of activation enthalpy of quenching (48 kJ mol^{-1} at 25) was very high (Rakhimberdieva *et al.*, 2007a). Thus light absorbed by carotenoid causes conformation changes of protein, inducing the protein-protein interaction typical for folding of water-soluble proteins (Oliveberg *et al.*, 1995).

The participation of PBS coupled with PSI or PSII and PBS coupled only with the membrane in carotenoid-induced quenching of PBS fluorescence in the cyanobacterium *Synechocystis* was discussed (Karapetyan, 2007; Rakhimberdieva *et al.*, 2007b). PSII coupled PBS fluorescence is quenched involving PSII light-harvesting Chl in the quenching. Fluorescence of PBS bound with membrane but energetically uncoupled from both photosystems is quenched by blue-green light (Rakhimberdieva *et al.*, 2007). If PBS is functionally coupled to PSI, energy absorbed by PBS migrates to PSI antenna Chls independently from the redox state of P700 since both P700 and $P700^+$ are the efficient quenchers. Thus fluorescence of PBS coupled with PSI is strongly quenched by PSI and therefore cannot be revealed. Hence blue light induced dissipation of excitation energy absorbed by PBS does not depend on the PBS status.

Carotenoid-induced quenching of PBS fluorescence diverts excitation away from both the reaction centers protecting the photosynthetic apparatus against photodestruction (Rakhimberdieva *et al.*, 2010).

The molecular mechanisms of carotenoid-induced fluorescence quenching of PBS is still unclear because the interaction of OCP with PBS as well as the interaction of PBS with PSII and PSI complexes are not well understood (Bald *et al.*, 1996; Sarcina *et al.*, 2001; Mullineaux, 2008). Several mechanisms of OCP-induced quenching of PBS fluorescence have been proposed to account for the role of conformational changes in energy dissipation (Karapetyan, 2007, 2008; Kirilovsky, 2007; Bailey and Grossman, 2008):

(i) The OCP binds to the PBS core and the carotenoid quenches excitation energy via direct interaction with the PBS.

(ii) The OCP binds to the PBS core, but a quenching center forms through re-orientation of PBS pigments.

(iii) The OCP acts indirectly in the development of quenching via formation of a PBS quenching complex with other yet unidentified components.

All these scenarios will lead to fluorescence quenching at room temperature. Recently shown light-induced changes in the OCP (Wilson *et al.*, 2008) back with the first (i) scenario; the long wavelength shift of carotenoid absorption probably allows quenching via low-energy S1 excited level of hydroxyechinenone OCP chromophores. Recently it was found that illumination of isolated OCP with blue-green light transforms reversibly the orange form of OCP into red form, which is considered "active" one essential for the induction of the photoprotective mechanism (Wilson *et al.*, 2008). A similar type of singlet–singlet Chl fluorescence quenching by zeaxanthin is known for LHCPII in higher plants (Ruban *et al.*, 2007).

The scheme of OCP-induced quenching is presented in Fig. 4.2. It is quite possible that the same carotenoid of OCP, serving as sensitizer, may function as a quencher. Absorption of blue-green light with carotenoid leads to conformation changes in protein that may regulate the binding of quenching-induced carotenoid-protein with PBS or make the OCP-PBS interaction easier. In this case the energy could transfer from APC to carotenoid and dissipated (Scott *et al.*, 2006; Rakhimberdieva *et al.*, 2007b). It may also be postulated that conformation change is important for increasing the overlap between S1

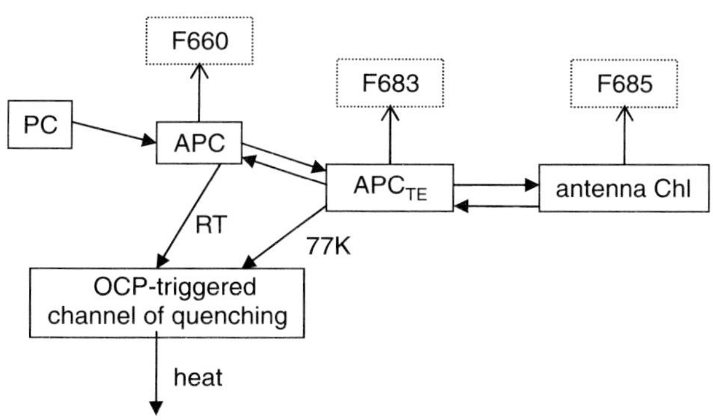

Fig. 4.2 Scheme of blue light induced energy dissipation in PBS of the cyanobacterium *Synechocystis* sp. PCC 6803. OCP triggers a heat dissipation of APC core excitation; APC_{TE} – terminal emitters of APC core. The origin of quencher is not yet known. F660, F683, F685 – 77K fluorescence bands peaked at 660, 683 and 685 nm, respectively; RT – room temperature (according to Karapetyan 2008).

levels of APC and OCP that stimulates resonance energy migration between them, even if the partners are located relatively far and have no direct contact. Other possibilities could also be mentioned: the conformational change in OCP induces a conformational change in a PBS core polypeptide, leading to a quenched conformation, or the conformational change in OCP may promote the coupling of PBS to another (unknown) quenching complex in the membrane.

P700$^+$-INDUCED FLUORESCENCE QUENCHING OF RED CHLS IN PSI

From an evolutionary standpoint, the PSI-type of the reaction center is accepted to be the most ancient (Blankenship, 1992). Primary cyanobacteria have formed the red Chls in PSI antenna, which as terminal energy acceptors may regulate the energy flow to open P700 and dissipation of excessively absorbed energy with closed P700. The PSI core antenna of the cyanobacterium *A. platensis* contains the red Chls absorbing at 740 and emitting at 760 nm at 77 K (Shubin *et al.*, 1991, 1993; Karapetyan *et al.*, 1997). This Chl considered the excitonically coupled Chls is more stable than the Chl emitting at 730 nm (Cometta *et al.*, 2000; Engelmann *et al.*, 2001; Schlodder *et al.*, 2007). Energy absorbed by red Chls of PSI trimers and monomers of *A. platensis* migrates uphill to P700 causing its oxidation (Shubin *et al.*, 1995). Energy migration delays on the way from bulk to long-wavelength Chls: energy equilibration within bulk antenna of PSI trimers of *A. platensis* at room temperature occurs within 4 ps, while equilibration within the heterogeneous antenna (including red Chls) - for15 ps. Energy migration to the most red Chl absorbing at 740 nm in PSI of *A. platensis* increases the trapping time up to 50 ps, while trapping time for PSI of *Synechocystis* that contain less red Chls is only 23 ps (Gobets *et al.*, 2001). Because of higher life-time of excited state, the red-most Chls function as terminal emitter participating in energy dissipation.

77 K fluorescence yield of the reddest Chl form in PSI of *A. platensis* [Shubin *et al.,* 1993] and *T. elongatus* (Byrdin *et al.*, 2000) highly depends on P700 redox state. P700$^+$ is an extremely efficient quencher of fluorescence at 760 nm of the reddest Chl (Fig. 4.3) (Shubin *et al.*, 1995; Karapetyan *et al.*, 1997). Under high light the reaction center of PSI is closed because of formation of cation radical or triplet state, dependent on redox state of the PSI acceptor side cofactors. The value of the overlap integral of the fluorescence band at 760 nm and the absorption band of P700$^+$ at 77 K is more than two orders of magnitude higher than that of fluorescence band of the same Chl and the absorption band of P700 [Karapetyan 1998, 2004a]. Although in core antenna the reddest Chl form is localized at a distance of 32-40 Å from P700 (Dorra *et al.*, 1998; Schlodder *et al.*, 2005), the high value of the overlap integral is the main reason of P700$^+$ as an efficient quencher. Red Chls in plant PSI emitting at 735 nm (77 K) are localized on Lhca1/Lhca4 (Jansson 1994) positioned on a large distance of about 65 Å from P700 (Dorra *et al.*, 1998), which makes the interaction of P700 with that of red Chls difficult. Therefore the fluorescence yield of red Chls in higher plants is independent of the P700 redox state.

The excitation absorbed by antenna Chls migrates to active P700 generating charge separation and electron transport, while transition of reaction center into inactive state leads to dissipation of excess energy, protecting against destruction. At room temperature, the reduced and oxidized P700 are equally efficient quenchers that may explain very low variable fluorescence of PSI (Karapetyan *et al.* 1973). P700$^+$ in PSI trimers of *A. platensis* at 77 K is much stronger quencher than P700 reduced (Shubin *et al.*, 1995; Karapetyan *et al.*, 1997; Schlodder *et al.*, 2005). When P700 and the cofactors of PSI acceptor side are prereduced, five decay components have been revealed in fluorescence kinetics of PSI trimers at 77 K (Dorra *et al.*, 1998; Karapetyan *et al.*, 2004a). The fastest component (9-12 ps) reflects the energy equilibrium within bulk antenna; 190 and 450 ps components correspond to the energy migration from the red Chls with emission maxima at 720 and 730 nm to the extreme red Chl emitting at 760 nm. An efficient quenching of fluorescence at 760 nm by P700$^+$ is accompanied by a decrease in decay lifetime components from 190 to 107 ps and from 450 to 350 ps (Holzwarth *et al.*, 1998). Accumulation of P700$^+$ leads to the formation of triplet state of antenna Chls; the decay half-times 10 and 80 ns at 77 K coincides exactly with the kinetics of formation of β-carotene triplet state is a result of direct triplet-triplet exchange (Schlodder *et al.*, 2001).

P700 in triplet state quenches the fluorescence efficiently at 760 nm emitted by red Chls in PSI trimers of *A. platensis* at reductive conditions (Fig. 4.3). When electron transport from A_0 to A_1 is blocked,

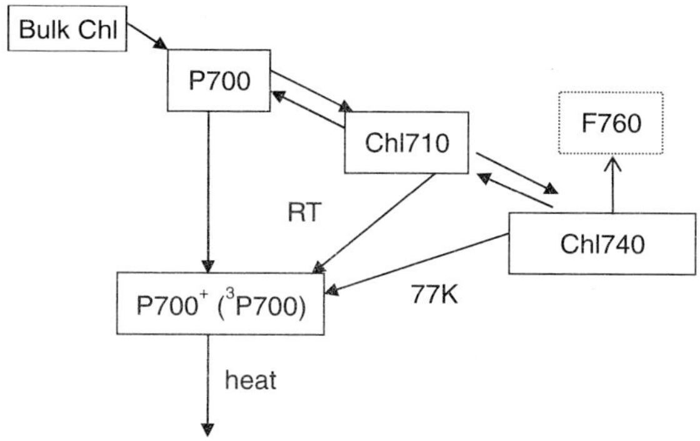

Fig. 4.3 Scheme of energy dissipation in core antenna of PSI trimers of the cyanobacterium *A. platensis*. P700$^+$ or P700 in triplet state quench the fluorescence at 760 nm, dependent on redox state of the PSI acceptor side cofactors. When P700 is reduced, energy from Chl740 migrates uphill to P700, causing its oxidation. F685, F730, F760 are the 77 K fluorescence bands peaked at 685, 730 and 760 nm, respectively (according to Karapetyan 2008).

since the PSI acceptor side cofactors are preliminary reduced by dithionite, recombination of separated charges in reaction center leads to the formation of P700 triplet state with half-times of 0.75 and 5.2 ms at 5 K (Schlodder *et al.*, 2005). The decay kinetics of P700 in triplet state coincide exactly with the kinetics of light-induced decrease of 760 nm fluorescence confirming that P700 triplet state quenches fluorescence of red Chls (Schlodder *et al.*, 2005).

Preferable organization of PSI complexes in cyanobacterial thylakoids in the form of trimers (Shubin *et al.*, 1993) is responsible for additional pathway of energy dissipation. When P700 in one of the monomer complexes within a PSI trimer of the cyanobacterium *A. platensis* is oxidized, dissipation of excess energy occurs as a result of interaction of antenna Chls via peripherally located most red Chl emitting at 760 nm (Karapetyan *et al.*, 1999a; Kruip *et al.*, 1999; Schlodder *et al.*, 2005). Thus, aggregation of pigment-protein complexes stimulates dissipation of excessively absorbed energy protecting photosynthetic membranes against photodestruction (Karapetyan, 2004b; Ihalainen *et al.*, 2005). Fluorescence quenching of the red Chl by P700 cation radical (or P700 triplet state) found at 77 K, may be considered as NPQ that occurs in cyanobacteria at physiological temperatures as well. Recently it was suggested that one of the functions of HLIPs associated with PSI trimers is to dissipate the excess energy from PSI trimers causing their stabilization on high light; deletion of HLIPs resulted in loss of PSI trimers in high light (Wang *et al.*, 2008).

The lack of electrons from PSII to PSI under high light is leading to accumulation of P700 in oxidized state. At high light stress conditions, up to 70-90% of P700 is accumulated as P700$^+$ in green leaves of different tropical plants, thus quenching the excess absorbed energy. Since the quantum yield of photodestruction of P700$^+$ is about three orders of magnitude lower than that of antenna Chls (Shubin *et al.*, 2008), it seems reasonable to postulate that PSI is resistant to high light stress *in vivo* because of the presence of P700$^+$. Under chilling stress a large part of P700 even under low light would be in its reduced form by electron flow from PSII, leading to the selective photoinhibition of PSI (Sonoike, 2006; Scheller and Haldrup, 2005). In these conditions, P700 cannot function as photochemical quencher because the flow-out of electrons from PSI is inhibited. It would be reasonable to assume that the absence of P700$^+$ is one of the reasons of the chilling-induced photoinhibition of PSI. Thus, the unique structure of PSI complex and the efficient P700$^+$-induced quenching of excess excitation energy are the basis of extremely high photostability of PSI. The extremely photostable P700$^+$ functions as an efficient quencher of singlet excitation energy in PSI for a long time, stabilizing the photosynthetic apparatus.

CONCLUSION

Photoprotection has appeared on the earliest phase of biological evolution since the ocean surface without ozone layer was irradiated by sunlight with intensity about two orders of magnitude higher than now. Therefore at high irradiance, the ancient heterotrophs could survive mainly in deep water layers of the ocean. Probably dissipative photochemistry could function in ancient cyanobacteria. The appearance of photosynthesis may be considered a transition from dissipative to productive photochemistry (Mulkidjanian and Junge, 1997) when coupling of energy transfer with electron transport in the reaction center has offered the ability to store the absorbed energy. The same pigments are involved in light-harvesting and dissipation of absorbed energy. Dissipation of excessively absorbed energy (NPQ) of pigment antenna is a general property of all photosynthetic biosystems to protect photosynthesis against destruction. Cyanobacteria, as an evolutionary oldest photosynthetic organisms, populated on water depth and could utilize photosynthesis light absorbed by carotenoids, phycobilins, and Chls. Oxygen evolved by cyanobacterial photosynthesis has formed the ozone screen, preventing the penetration of rigid short-wavelength UV on the earth surface. In course of evolution, the photosynthetic organisms created several strategies to dissipate the excessively absorbed energy thus stabilizing the photosynthetic apparatus. Under high light illumination, due to the imbalance of absorbed energy and CO_2 assimilation, energy dissipation processes were involved in protection of the photosynthetic apparatus against the damaging effect of excessively absorbed energy. The OCP-induced quenching of PBS fluorescence was probably the oldest protection mechanism of the photosynthetic apparatus of cyanobacteria. OCP, which is localized outside the membrane, may be involved in dissipation of energy absorbed by PBS. Interaction of water-soluble carotenoid-binding protein OCP with extra-membrane PBS in cyanobacteria was replaced in course of evolution by energy dissipation via membrane-bound carotenoid in LHCII complex. Although the discussed pathways of energy dissipation differ in mechanisms, in both cases energy arriving to the terminal energy acceptors in PBS or in PSI antenna of cyanobacteria dissipates.

ACKNOWLEDGEMENTS

This work has been supported by the Russian Academy of Sciences, program "Molecular and cell biology" and the Russian Foundation of Basic Research, grant 08-04-00143a.

REFERENCES

Adir N (2005) Elucidation of the molecular structures of components of the phycobilisome: reconstructing a giant. Photosynth Res 85: 15-32

Allen JF, Holmes NG (1986) A general model for regulation of photosynthetic unit function by protein phosphorylation. FEBS Lett 202: 175-181

Arteni AA, Ajlani G, Boekema EJ (2009) Structural organisation of phycobilisomes from *Synechocystis* sp. strain PCC 6803 and their interaction with the membrane. Biochim Biophys Acta 1787: 272-279

Bailey S, Mann N, Robinson C, Scanlan DJ (2005) The occurrence of rapidly reversible non-photochemical quenching of chlorophyll fluorescence in cyanobacteria. FEBS Lett 579: 275-280

Bailey S, Grossman A (2008) Photoprotection in cyanobacteria: Regulation of light harvesting. Photochem Photobiol 84: 1410-1420

Bald D, Kruip J, Roegner M (1996) Supramolecular architecture of cyanobacterial thylakoid membranes: How is the phycobilisome connected with the photosystems? Photosynth Res 49: 103-118

Biggins J, Bruce D (1989) Regulation of excitation energy transfer in organisms containing phycobilisomes. Photosynth Res 20: 1-34

Bittersmann E, Vermaas WFJ (1991) Fluorescent lifetime studies of cyanobacterial photosystem II mutants. Biochim Biophys Acta 1098: 105-116

Blankenship RE (1992) Origin and early evolution of photosynthesis. Photosynth Res 33: 91-111

Bouley C, Abasova L, Six C, Vass I, Kirilovsky D (2008) Occurrence and function of the orange carotenoid protein in photoprotective mechanisms in various cyanobacteria. Biochim Biophys Acta 1777: 1344-1354

Byrdin M, Rimke I, Schlodder E, Stehlik D, Roelofs TA (2000) Decay kinetics and quantum yields of fluorescence in photosystem I from *Synechococcus elongatus* with P700 reduced and oxidized state: are the kinetics of excited state decay trap-limited or transfer-limited? Biophys J 79: 992-1007

Cadoret J-C, Demouliere R, Lavand J, van Gorkom H, Houmard J, Etienne A-L (2004) Dissipation of excess energy triggered by blue light in cyanobacteria with CP43' (IsiA). Biochim Biophys Acta 1659: 100-104

Capuano V, Braux AS, Tandeau de Marsac N, Houmard J (1991) The "anchor polypeptide" of cyanobacterial phycobilisomes. Molecular characterization of the *Synechococcus* sp. PCC 6301 *apcE* gene. J Biol Chem 266: 7239-7247

Cometta A, Zucchelli G, Karapetyan NV, Engelmann E, Garlaschi FM, Jennings RC (2000) Thermal behavior of long wavelength absorption transitions in *Spirulina platensis* photosystem I trimers. Biophys J 79: 3235-3243

Dorra D, Fromme P, Karapetyan NV, Holzwarth AR (1998) Fluorescence kinetics of photosystem I: multiple fluorescence components. In Garab G (ed) Photosynthesis: Mechanisms and Effects. Kluwer Acad Publ, Dordrecht, Vol 1, pp 587-590

El-Bissati K, Delphin E, Murata N, Etienne A-L, Kirilovsky D (2000) Photosystem II fluorescence quenching in the cyanobacterium *Synechocistis* sp. PCC 6803: involvement of two different mechanisms. Biochim Biophys Acta 1457: 229-242

Engelmann E, Tagliabue T, Karapetyan NV, Garlaschi FM, Zucchelli G, Jennings RC (2001) CD spectroscopy provides evidence for excitonic interactions involving red-shifted chlorophyll forms in photosystem I. FEBS Lett 499: 112-115

Fujita J (1997) A study of the dynamic features of photosystem stoichiometry: Accomplishments and problems for future studies. Photosynth Res 53: 83-93

Funk C, Vermaas WFJ (1999) A cyanobacteria gene family coding for single-helix proteins resembling part of the light-harvesting proteins from higher plants. Biochemistry 38: 9397-9404

Gobets B, van Stokkum IHM, Roegner M, Kruip J, Schlodder E, Karapetyan NV, Dekker JP, van Grondelle R (2001) Time-resolved fluorescence emission measurements of photosystem I particles of various cyanobacteria: a unified compartmental model. Biophys J 81: 407-424

Grimm B, Kruse E, Kloppstech K (1989) Transiently expressed early light-inducible thylakoid proteins share transmembrane domains with light-harvesting chlorophyll-binding proteins. Plant Mol Biol 13: 583-593

Henderson JN, Zhang J, Evans B, Redding K (2003) Disassembly and degradation of photosystem I in an *in vitro* system are multievent, metal-dependent processes. J Biol Chem 278: 39078-39086

Holzwarth AR, Dorra D, Mueller MG, Karapetyan NV (1998) Structure-function relationship and excitation dynamics in photosystem I. In: Garab G (ed) Photosynthesis: Mechanisms and Effects. Kluwer Academic Publishers, Dordrecht Vol 1, pp 497-502

Horton P, Ruban A, Walters RG (1996) Regulation of light harvesting in green plants. Ann Rev Plant Physiol Plant Mol Biol 47: 655-684

Horton P, Ruban AV, Wentworth M (2000) Allosteric regulation of the light-harvesting system of photosystem II. Phil Trans R Soc London 355B: 1361-1370

Ihalainen JA, D'Haene S, Yeremenko N, van Roon H, Arteni AA, Boekema EJ, van Grondelle R, Matthijs HCP, Dekker JP (2005) Aggregates of the chlorophyll-binding protein IsiA (CP43') dissipate energy in cyanobacteria. Biochemistry 44: 10846-10853

Isono T, Katoh T (1987) Subparticles of *Anabaena* phycobilisomes II. Molecular assembly of allophycocyanin cores in reference to "anchor" protein. Arch Biochem Biophys 256: 317-324

Jansson S (1994) The light-harvesting chlorophyll a/b binding proteins. Biochim Biophys Acta 1184: 1-19

Jordan P, Fromme P, Klukas O, Witt HT, Saenger W, Krauss N (2001) Three-dimensional structure of cyanobacterial photosystem I at 2.5 Å resolution. Nature 411: 909-917

Joshua S, Bailey S, Mann NH, Mullineaux CW (2005) Involvement of phycobilisome diffusion in energy quenching in cyanobacteria. Plant Physiol 138: 1577-1585

Karapetyan NV (1998) Organization and the role of long-wave chlorophylls in the photosystem I of the cyanobacterium *Spirulina*. Membr Cell Biol 12: 571-584

Karapetyan NV (2004a) The dynamics of excitation energy in photosystem I of cyanobacteria: transfer in the antenna, capture by the reaction center and dissipation. Biophysika (Russ) 49: 212-226

Karapetyan NV (2004b) Interaction of pigment-protein complexes within aggregates stimulates dissipation of excess energy. Biochemistry (Moscow) 69: 1598-1605

Karapetyan NV (2007) Non-photochemical quenching of fluorescence in cyanobacteria. Biochemistry (Moscow) 72: 1127-1135

Karapetyan NV (2008) Protective dissipation of excess absorbed energy by photosynthetic apparatus of cyanobacteria: role of antenna terminal emitters. Photosynth. Res. 97: 195-204

Karapetyan NV, Klimov VV, Krasnovsky AA (1973) Light-induced changes in the fluorescence yield of particles obtained by digitonin fragmentation of chloroplasts. Photosynthetica 7: 330-337

Karapetyan NV, Dorra D, Schweitzer G, Bezsmertnaya IN, Holzwarth AR (1997) Fluorescence spectroscopy of the longwave chlorophylls in trimeric and monomeric photosystem I core complexes from cyanobacterium *Spirulina platensis.* Biochemistry 36: 13830-13837

Karapetyan NV, Holzwarth AR, Roegner M (1999a) Photosystem I trimer of cyanobacteria: molecular organization, excitation dynamics and physiological significance. FEBS Lett 460: 395-400

Karapetyan NV, Shubin VV, Strasser RJ (1999b) Energy exchange between the chlorophyll antennae of monomeric subunits within the photosystem I trimeric complex of cyanobacterium *Spirulina*. Photosynth Res 61: 291-301

Karapetyan NV, Schlodder E, van Grondelle R, Dekker JP (2006) Long-wavelength chlorophylls of photosystem I. In: Golbeck JH (ed) Photosystem I: The Light-Driven Plastocyanin: Ferredoxin Oxydoreductase, series Advances in Photosynthesis and Respiration, Springer, Dordrecht, Vol 24, pp 177-192

Kerfeld CA (2004) Water-soluble carotenoid proteins of cyanobacteria. Arch Biochim Biophys 430: 2-9

Kerfeld CA, Sawaya MR, Brahmandam V, Cascio D, Ho KK, Trevithick-Sutton CC, Krogmann DW, Yeates TO (2003) The crystal structure of a cyanobacterial water-soluble protein. Structure 11: 56-65

Kirilovsky D (2007) Photoprotection in cyanobacteria: the orange carotenoid protein (OCP)-related non-photochemical-quenching mechanism. Photosynth Res 93: 7-16

Kruip J., Karapetyan NV, Terekhova I.V., Roegner M (1999) *In vitro* oligomerisation of a membrane protein complex: liposome based reconstitution of trimeric photosystem I from isolated monomers. J Biol Chem 274: 18181-18188

Lundell DJ, Glazer AN (1981) Allophycocyanin B: A common β subunit in *Synechococcus* allophycocyanin B (A_{max} 670 nm) and allophycocyanin (A_{max} 650 nm). J Biol Chem 256: 12600-12606

McColl R (1998) Cyanobacterial phycobilisomes. J Struct Biol 124: 311-334

Mimuro M (2004) Photon capture, exciton migration and trapping and fluorescence emission in cyanobacteria and red algae. In: Papageprgiou G, Govindjee (eds) Chlorophyll a Fluorescence: A Signature of Photosynthesis, Springer, pp 173-195

Mimuro M, Lipschultz CA, Gantt E (1986) Energy flow in the phycobilisomes of *Nostoc* sp. (Mac): Two independent terminal pigments. Biochim Biophys Acta 852: 126-132

Mimuro M, Kikuchi H, Murakami A (1999) Structure and function of phycobilisomes. In: Singhal GS, Renger G, Sopory SK, Irgang K-D, Govindjee (eds), Concepts in Photobiology: Photosynthesis and Photomorphogenesis, Narosa Publ House, New Delhi, India, pp 104-135

Mulkidjanian AY, Junge W (1997) On the origin of photosynthesis as inferred from sequence analysis. Photosynth Res 51: 27-42

Mullineaux CW (2008) Phycobilisome-reaction centre interaction in cyanobacteria. Photosynth Res 95: 175-182

Mullineaux CW, Holzwarth AR (1991) Effect of photosystem II reaction centre closure on fluorescence decay kinetics in a cyanobacterium. Biochim Biophys Acta 1098: 68-78

Mullineaux CW (1992) Excitation energy transfer from phycobilisomes to photosystem I in a cyanobacterium. Biochim Biophys Acta 1100: 285-292

Mullineaux CW, Pascal AA, Horton P, Holzwarth AR (1993) Excitation-energy quenching in aggregates of the LHC-II chlorophyll-protein complex – a time-resolved fluorescence study. Biochim Biophys Acta 1141: 23-28

Niyogi KK (1999) Photoprotection revisited: Genetic and molecular approaches. Ann Rev Plant Physiol Plant Mol Biol 50: 333-359

Oliveberg M, Tan Y-J, Fersht AR (1995) Negative activation enthalpies in the kinetics of protein folding. Proc Natl Acad Sci USA 92: 8926-8929

Polivka T, Kerfeld CA, Pascher T, Sundstroem V (2005) Spectroscopic properties of the carotenoid 3'-hydroxyechinenone in the orange carotenoid protein from the cyanobacterium *Arthrospira maxima*. Biochemistry 44: 3994-4003

Rakhimberdieva MG, Boichenko VA, Karapetyan NV, Stadnichuk IN (2001) Interaction of phycobilisomes with photosystem 2 dimers and photosystem 1 monomers and trimers of the cyanobacterium *Spirulina platensis*. Biochemistry 40: 15780-15788

Rakhimberdieva MG, Stadnichuk IN, Elanskaya IV, Karapetyan NV (2004) Carotenoid-induced quenching of the phycobilisome fluorescence in photosystem II-deficient mutant of *Synechocistis* sp. FEBS Lett 574: 85-88

Rakhimberdieva MG, Elanskaya IV, Vavilin DV, Vermaas WFJ, Karapetyan NV (2006) Blue light-induced fluorescence quenching of phycobilisomes in the cyanobacterium *Synechocystis* sp. PCC 6803 in the absence of the IsiA protein. Biochim Biophys Acta 1757 Supl 1, 286-287

Rakhimberdieva MG, Bolychevtseva YV, Elanskaya IV, Karapetyan NV (2007a) Protein-protein interactions in carotenoid triggered quenching of phycobilisome fluorescence in *Synechocystis* sp. PCC 6803. FEBS Lett 581: 2429-2433

Rakhimberdieva MG, Vavilin DV, Vermaas WFJ, Elanskaya IV, Karapetyan NV (2007b) Phycobilin/chlorophyll excitation equilibration upon carotenoid-induced non-photochemical fluorescence quenching in phycobilisomes of the cyanobacterium *Synechocystis* sp. PCC 6803. Biochim Biophys Acta 1767: 757-765

Rakhimberdieva MG, Elanskaya IV, Vermaas WFJ, Karapetyan NV (2010) Carotenoid-triggered energy dissipation in phycobilisomes of *Synechocystis* sp. PCC 6803 diverts excitation away from reaction centers of both photosystems. Biochim Biophys Acta 1797: 241-249

Ruban AB, Berera R, Ilioaia C, van Stockkum IHM, Kennis JTM, Pascal AA, van Amerongen H, Robert B, Horton P, van Grondelle R (2007) Indentification of a mechanism of photoprotective energy dissipation in higher plants. Nature 450: 575-578

Sarcina M, Tobin MJ, Mullineaux CW (2001) Diffusion of phycobilisomes on the thylakoid membranes of the cyanobacterium *Synechococcus* 7942: Effect of phycobilisome size, temperature, and membrane lipid composition. J Biol Chem 276: 46830-46834

Scheller HV, Haldrup A (2005) Photoinhibition of photosystem I. Planta 221: 5-8

Schlodder E, Paul A, Cetin M (2001) Triplet states in photosystem I complexes from *Synechococcus elongatus*, in: Proc of the 12th Internat Congress on Photosynthesis, CSIRO; Publishing: Melbourne, Australia, www. publishcsiro.au/ps2001; contribution S6-015

Schlodder E, Cetin M, Byrdin M, Terekhova IN, Karapetyan NV (2005) P700^{+}- and ^{3}P700-induced quenching of the fluorescence at 760 nm in trimeric photosytem I complexes from the cyanobacterium *Arthrospira platensis*. Biochim Biophys Acta 1706: 53-67

Schlodder E, Shubin VV, El-Mohsnawy E, Roegner M, Karapetyan NV (2007) Steady-state and transient polarized absorption spectroscopy of photosytem I complexes from the cyanobacteria *Arthrospira platensis* and *Thermosynechococcus elongatus.* Biochim Biophys Acta 1767: 732-741

Scott M, McCollum C, Vasiliev S, Croizer C, Espie GS, Krol M, Huner NPA, Bruce D (2006) Mechanism of the down regulation of photosynthesis by blue light in the cyanobacterium *Synechocystis* sp. PCC 6803. Biochemistry 45: 8952-8958

Shen G, Boussiba S, Vermaas WFJ (1993) *Synechocystis* sp. PCC 6803 strains lacking photosystem I and phycobilisome function. Plant Cell 5: 1853-1863

Shubin VV, Murthy SDS, Karapetyan NV, Mohanty P (1991) Origin of the 77K variable fluorescence at 758 nm in cyanobacterium *Spirulina platensis.* Biochim Biophys Acta 1060: 28-36

Shubin VV, Bezsmertnaya IN, Karapetyan NV (1992) Isolation from *Spirulina* membranes of two photosystem I-type complexes one of which contains chlorophyll responsible for the 77 K fluorescence band at 760 nm. FEBS Lett 30: 340-342

Shubin VV, Tsuprun VL, Bezsmertnaya IN, Karapetyan NV (1993) Trimeric forms of the photosystem I reaction center complex pre-exist in the membranes of the cyanobacterium *Spirulina platensis.* FEBS Lett 334: 79-82

Shubin VV, Bezsmertnaya IN, Karapetyan NV (1995) Efficient energy transfer from the longwave antenna chlorophylls to P700 in photosystem I complexes from *Spirulina.* J Photochem Photobiol 30B: 153-160

Shubin VV, Terekhova IV, Kirillov BA, Karapetyan NV (2008) Quantum yield of $P700^+$ photodestruction in isolated photosystem I complexes of the cyanobacterium *Arthrospira platensis.* Photochem Photobiol Sci 7: 956-962

Sonoike K (2006) Photoinhibition and protection of photosystem I. In: Golbeck JH (Ed) Photosystem I: The Light-Driven Plastocyanin:Ferredoxin Oxydoreductase, series Advances in Photosynthesis and Respiration, Springer, Dordrecht, Vol 24, pp 657-668

Van der Weij-de Wit CD, Ihalainen JA, van der Vijver E, D'Haene S, Matthijs HCP, van Grondelle R, Dekker JP (2007) Fluorescence quenching of IsiA in early stage of iron deficiency and at cryogenic temperatures. Biochim Biophys Acta 1767: 1393-1400

Vavilin DV, Hu S, Lin S, Vermaas WFJ (2003) Energy and electron transfer in photosystem II of a chlorophyll *b*-containing *Synechocystis* sp. 6803 mutant. Biochemistry 42: 1731-1746

Wang Q, Jantaro S, Lu B, Majeed W, Bailey M, He Q (2008) The high light-inducible polypeptides (HLIP) stabilize trimeric photosystem I complex under high light conditions in *Synechocystis* PCC 6803. Plant Physiol 147: 1239-1250

Wilson A, Ajlani G, Verbavatz J-M, Vass I, Kerfeld CA, Kirilovsky D (2006) A soluble carotenoid protein involved in phycobilisome-related energy dissipation in cyanobacteria. Plant Cell 18: 992-1007

Wilson A, Boulay C, Wilde A, Kerfeld CA, Kirilovsky D (2007) Light-induced energy dissipation in iron-starved cyanobacteria: Roles of OCP and IsiA proteins. Plant Cell 19: 656-672

Wilson A, Punginelli C, Gall A, Bonetti C, Alexandre M, Routaboul J-M, Kerfeld CA, van Grondelle R, Robert B, Kennis JTM, Kirilovsky D (2008) A photoactive carotenoid protein acting as light intensity sensor. Proc Natl Acad Sci USA 33: 12075-12080

CHAPTER 5

The Functional Sites of Electron Transfer Cofactors in Photosystem II Studied by Pulsed EPR

Asako Kawamori[1] and Hiroyuki Mino[2]

[1]*Agape-Kabutoyama Institute of Medicine, Nishinomiya 662-0001, Japan*

[2]*Graduate School of Science, Nagoya University, Nagoya 464-8602, Japan*

ABSTRACT

Dipolar interactions between radical pairs are detected by Pulsed ELDOR and Spin Polarized ESEEM to measure distances between electron transfer cofactors in Photosystem I and II. The distances from reaction centers P700 and P680 to the primary acceptor quinone A_1 and Q_A were determined accurately using Spin Polarized ESEEM. Pulsed electron double resonance (PELDOR) is the most accurate method and provides the direct distance information of radical pairs formed by light illumination. Chlorophyll Z in the inhibited electron transfer path was determined to be in D2-protein in spinach. The basic assumption for these methods is a point dipole interaction between a radical pair. This assumption is not applicable to the cluster of paramagnetic molecules or ions which are usually found in the biological enzyme systems. To draw the correct distance information, dipole interactions between $Y_D^{\bullet}$ radical and each manganese were calculated using spin projections of each manganese for valence state of Mn(III) $Mn(IV)_3$ in the S_2-state and $Mn(III)_3Mn(IV)$ in the S_0-state and X-ray data for coordinates of Y_D and four manganese. The simulations over the assigned spin projections of four manganese fit well the experiential PELDOR time profile of $Y_D^{\bullet}$- Mn_4 cluster in the S_2 and S_0-states.

Keywords: Photosystem, Pulsed ELDOR, ESEEM, Electron transfer cofactors.

INTRODUCTION

Electron paramagnetic resonance (EPR) has been one of the important means to study electron transfer in photosystems.

Among several pioneering works Feher *et al.* (1975) has proven that the reaction center of a bacterial photosystem consists of dimer chlorophyll by ENDOR (Electron Nuclear Double Resonance). Since then other photosystems, PSI and II in plants and cyanobacteria, have been found to have a dimer reaction center chlorophylls (Feher *et al.*, 1975). The hyperfine structure constant of reaction center chlorophylls is a half of that in a monomer chlorophyll. In that time analysis of crystal structure had been difficult, because of the difficulty in single crystal preparation. The first crystal structure analysis of bacterial reaction center was reported by Deisenhofer *et al.* (1985) and the dimer structure has been confirmed.

The crystal structure of cyanobacterial PSI was reported by Shubert *et al.* (1997). The crystal structure of cyanobacterial PSII was reported for the first time by Zouni *et al.* (2001). The distance relation between electron transfer cofactors is essential to understand electron transfer mechanism in a photo system. Before the crystal structure data have been available, several workers estimated the distance between cofactors by dipolar broadening and spin lattice relaxation measurement of EPR in a radical such as $Y_D^{\bullet}$ affected by non-heme iron at the acceptor side (Inne *et al.*, 1989 and Kodera *et al.*, 1992), which gave rather indirect methods.

Pulsed EPR was applied to distance estimation of photosystem II for the first time by Kodera *et al.* (1994) to estimate the distance from Y_D to the Mn-cluster and to the non-heme iron. This method was called selective hole burning based on a similar principle as spin lattice relaxation measurement.

Later spin polarized ESEEM (Dzuba *et al.*, 1997) and ELDOR (Hara *et al.*, 1996) methods gave accurate methods for distance determination with 0.5 to 1 Å resolution less than 2.8 Å resolution (Loll *et al.*, 2005) in the crystal structure analysis.

After crystal structure analysis has been developed, these data now will be available for determination of electron transfer path in D1 or D2 proteins. In this report, we will describe how spin polarized ESEEM and PELDOR were used to determine the active site in electron transfer.

CRYSTAL STRUCTURE AND ELECTRON TRANSFER PATH IN PSII

One of the recent crystal structure data of photosystem II (Kamiya and Shen, 2003) is shown in Fig. 5.1. A reaction center protein of photosystem II composes of D1 and D2 proteins for main electron transfer path and other extrinsic proteins. By light illumination an electron transfers from water to acceptor quinine Q_A via water oxidase (Mn_4-cluster) and Y_Z in D1 protein and then to P680 belonging to D1 and D2 proteins via pheophitin to Q_A in D2-protein (Wydrzynski and Satoh, 2005). There are several pathways presented in inhibited PSII at low temperatures from cyt559 to P680 (Hanley *et al.*, 1999).

X-ray cannot determine the position of Q_A directly, because it is a small molecule. The X-ray data have identified positions of chlorophylls, Mn_4-cluster and irons and large protein helixes. Tyrosine molecule is also difficult to identify because of low resolution of 3.0 Å. Spin polarized ESEEM and PELDOR can determine the position of the small molecules.

There are four monomer chlorophylls in PSII reaction center proteins. Two, Chl_{D1} and Chl_{D2}, reside next to P680. The other two are situated far from P680, $ChlZ_{D1}$ and $ChlZ_{D2}$, and the one or both contribute to electron transfer directly or indirectly to P680. Which chlorophyll contributes to electron transfer could not be determined by optical or X-ray method only. However, ELDOR could determine the active position referred to X-ray data.

PULSE EPR FOR MEASUREMENT OF DIPOLAR INTERACTION

Generally, electron transfer gives rise to a radical pair, donor (oxidized reaction center $P680^+$) and acceptor (reduced quinine Q_A^-), in Photosystem II.

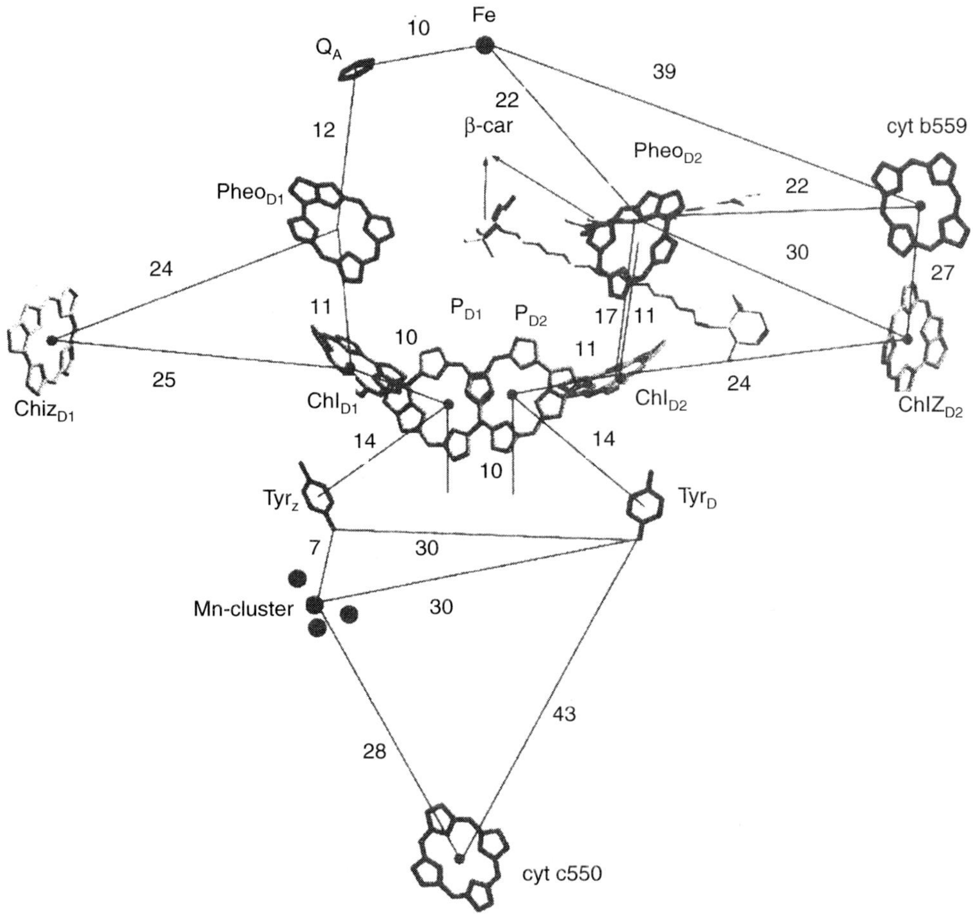

Fig. 5.1 Side view of configurations of electron transfer components depicted by the X-ray analysis from Kamiya and Shen (2003). The axis from the center of reaction center chlorophylls (P_{D1} and P_{D2}) to Fe atom is parallel to crystalline *c*-axis.

Figure 5.2 shows a radical pair system with magnetic dipolar interaction and two methods of measurement by pulsed EPR.

Dipolar Interaction

Figure 5.2(a) shows dipolar interaction energy referred to *A*-spin, where *B*-spin exerts a dipolar field of the form ΔH, r is the distance between *A* and *B* spins and θ is the angle of applied magnetic field from distance axis.

$$\varepsilon(r, \theta) = \gamma_A \, \Delta H = \frac{\gamma_A \, \mu_B \, (3\cos^2\theta - 1)}{r^3} \tag{1}$$

The radical pair $P680^+$-Q_A^- is produced by laser irradiation as shown in Fig. 5.2(b). When Q_A is pre-reduced, oxidized $P680^+$ and Q_A^- form a triplet state and can be detected by two-pulse echo before the electron comes back to the ground state after about 100 μs. The radical pair state is polarized 100% and

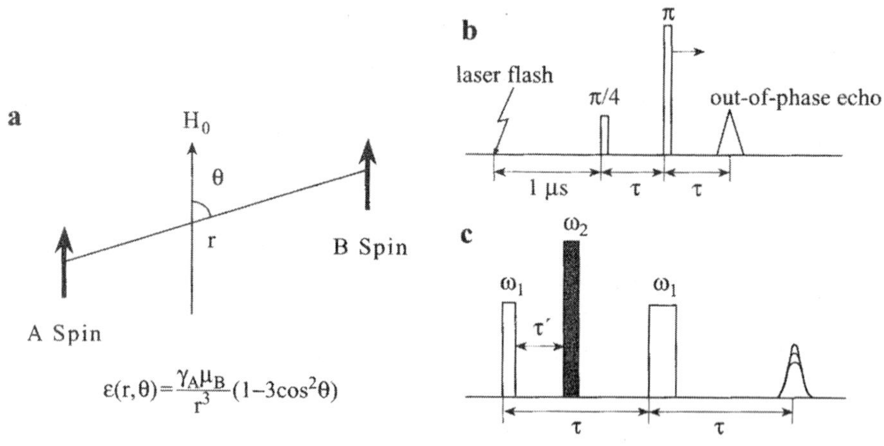

Fig. 5.2 (a) Dipolar interaction and pulse sequences for the pulse methods for distance measurement. (b) Spin polarized ESEEM. (c) PELDOR. '2 + 1' is a special case of $\omega_1 = \omega_2$.

yields a strong oscillating signal of dipolar interaction frequency. In non-oriented membranes the angle θ takes all directions and averaging over all directions gives the frequency of $\gamma_A \mu_B/r^3$. Thus, we will be able to determine the distance *r* between the radical pair. In oriented membrane or single crystal we will be able to determine the angle θ_n, the direction of distance vector ***r*** relative to the membrane normal ***n***, also as shown later.

The second method PELDOR (Pulsed Electron Double Resonance) shown in Fig. 5.2(c) is applied to trapped or stable paramagnetic pairs such as $Y_D^{\bullet} - Q_A^{-}$, $Y_D^{\bullet} - Chl_Z^{+}$ and $Y_D^{\bullet} - Mn_4$ cluster.

The equation (1) shows a point dipole approximation. In fact molecules Y_D and Q_A have a size of 2-3 Å much smaller compared to the distance 20-30 Å between electron transfer cofactors. However, the Mn-cluster is larger with the size of about 5^3 Å^3 and carotenoid is as long as about 20 Å.

Then we have to take spin distribution over the molecule as shown later for Y_D–Mn_4-cluster.

Pulse EPR Measurements

ESEEM measurements were carried out using Bruker ESP380 X-band spectrometer equipped with a cylindrical dielectric resonator (ER4117DHQ-H). A He gas flow system (CF935 Oxford Instrument) was used to control temperature at 80 K. Light illumination was done by N_D-YAG laser (Continuum Surelite 1) with a 5 ns FWHM at 532 nm, a power of 120 mJ and repetition rate of 10 Hz.

PELDOR consists of three microwave pulses (π/2, π and π) as shown in Fig. 5.2(c). The first and third pulses with frequency ω_1 form the spin echo of *A* spin, and the second pulse with frequency ω_2 rotates *B* spin by π. Now Bruker ESP 580 pulse EPR provides the second frequency oscillator. We have equipped a synthesizer sweeper HP83751A (Hewlett-Packard).

For the measurements of ESEEM and PELDOR, we have a oscillating time profile as given by

$$V(\tau, \tau') \propto 1 - p\{\cos(2\pi D\tau') - 1\} \qquad \begin{array}{l}\tau = \tau' \text{ for ESEEM}\\ \tau < \tau' \text{ for PELDOR}\end{array} \tag{2}$$

where $D = D_0(1 - 3\cos^2\theta)$ with $hD_0 = \gamma_A \mu_B/r^3$ and $0 < p \leq 1$

Samples

PSI particles were isolated by the digitonin–treatment of spinach according to [14]. Reaction mixture at the concentration of 20 μM P700 in Tris-HCl buffer (pH 7.5) contained 60% glycerol to improve illumination efficiency. The samples were loaded into suprasil quartz tubes with inner diameter of 4 mm.

The oxygen-evolving PSII particles were prepared from spinach according to Kuwabara and Murata (1982). The Tris-treatment was carried out by incubation of the core complexes in a buffer containing 0.8 M Tris. Removal of the non-heme iron from the core complex and its substitution by Zn^{2+} were carried out by a slightly modified method of Klimov *et al.* (1980). The Zinc substituted PSII particles were suspended in a buffer of 0.4 M sucrose, 10 mM NaCl and 50 mM MES.

For measurement of oriented PSI and PSII particles, the membranes were dried on pieces of myler sheets at 0°C for about 16 hours using nitrogen gas under controlled humidity of 90%. Five myler sheets with 3 × 20 mm^2 were put into quartz tube with inner diameter of 4 mm. The definition of angle θ is shown in Fig. 5.3 (A) and (B). For determination of the angle θ relative to the membrane normal, the distance *r* should be measured for non-oriented membranes.

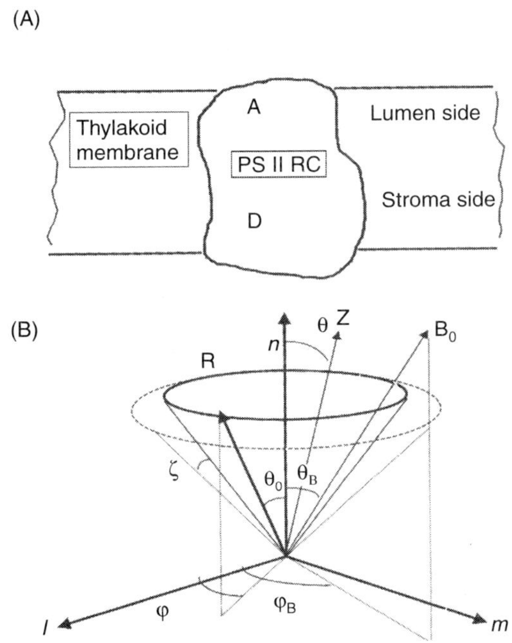

Fig. 5.3 (A) Schematic drawing showing PSII RC in a fragment of a thylakoid membrane. D and A show the donor and acceptor sides of PSII. (B) Orientation of a PSII RC and the magnetic field direction $\mathbf{B}_0$ with respect to the membrane coordinate system (***l, m, n***), where ***n*** is the membrane normal, while ***l*** and ***m*** are arbitrarily orthogonal axes in the membrane plane. Axis **Z** is associated with the PSII RC, whose deviation from ***n***-axis is defined by ζ. All orientations are obtained by a rotation of PSII RC around Z axis. φ is assumed to have equal probability.

Otherwise it can be found from the lowest frequency of the angular dependence. Actually there is some mis-orientation and the observable time profile is given by, using a distribution function $G(\theta - \theta_B)$

$$\langle V(\tau, \tau')\rangle_{\text{oriented}} \propto \int_0^{\pi}\int_0^{2\pi}\int_0^{2\pi} \langle V(\tau, \tau')\rangle \, d\varphi_B \, d\varphi \, G(\theta - \theta_B) \sin\theta \, d\theta \tag{3}$$

For non-oriented particles PELDOR time profile Eq. (2) will be averaged over all orientations of θ as given by Eq. (2) with $D = D_0 = \gamma_A \mu_B/hr^3$.

THE DISTANCE AND DIRECTION FROM THE REACTION CENTER CHLOROPHYLLS TO THE ACCEPTOR QUINONE IN PS I AND PS II

To reduce A_1 in PSI the oriented membranes have been illuminated at 200 K.

Figure 5.4 (A) shows the time profile of spin polarized ESEEM observed in oriented PSI membranes for the direction of the magnetic field parallel and perpendicular to the membrane normal. Figure 5.4(B) shows Fourier transforms from the time profile (A) (Yoshii *et al.*, 1999). Detailed angular variation gave best fitted angle of θ of P700–A_1 distance vector shown in the upper-right in Fig. 5.4(B).

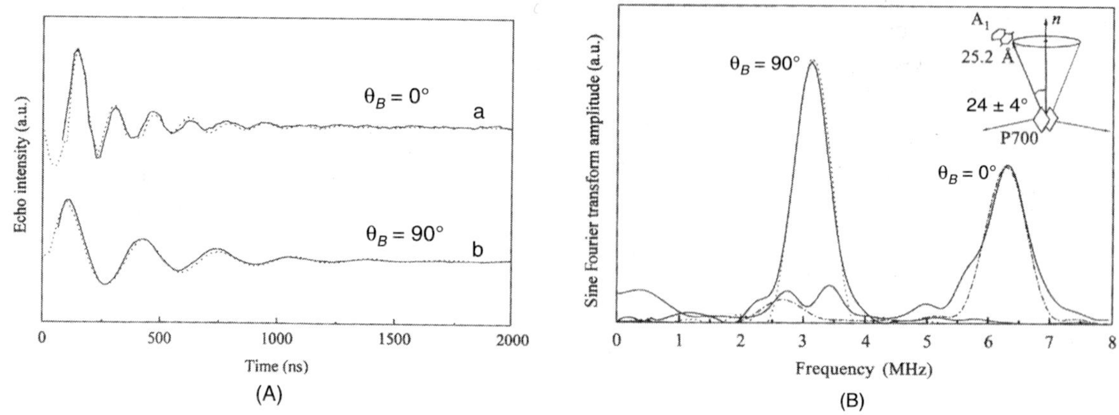

Fig. 5.4 (a) Spin polarized ESEEM spectra observed in oriented PSI particles at the field direction relative to the membrane normal θ_B = 0° (a) and 90° (b) were shown with corrections for dead time regions. Dotted lines were calculated for D = 1.73 G, by assuming the angle θ_0 between the radius vector joining P700 and A_1 ***R***(P700-A_1) and ***n*** to be 24°. (b) Results of sine Fourier transformation of time domain spectra shown above. Solid curves show the experimental data observed at 80 K in PSI particles and dotted lines show the simulated curves with parameters shown in the inset that shows a geometry of the reaction center chlorophylls and phylloquinone (A_1) based on the simulation.

To observe spin polarized ESEEM of P680$^+$-Q_A^- in PSII, we have to reduce Q_A in Zn-substituted PSII membranes by dithionite. Figure 5.5 (a) shows the time profiles of P680$^+$-Q_A^- for the magnetic field direction parallel and perpendicular to the membrane normal (Yoshii *et.al.*, 1999). Fourier transforms and the derived distance and directions using the distance obtained from non-oriented membrane, are shown in Fig. 5.5(b).

Using the value of r obtained in non-oriented membrane, we have determined the angle between distance vector and membrane normal 21°.

Figure 5.6 shows schematic drawings of orientations of distance vector from primary electron donors (P700 and P680) to the acceptor quinones (A_1 and Q_A) which are compared with *X*-ray data.

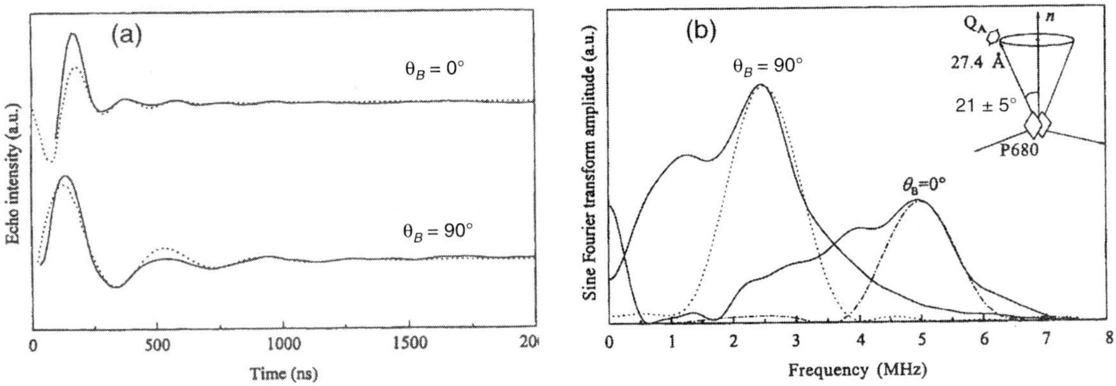

Fig. 5.5 (a) Spin-polarized ESEEM spectra observed in oriented Zn-substituted PSII core particles at the field directions $\theta_B = 0°$ and 90°. (b) Results of sine Fourier transformation of the time domain spectra obtained from (a)

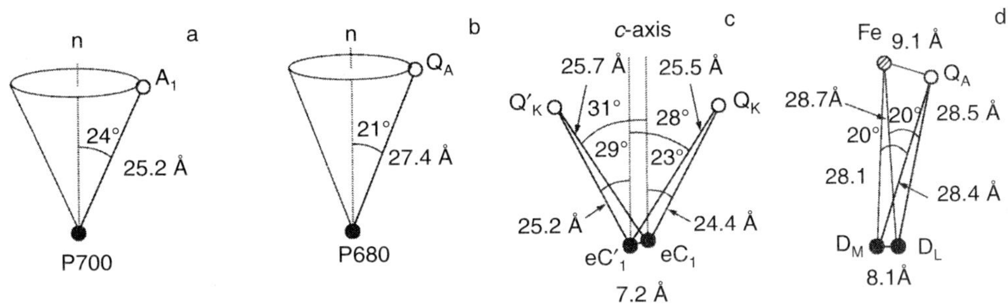

Fig. 5.6 Schematic drawing of orientation of primary electron donor (P700 and P680) and the acceptor quinones (A_1 and Q_A) in reaction center complexes PSI (a) and PSII(b). a and b show distances and angles P700-A_1 and P680-Q_A determined by ESEEM. (c) Location of cofactors in the cyanobacterial PSI complex determined by X-ray. eC1 and eC2 are chlorophyll molecules constituting P700, and Q_K and $Q_{K'}$ are proposed positions of phyloquinones. (d) Location of cofactors in the Rb.spaeroides reaction center calculated by X-ray data. Fe is non-heme iron, DL and DM bacteriochlorophylls constituting P870. At present, PSII X-ray data show similar arrangement.

ELDOR FOR $Y_D^{\bullet}$ AND Chl_Z^+

Chl_Z^+ radical was trapped by illumination of the Mn-depleted PSII at 200 K (Tonaka *et al.*, 2000). As $Y_D^{\bullet}$ and Chlorophyll radical overlap, '2 + 1' method, a special case of ELDOR with the same frequency was applied. In the membrane oriented Mn-depleted preparation of PSII, the angular variation of time profile is shown in Fig. 5.7(a), that shows the maximum frequency oscillation at the angle θ of about 60°. Detailed angular dependence shows the maximum frequency 2 MHz at 50°. Using the determined distance $r = 29.4 \pm 0.5$ Å in the non-oriented PSII, the obtained angle θ from the membrane normal was determined to be 50.0 ± 5°.

There are two Chlorophyll Z in PSII, the one is situated in D2 protein rather close to Y_D, and the other one is in D1 protein close to Y_Z (see Fig. 5.1).

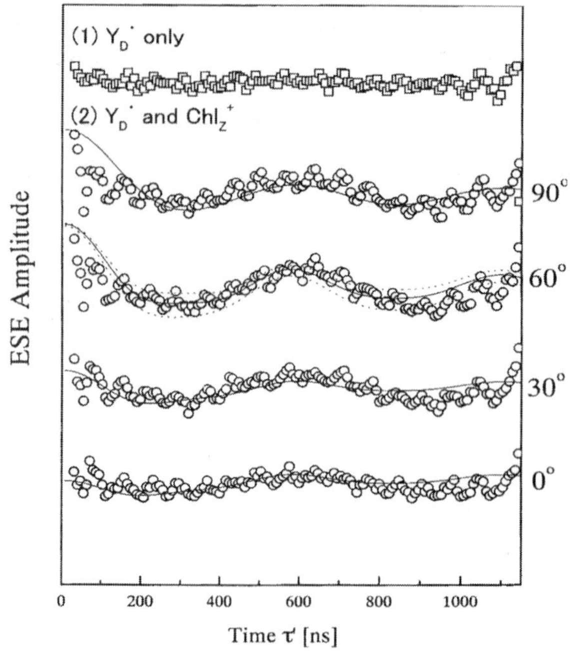

Fig. 5.7 '2 + 1' ESE traces recorded in the illuminated Mn-depleted PSII membranes for orientations from 0° to 90°(Y_D• and Chl_Z^+). Trace 1 shows the time profile of the dark adapted PSII for 90 direction (Y_D• only). The broken lines of 60 time profile show fitting lines for 55.0 and 45.0.

We have to prove which of $ChlZ_{D1}$ and $ChlZ_{D2}$ shows the observed ELDOR profile (Kawamori *et al.*, 2002b and Lakshmi *et al.*, 2003). Y_D-less mutant of *Chlamidomonas reinhardti* was investigated by PELDOR to measure the distance between $Y_Z^{\bullet}$-Q_A^- radical pair, which was trapped by 253 K illumination of Tris-treated and Zn-substituted PSII. The Tris-treated and non-heme iron retained PSII was illuminated at 253 K and further illuminated at 200 K to trap Chl_Z^+ and to observe $Y_Z^{\bullet}$-Chl_Z^+ PELDOR. We could observe the signal for $Y_Z^{\bullet}$-Q_A^- in Zn-substituted PSII and determined the distance 34.5 Å for $Y_Z^{\bullet}$-Q_A^-. However, we could not observe PELDOR for $Y_Z^{\bullet}$-Chl_Z^+ in 200 K illuminated PSII, even if we observed $Y_Z^{\bullet}$-Chl_Z^+ EPR signal (Kawamori *et al.*, 2002a). Then we could determine that the Chl_Z^+ is not in D1 but is in D2 protein. Therefore, the side-path from Cyt b559 to P680 in spinach PSII is confirmed on D2-side. (Hanley *et al.*, 1999).

As for the electron transfer to P680 at low temperatures in Mn-depleted PSII carotenoid is also one of the candidates of electron career. In cyanobacteria this radical is produced by 200 K illumination. Therefore, the electron transfer path was suggested on D1 side (Lakshmi *et al.*, 2003). In spinach it is produced by 20-5 K illumination (Hanley *et al.*, 1999). The problem in the inhibited PSII is the electron transfer path have not yet been completely clarified. It will be interesting to determine the functional carotenoid position, though this molecule is very long and point dipole approximation will not be applicable.

PELDOR OF Y_D AND Mn_4-CLUSTER

In oxygen evolving photosystem II the distance between Y_D and the Mn-cluster in the S_2-state was derived by various methods. Latest PELDOR data in oriented membranes has shown the distance *r* of

27.4 Å and the angle θ of 110° (Astashkin *et al.*, 1998). The relative position derived by X-ray data (Zouni *et al.*, 2001) show the average distance from four manganese to the center of Y_D molecule is about 30 Å the angle from 118° to 128°. Evidently these deviations from the crystal structure data show that the point dipole approximation was not correct. However, for the deviation in the distance direction we

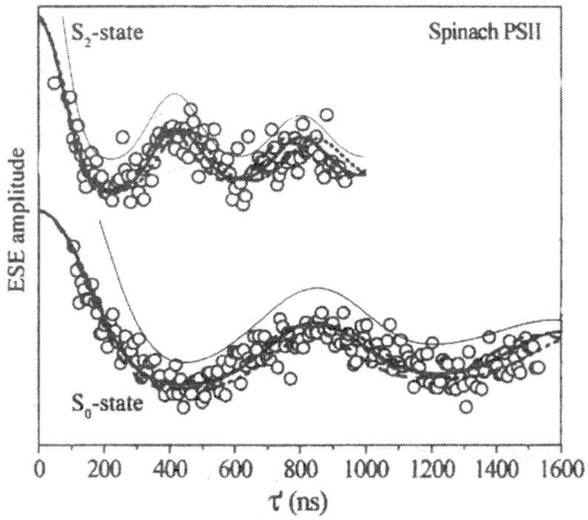

Fig. 5.8 (a) PELDOR time profiles observed on the manganese multiline in the S_2-state (upper) and S_0-state (lower) with excitation of the dipolar interacting second spin of Y_D radical. Curves on the data points were drawn by assuming point dipole interaction. Each slim curves right above are drawn taking spin projections in Table 5.4 in S_2 and S_0-states into account.

have to take it into account that the crystalline *c*-axis is inclined about 10° from the membrane normal (Zouni *et al.*, 2001), and this deviation can be considered to be a minor one.

After the manganese multiline in the S_0-state had been observed (Messinger *et al.*, 1997), PELDOR for Y_D-Mn_4 cluster was also observed in the NH_2OH reduced S_0-state PSII (Kawamori *et al.*, 2002a, b). Figure 5.8(a) shows PELDOR profiles observed in S_2 and S_0 states of oxygen evolving non-oriented photosystem II. The S_2 state was prepared by illumination of PSII at 200 K about 10 min. The S_0 state was prepared by reduction of S_1-state PSII by hydroxyl amine (Messinger *et al.*, 1997). These profiles indicate the S_2-state and S_0-state showed different PELDOR frequency oscillation. This means the different distances for S_2 and/or S_0-states or change in spin density for these states.

Figure 5.8(b) shows a model tentatively suggesting distances and directions of Mn_4-cluster-Y_D derived by a point dipole approximation. This model shows the center of spin density transfers from the S_0 to the S_2-state. In fact so much change in manganese position have not yet reported by EXAFS (Messinger *et al.*, 2001).

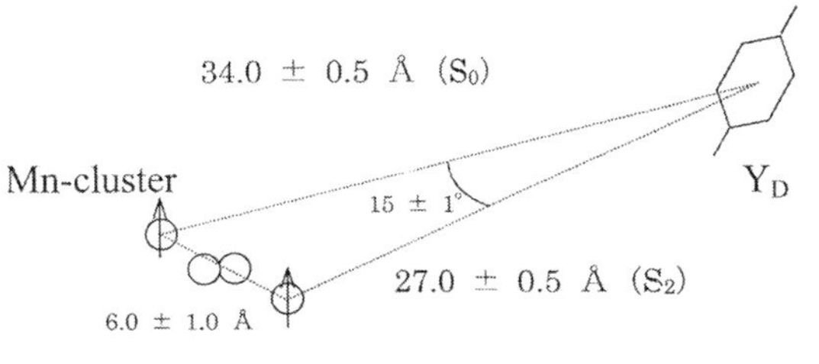

Fig. 5.8 (b) Point dipolar approximation.

In the X-ray data shown in Fig. 5.1 the size of the Mn-cluster is about 5^3 Å^3. Point dipolar interaction should be considered for each pair of Y_D-Mn_i with *i* from 1 to 4, as the following. It is interesting to remark the change of spin center about 15° is comparable to the diference of Mn_i-Y_D vector direction studied by X-ray.

Dipolar interaction (1) should be summed over 4 manganese

$$M_y(t) = \int\int \sin\theta_B\, d\theta_B\, d\phi_B \cos[\omega(\theta_B)\, t] \tag{3}$$

$$\hbar\omega(\theta_B) = \sum_{i=1,4} P_i (3\cos^2\theta_i - 1)/r_i^3 \tag{4}$$

where r_i is distance between Y_D and *i*th manganese.

The Y_D molecule is small about 2^3Å^3 and the spin center is assumed to be on the molecular center.

P_i = Spin projection on *i*th manganese

Peloquin *et al.* (2000) derived spin projection based on Pulsed ENDOR. For simplicity, the anisotropy in projection approximated by an isotropic projection is

$$P_{iso} = [P// + 2P_\perp]/3 \tag{5}$$

The valence for each Mn(III)Mn(IV)$_3$ has been suggested for S_2-state by EXAFS study (Messinger *et al.*, 2001). The valence Mn(III)$_3$Mn(IV) was suggested by Peloquin *et. al.* (2000).

Using the spin projection given in Table 5.1, which has been given by Eq. (5) using pulsed ENDOR data for the S_2-state and r_i derived for Y_D-Mn distance based on X-ray analysis [9] as shown in Table 5.2, the ELDOR profiles are well fitted on the experimental data. Spin distribution tentatively assigned using ENDOR data (Kulik *et al.*, 2007) on the manganese atoms as shown in Fig. 5.9.

Table 5.1 The values of spin projections and valence shown in Peloquin *et. al.*, (2000).

	Mn(1)	*Mn(2)*	*Mn(3)*	*Mn(4)*
S_2-state	1.705 (III)	−0.99(IV)	−1.0(IV)	1.275(IV) -
S_0-state	−1.8(III)	2.0(III)	−1.1(IV)	1.9 (III)

For S_2-state the values are given by Peloquin *et al.* (2000) and for S_0-state tentative values are assigned.

Table 5.2 The derived values of distances along crystalline axes (*a, b, c*) from Y_D-Mn_i with i = 1 to 4 in unit of Å taken from Lolls *et al.* (2005).

i	a	b	c	*r*	*θ(°)*
Mn1	22.52	−5.332	18.03	29.34	128
Mn2	24.71	−5.476	19.51	31.96	128
Mn3	25.23	−7.075	17.39	31.45	124
Mn4	25.60	−7.694	14.23	30.28	118

The values of θ show the direction of distance vector from the crystalline *c*-axis.

We have shown the best fitted curves on S_0 and S_2 states by slim line in Fig. 5.8(a). From Table 5.1, the high spin density is on Mn(1) for the S_2 state and is on Mn(4) for the S_0-state with the movement of spin density shown in Fig. 5.8(b).

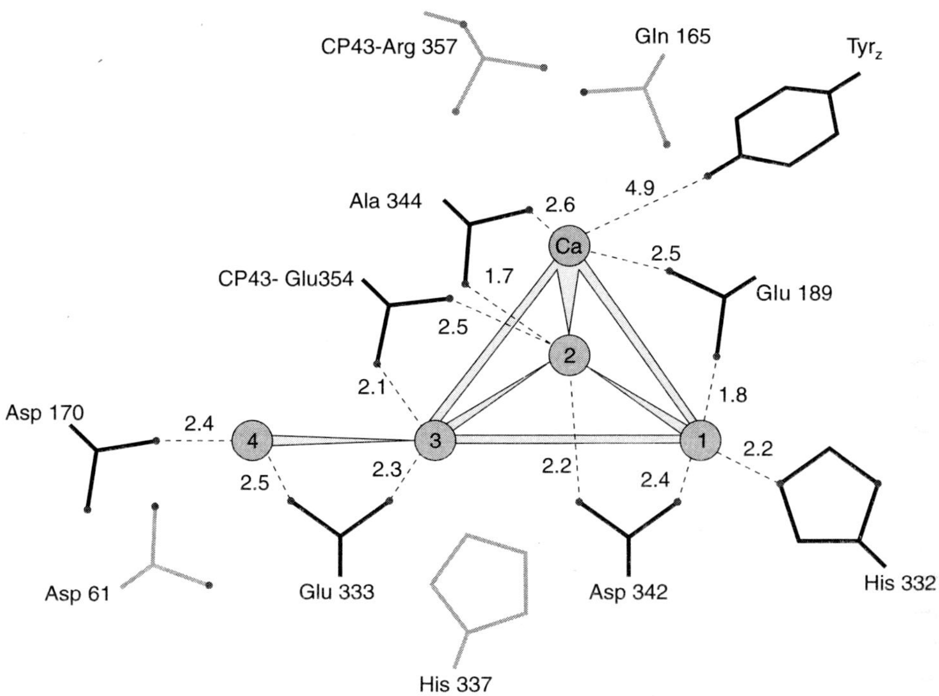

Fig. 5.9 Detailed structure of Mn_4-cluster presented by Loll et al. [10]. The manganese atoms are numbered from 1 to 4.

From the PELDOR for $Y_D^{\bullet}$-Mn_4 cluster we can conclude Mn(1) is trivalent in S_2-state. Mn(3) takes tetra-valence in the S_0 state.

The order of oxidation in water splitting and the different distribution may have been suggested (Kulik *et al.*, 2007).

$Yz^{\bullet}$–Mn_4-cluster for the S_2 state will give more accurate spin distribution over the four manganese atoms. We are now in progress of analyzing PELDOR data on $Yz^{\bullet}$–Mn_4-cluster for the S_2 state.

REFERENCES

Astashkin AV, Hara H, Kawamori A (1998) The pulsed electron-electron double resonance and '2+1' electron spin echo study of the oriented oxygen-evolving and Mn-depleted preparations of photosystem II. J Chem Phys 108: 3805-3812

Deisenhofer J, Epp O, Miki K, Huber R, Michel H* (1985) Structure of the protein subunits in phtotosynthetic reaction enters of the *Rhodopseudomonas viridis* at 3 Å. Nature 318 618-624

Dzuba SA, Hara H. Kawamori A, Iwaki M, Itoh, S, Tsvetkov Yu, D (1997) Electron spin echo of spin-polarised radical pairs in intact and quinone-reconstituted plant photosystem I reaction centres. Chem Phys Lett 264: 238-244

Feher G, Hoff AJ, Isaacson RA, Ackerson LC (1975) ENDOR Experiments on chlorophyll and bacteriochlorophyll *in vitro* and in photosynthetic unit. Ann N Y Acad Sci 244: 260-280

Hanley J, Deligiannakis Y, Pascal A, Faller P, Rutherford AW (1999). Carotenoid oxidation in photosystem II. Biochemistry 38: 8189-8195

Hara H, Kawamori A, Astashkin AV, Ono T-A (1996) The distances from tyrosine D to redox-active components on the donor side of photosystem II Determined by pulsed electron-electron double resonance. Biochim Biophys Acta 1276: 140-146

Inne JB and Brudvig GW (1989) Location and magnetic relaxation properties of the stable tyrosine radical in photosystem II. Biochemistry 28: 1116-1125

Itoh S., Iwaki M, Ikegami I, (1987) Extraction of vitamin K1 from photosystem I particles by treatment with diethyl ether and its effects on the A1 electron paramagnetic resonance signals and system I photochemistry. Biochim Biophys Acta 893: 508-516

Kamiya N, Shen J-R (2003) Crystal structure of oxygen evolving photosystem II from *Thermo synechococcus vulcanus* at 3.7-Å resolution. Proc Natl Acad Sci USA 100: 98-103

Kawamori A, Katsuta N, Arao S, Ishi A, Minagawa J, Mino H, Ono T-A (2002a) Three-dimensional structure of photosystem II studied by pulsed EPR. Proceedings of 12th International Congress of Photosynthesis CSIRO S5-4

Kawamori A, Katsuta, N Mino, H. Ishii, A Minagawa J, Ono, T-A (2002b) Positions of Q_A and Chl_Z relative to tyrosine Y_Z and Y_D in photosytem II studied by pulsed EPR. J Biol Phys 28: 413-426

Kawamori A, Ono T-A, Ishii A, Nakazawa S, Hara H, Tomo T, Minagawa J, Bittl R, Dzuba SA (2005) The functional sites of chlorophylls in D1 and D2 subunits of photosystem II identified by pulsed EPR. Photosynthesis Research 84: 187-192

Klimov VV, Dolan E, Shaw ER, Ke B (1980) Interaction between the intermediary electron acceptor (pheophytin) and a possible plastoquinone-iron complex in photosystem II reaction centers. Proc Natl Acad Sci USA 77: 7227-7231

Kodera Y, Takura, K, and Kawamori A (1992) Distance of P680 from the manganese complex in photosystem II studied by time resolved EPR. Biochim Biophys Acta 1101: 23-32

Kodera,Y, Dzuba SA, Hara H, Kawamori A (1994) Distances from tyrosine D^+ to the manganese cluster and the acceptor iron in photosystem II as determined by selective hole burning in EPR spectra. Biochem Biophys Acta 1186: 91-99

Kulik LV, Epel B, Lubitz W, Messinger J (2007) Electronic structure of the Mn_4O_xCa cluster in the S_0 and S_2 states of the oxygen-evolving complex of phtotosystem II based on pulse ^{55}Mn-ENDOR and EPR spectroscopy. J Am Chem Soc 129: 13421-13435

Kuwabara T, Murata N (1982) Inactivation of photosynthetic oxygen evolution and concomitant release of three polypeptides in the photosystem II particles of spinach chloroplasts. Plant Cell Physiology 23: 533-540

Lakshmi KV, Poluektov OG, Reifler MJ, Wagner AM, Thurnauer MC, Brudvig GW: (2003) Pulsed high-frequency EPR study on the location of carotenoid and chlorophyll cation radicals in photosystem II. J Am Chem Phys 125: 5005-5014

Loll B, Kern J, Saenjer W Zouni A, Biesiadka J (2005) Towards complete cofactor arrangement in the 3.0 Å resolution structure of phtotosystem II. Nature 438: 1040-1044

Messinger J, Nugent JHA, Evans MCW (1997) Detection of an EPR multiline signal for the S_0* state in phtotosystem II. Biochemistry 36: 11055-11060

Messinger J, Robblee JH, Bergmann U, Fernandez C, Glatzel P, Visser H, Cinco RM, McFarlane KI, Bellacchio E, Pizarro SA, Cramer SP, Sauer K, Klein MP, Yachandra VK (2001) Absence of Mn-centered oxidation in the S2 to S3 transition implication for the mechanism of photosynthetic water oxidation. J Am Chem Soc 123: 7804-7820

Peloquin JM, Cambell KA, Randall DW, Evanchic MA, Pecoraro VL, Armstrong WH, Britt RD (2000) ^{55}Mn ENDOR of the S_2-state multiline EPR signal of photosystem II: Implications on the structure of tetranuclear Mn cluster. J Am Chem Soc 122: 10926-10942

Schubert WD, Klukas O, Krauß N, Saenger W, Fromme P, Witt. HT (1997) Photosystem I of *Synechococcus elongatus* at 4 Å resolution: A comparable structure analysis. J Mol Biol 272: 741-769

Tonaka M, Kawamori A, Hara H, Astashkin AV (2000) "Three-dimensional structure of electron transfer components in photosystem II: '2+1' ESE of chlorophyll Z and tyrosine D". Appl Magn Reson 19: 141-150

Wydrzynski J, Kimiyuki S, (ed) (2005) Photosystem II: The Light driven Water Plastoquinone Oxidoreductase by Springer.

Yoshii T, Hara H, Kawamori A, Akabori K, Iwaki M, Itoh S (1999) ESEEM study of the location of spin-polarized chlorophyll-quinone radical pair in membrane-oriented spinach photosystem I and II complexes Appl Magn Reson 16: 565-580

Zouni A, Witt, J, Kern J, Fromme P, KraußW, Saenger W, Orth P (2001) Crystal structure of photosystem II from *Synechococcus elongatus* at 3.8 Å resolution. Nature 409: 739-743

CHAPTER 6

Delayed Fluorescence of Chlorophyll *d* and Chlorophyll *a*; Emission/Excitation Spectra and Activation Energy in Photosystem II of *Acaryochloris marina* and *Synechocystis* sp. PCC 6803

Yoshimasa Fukushima and *Shigeru Itoh

Photo-bioenergetics Laboratory, Division of Material Science (Physics), Graduate School of Science, Nagoya University, Furocho, Chikusaku, Nagoya 464-8602, Japan

ABSTRACT

The intact cells of cyanobacterium *Acaryochloris marina* contain Chl *d* as a major pigment with a few percentages of Chl *a*. Delayed fluorescence (DF) emitted at 0.1-2 ms by intact cells after the light illumination showed 1) an emission peak at 719 nm as does prompt fluorescence, indicating Chl *d* to be the emitter pigment, and 2) the excitation spectrum with peaks at 620 and 702 nm indicating almost equal contributions of phycocyanin and Chl *d* as the PS II antenna pigments. These features are different from *Synechocystis* sp. PCC 6803 cells, which showed a 681-nm emission peak of Chl *a*-DF and a 625-nm excitation peak of phycocyanin and a 680-nm shoulder of Chl *a*. Temperature dependencies of the yield and decay rate of the Chl *d*-DF at 0.1-10 s, which were measured in the presence of DCMU, gave the activation enthalpies larger than the corresponding ones for the Chl *a*-DF in *Synechocystis*, suggesting different energy levels of PS II redox cofactors. The features of DF/PF of PS II Chl *d* in *Acaryochloris marina* were however similar to those of DF/PF of PS II Chl *a* of *Synechocystis*, indicating that the oxygenic photosynthesis in *Acaryochloris* with Chl *d* based PS II and PS I is performed efficiently with low charge recombination probabilities.

Keywords: Charge recombination, Chlorophyll *d*, Cyanobacteria, Delayed fluorescence, Electron transfer, Excitation energy transfer, Photosystem II.

Abbreviations

Chl: chlorophyll, *DCMU:* 2,3-dichloro-1, 1-dimethlyurea, *DF:* delayed fluorescence, *H:* photosystem II primary electron acceptor pheophytin *a*, *PF:* prompt fluorescence, *PS:* photosystem, *P:* primary electron donor chlorophyll, Q_A *and* Q_B *:* photosystem II electron acceptor plastoquinone A and B.

INTRODUCTION

Oxygenic photosynthesis in plants and cyanobacteria has been known to use chlorophyll (Chl) *a* ubiquitously as an essential pigment for the light harvesting and light-induced charge separation. Other photosynthetic pigments, such as Chls *b* and *c*, carotenoids and phycobilins have been known to function only as the antenna pigments that absorb light at 400-650 nm and transfer the excitation energy to the red absorption bands of Chl *a* at 670-680 nm. The electron donors named P680 and P700 of PS II and PS I reaction center complexes are known to be a special pair of Chl *a* (Loll *et al.*, 2005; Kamiya and Shen, 2003) and a heterodimer of Chl *a* and Chl *a'* (Kobayashi *et al.*, 1988; Jordan *et al.*, 2001), respectively. Some cyanobacteria use divinyl Chl *a* and divinyl Chl *a'* that have spectral and chemical properties very similar to those of Chl *a* (Chisholm *et al.*, 1992). *Acaryochloris marina,* which is a newly identified species of cyanobacteria (Miyashita *et al.*, 1996; Miyashita *et al.*, 2003) uses Chl *d* that absorbs light at 700-740 nm as the major pigment with a few percentages of Chl *a* (Miyashita *et al.*, 1997). The special pair Chl in PS I of this organism was shown to be a Chl *d* dimer (presumably a heterodimer of Chl *d* and Chl *d'* in analogy to P700 (Kobayashi *et al.*, 1988)) and was named P740 (Hu *et al.*, 1998). Although P740 absorbs the 740 nm light, it seems to produce the reducing power, comparable to that of Chl *a*-type PS I of other organisms, to reduce iron sulfur centers (Hu *et al.*, 1998). On the other hand, in PS II of *Acaryochloris*, the electron donor has been assumed to be either the special pair of Chl *d* (Itoh *et al.*, 1998; Chen *et al.*, 2005a, 2005b), or a Chl *a* dimer P680 as in all other oxygenic organisms as shown in Fig. 6.1. The function of uphill energy transfer from the antenna Chl *d* to P680, as proposed by Mimuro *et al.* (1999, 2000, 2004, 2007), was rather difficult (Nieuwenburg *et al.*, 2003).

Chl *d* functions as the major antenna pigment in PS II of *Acaryochloris* as shown by the high efficiency of 720 nm light for the oxygen evolution (Miyachi *et al.*, 1997), the Chl *d* peak in the excitation spectrum for the oxygen evolution, and the Chl *d* fluorescence induction in response to the reduction of the PS II electron acceptor plastoquinone Q_A (Schiller *et al.*, 1997). Itoh *et al.* (1998) proposed the PS II special pair Chl (P) to be a Chl *d* dimer absorbing at 725 nm based on the flash-induced absorption changes around 725 nm in the ferricyanide-oxidized thylakoid membranes. Chen *et al.* (2005a, 2005b) isolated the crude PS II RC complex and detected the photoreduction of the acceptor pheophytin *a* (H) and the absorption change of bulk Chl *d* (Chen *et al.*, 2005b). Mimuro and his colleagues (Mimuro *et al.*, 1999, 2000, 2004, 2007) assumed P680 to function in PS II of this organism based on the long-lived Chl *a*-like minor fluorescence component at 680 nm, which was accompanied by a major 720 nm Chl *d* fluorescence band. The former was proposed to be the delayed fluorescence (DF) (Mimuro *et al.*, 1999, 2000, 2004, 2007). The argument about the PS II electron donor still continues in recent studies (Itoh *et al.*, 2007; Tomo *et al.*, 2007; Schlodder *et al.*, 2007). Schlodder *et al.* (2007) proposed a Chl *a*/Chl *d* heterodimer (or a Chl *a* dimer) to be the special pair, and the accessory pigments to be Chl *d* based on the absorption changes in membranes similar to those reported by Itoh *et al.* (1998). Itoh *et al.* (2007) proposed the electron donor and accessory pigments to be Chl *d* based on the absorption changes in the ferricyanide-oxidized thylakoid membranes. Tomo *et al.* (2007) proposed the donor to be a Chl *d* special pair based on the FTIR measurement in a PS II RC preparation that had 22 Chl *d* and 4 Chl *a* per 2 pheophytin *a*. Tomo *et al.* (2007) detected the 15-ns fluorescence at 680 nm and re-assigned it to be emitted from the Chl *a* adjacent to the special pair of Chl *d*. On the other hand, we reported the Chl *d*-type emission spectrum of the millisecond-delayed fluorescence (DF) in *Acaryochloris* cells (Itoh *et al.*, 2002). The mechanism of energy conversion and electron transfer in *Acaryochloris* is, thus, still not fully clear, although it is interesting how the electron transfer system in PS II in *Acaryochloris* enables the oxygen evolution reaction with the lower quantum energy of photon absorbed by Chl *d*.

In the Chl *a*-based PS II of plants and cyanobacteria, DF has been known to be almost exclusively emitted from antenna Chl *a* molecules that are re-excited in the dark by the charge recombination reactions (Strehler and Arnold, 1951; Itoh *et al.*, 1971; Wraight and Crofts, 1971; Grabolle and Dau, 2005). DF intensity/time course is known to vary depending on the energy levels of the intermediate radical pair states that give rise to the charge recombination reactions (Grabolle and Dau, 2005), and also on the electrochemical potential gradient across the thylakoid membranes (Itoh *et al.*, 1971; Wraight and Crofts, 1971). In this article we have compared the emission and excitation spectra of DF in *Acaryochloris* cells to those of *Synechocystis* to know the mechanisms of this unique Chl *d*-based PS II. The results indicated that DF in *Acaryochloris* is emitted from antenna Chl *d* molecules, and the energy levels of the PS II intermediate states are somewhat different from those in Chl *a*-based PS II.

FLUORESCENCE MEASUREMENTS

Cells of *A. marina* (an MBIC 11017 strain) were grown under white fluorescent lamps in a MK medium for about a week as described previously (Hu *et al.*, 1998). Cells were harvested and re-suspended in the same fresh medium.

DF emitted at 0.1-2 ms during the dark period after the 2 ms excitation period (with a light-dark cycle time of 5 ms) was measured repetitively with a home-built Becquerel-type phosphoroscope that has co-axial two disks with 20 cm diameters, rotating at 3000 rpm on both sides of the sample cuvette (Fig. 6.2). The openings of the two disks are displaced from each other to isolate only DF (Itoh *et al.*, 1971; Wraight and Crofts 1971). DF was excited by light from a 150W tungsten-iodine lamp and was measured with a photomultiplier (Hamamatsu R-490, Hamamatsu Photonics, Hamamatsu) connected to a lock-in amplifier (L1-570, NF circuits Co., Tokyo, Japan). A stepping-motor-driven filter monochromator (*fm*) (7155, Oriel, Stratford, CT) was inserted into a light path, either between the excitation light and the sample cuvette (position 2) or between the photomultiplier and the sample (position 1) to measure the excitation or emission spectrum, respectively. The filter monochromator had spectral resolution of an 11 nm full width at a half maximum and was scanned at a rate of 0.15 nm/sec.

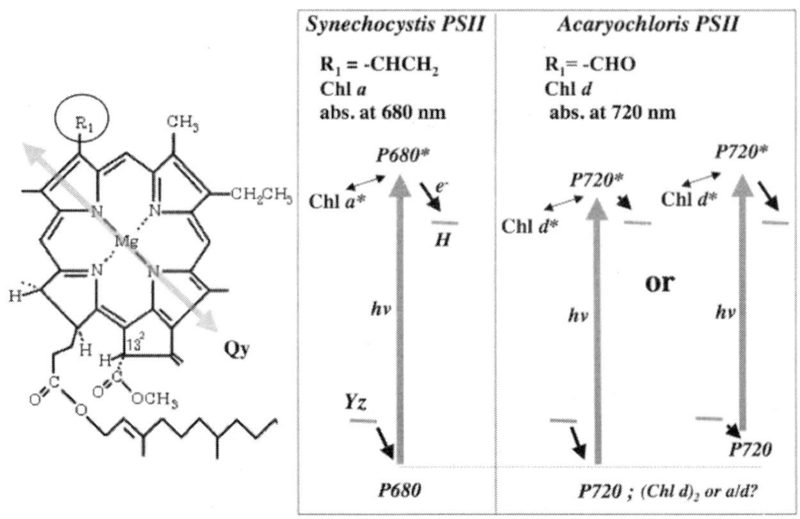

Fig. 6.1 Left; Chemical structure of chlorophyll *d* and chlorophyll *a*. Right; Energy diagram of Chl *a*-based PS II of cyanobacteria (plants) and expected energy diagrams of Chl *d*-based PSII of *Acaryochloris*.

Emission spectra of 0.1-2 ms DF were measured with a monochromatic excitation light obtained from the tungsten-iodine lamp through a heat-cut filter at an intensity of 15-50 mW/cm^2 on the sample surface. Excitation spectra of *Acaryochloris* and *Synechocystis* cells were obtained from the DF intensities at 720 nm and 680 nm peaks, respectively. These spectra were calibrated for the sensitivity of the detection/ excitation system by the use of a standard tungsten lamp. Prompt fluorescence (PF) emitted during the excitation period was measured simultaneously with a silicon photodiode (S639, Hamamatsu Photonics, Hamamatsu) coupled through a long-pass filter. The emission spectrum of PF excited at 590 nm was measured with a spectrofluorimeter (F4500, Hitachi, Hitachi).

DF decay kinetics at 0.1-10 s after a single turnover flash excitation was measured by using a photon counting system (C1230, Hamamatsu Photonics, Hamamatsu). Samples were excited with a Xenon flash (30 μs duration) from a stroboscope (PS-240A, Sugawara, Tokyo) through a blue glass filter passing the 330-480 nm light. DF was measured with a cooled photomultiplier through a 660-nm long-pass filter.

Absorption Spectrum of *Acaryochloris* Cells

Acaryochloris cells contain Chl *d* as the major pigment with a few percents of Chl *a* together with phycobilisomes, which contain phycocyanin and allophycocyanin as reported elsewhere (Miyashita *et al.*, 1997). In the absorption spectra (see the action spectrum with a dotted line in (Fig. 6.5B), phycobilisome showed peaks at 620-635 nm, both in *Acaryochloris* and *Synechocystis*. *Synechocystis* cells showed an absorption peak of Chl *a* at 678 nm, while *Acaryochloris* cells showed almost no peak of Chl *a* at 670-680 nm but showed a large peak at 709 nm of Chl *d* (Fig. 6.5A). The two organisms, therefore, have different chlorophyll species together with phycobilins (Hu *et al.*, 1999; Marquardt *et al.*, 2000).

Time Courses of Millisecond Delayed Fluorescence and Prompt Fluorescence in Cells of *Acaryochloris* and *Synechocystis* PCC6803

DF emitted at 0.1-2 ms after the 2 ms repetitive flash excitations of intact cells of *Acaryochloris* was continuously measured with a Becquerel type phosphoroscope (Fig. 6.2). Prompt fluorescence (PF) emitted during the 2-ms excitation periods was also monitored simultaneously. Time course of DF intensity was measured (Fig. 6.3 curve a) together with that of PF (curve b). Both time courses were measured at their emission peaks of Chls (Fig. 6.4A). The intensity of DF increased from zero to an initial peak within 1 s and then slowly to the steady state level. PF intensity, on the other hand, rose from the initial F_0 level to the maximum F_m level indicating the reduction of the electron acceptor Q_A in PS II as is well known (curve b). The F_0 level could be identified more clearly in the trace with an expanded time scale (not shown). Figure 6.3 also shows the time courses of PF (curve d) and DF (curve c) of Chl *a* in *Synechocystis* sp. PCC 6803 cells. The time course of PF was similar to that in *Acaryochloris* cells. The time course of DF was a little different from that in *Acaryochloris* showing the faster decrease during the illumination. The intensities of DF were not very different between these two organisms apparently, though more accurate comparison is required to evaluate the effects of the different absorption and emission spectra, and PS I/II ratio of the cells.

Addition of DCMU, which inhibits the electron transfer from Q_A to Q_B, modulated the time courses of DF and PF (not shown) and accelerated the rise of PF indicating the fast accumulation of Q_A^-. DF did not show the slow rising phase after the initial rise both in *Acaryochloris* and *Synechocystis* in the presence of DCMU, and depressed the steady state level. These effects indicate that DF is emitted from PS II

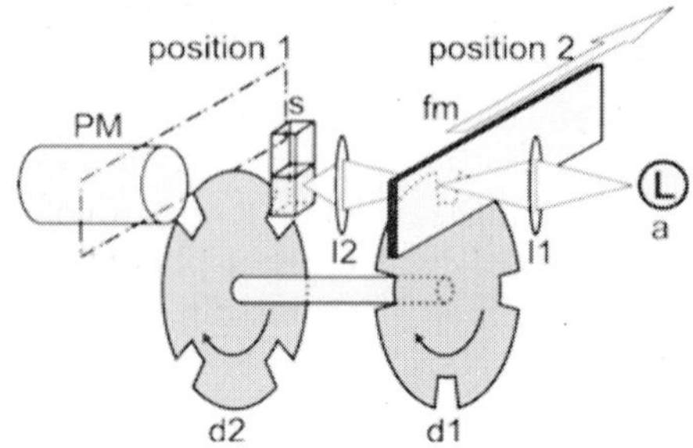

Fig. 6.2 A schematic diagram of a Becquerel type phosphoroscope equipped with a filter monochromator used for the simultaneous measurements of prompt (PF) and delayed fluorescence (DF). Light from the excitation tungsten lamp passes through a glass cutoff filter (f1), a filter monochromator (fm) (at position 2), a rotating disk (disk-1), reflected by mirrors (m) and then illuminates a sample in a quartz or acrylic cuvette (S). DF emitted from the sample passes through another rotating disk (disk-2), a glass filter (f2) and a filter monochromator (fm) (at position 1), and then, detected by a photomultiplier (pm). PF from the sample was also measured by a photodiode (pd) that is set at the opposite side through a long-pass filter. The signal from a photomultiplier was amplified with a preamp and by a lock-in amplifier with a reference signal from another photodiode monitoring the disk opening, and then, fed to a computer or a strip chart recorder.

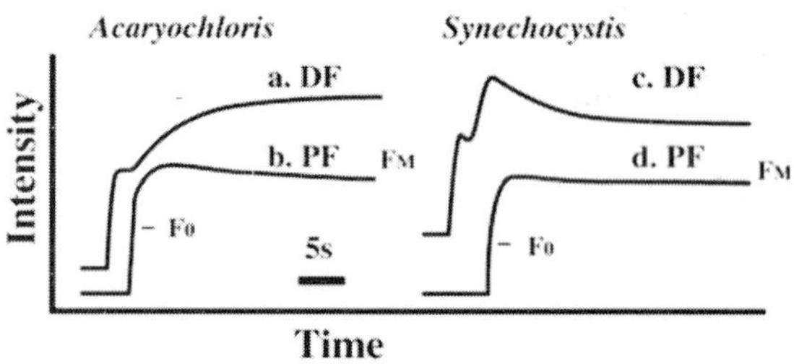

Fig. 6.3 Time courses of PF and DF in intact cells of *Acaryochloris* and *Synechocystis* sp. PCC 6803 under intermittent illumination. a and b represent the time courses of PF and DF in *Acaryochloris*, respectively, measured simultaneously with a system shown in Fig. 6.2. c and d, PF and DF in *Synechocystis* PCC 6803 cells. The PF and DF were monitored at every 0-1.9 ms illumination period and 2.0-3.9 ms dark period, respectively, under the continuous repetitive illumination.

in *Acaryochloris*, as reported in other cyanobacteria and plants (Itoh *et al.*, 1971; Wraight and Crofts, 1971; Grabolle and Dau, 2005).

Emission Spectra of PF and DF

The spectrum of PF that was measured separately with a fluorimetry with the excitation at 590 nm, showed a small emission peak at 648 nm (with a shoulder at 665 nm) of phycocyanin and a large distinct peak at 719 nm of Chl *d* (Fig. 6.4A dotted line). We also measured the emission spectrum of DF by scanning the filter-monochromator that was inserted between the photomultiplier and the sample cuvette. The emission spectrum of DF measured by the phosphoroscope, on the other hand, showed a single peak at 720 nm, at the same wavelength as the larger peak of PF, indicating that DF is almost exclusively emitted from Chl *d*. The DF emission spectrum revealed no significant contributions of phycocyanin and Chl *a* at around 650 and 680 nm, respectively.

The emission spectra of PF and DF in *Synechocystis* cells (Fig. 6.4B) showed features somewhat different from those of *Acaryochloris* (Fig. 6.4A). PF showed a major peak at 653 nm of phycocyanin and a small shoulder peak at 680 nm of Chl *a*, respectively (dotted line). DF, on the other hand, showed

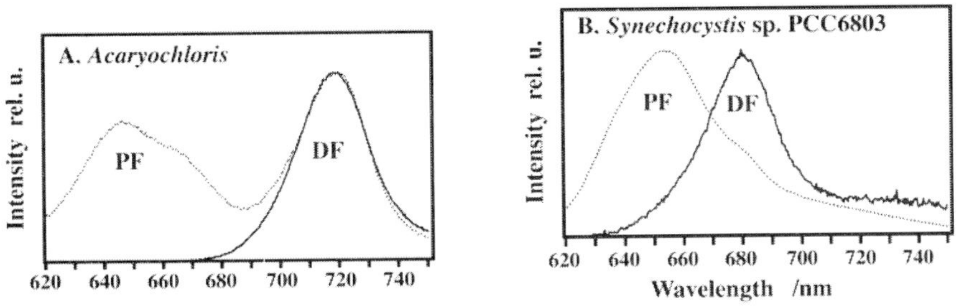

Fig. 6.4 Emission spectra of PF and DF of *Acaryochloris* (A) and *Synechocystis* 6803 cells (B). Emission spectra of PF (broken lines) and DF (solid lines) were measured by scanning a filter monochromator. The spectra were normalized at the peak wavelengths of chlorophyll emissions. Excitation light for DF measurement was obtained from the 150 W tungsten lamp through a heat-cut filter. Other conditions were similar to those in Fig. 6.3.

a single peak at 680 nm (solid line) indicating the emission solely from Chl *a* in PS II as reported (Hideg *et al.*, 1991)

The results in Fig. 6.4 clearly indicate that *Acaryochloris* emits DF from Chl *d* but not from Chl *a* or phycobilins. *Synechocystis* cells, on the other hand, emit DF exclusively from Chl *a*. In both the organisms, phycocyanin gave only PF.

Excitation Spectrum of DF

The excitation spectrum for the millisecond DF was measured in *Acaryochloris* by scanning the filter monochromator placed between the excitation light and the sample as depicted in Fig. 6.2 (position 2). The phosphoroscope enables the physical separation of DF emission from the excitation beam. The method provides a clear excitation spectrum even in the wavelength region overlapping with the DF emission band.

The DF excitation spectrum in *Acaryochloris* showed a broad peak at around 620 nm and a narrow peak at 702 nm, respectively (a solid line in Fig. 6.5A). The two peaks gave almost the same height. The 620 nm band agrees well with the absorption band (dotted line) of phycocyanin and indicates that phycocyanin efficiently transfers the excitation energy to PS II RC. The 702 nm excitation peak almost agrees with the absorption band of chlorophyll *d* with an extent at a half of the latter when normalized at the phycocyanin peaks. It suggests that about a half of Chl *d* transfers excitation energy to PS II and the other half to PS I. No peak of Chl *a* at 670-80 nm was detected in the excitation spectrum.

Figure 6.5B compares the two spectra after normalization at the Chl *d* peaks. The excitation spectrum gave a peak at 702 nm in contrast to the absorption peak of Chl *d* at 709 nm. Chl *d* in PS II, thus, seems to be blue shifted compared to the absorption band of total or PS I Chl *d*. The result agrees with the action spectra of the PS II oxygen evolution and the PS I oxygen absorption reported in this organism (Boichenko *et al.*, 2000).

The excitation spectrum for the 680 nm DF of Chl *a*, measured in *Synechocystis* cells (Fig. 6.5C solid line), showed a large phycocyanin peak at 620 nm with a small shoulder at 680 nm showing the small contribution of Chl *a* to PS II. The spectrum is different from the absorption spectrum that gave a large distinct peak of Chl *a* at 678 nm and suggests the abundance of PS I. It agrees with the report that *Acaryochloris marina* cells show a high PS II/PSI ratio of 1.8 under similar growth conditions (Itoh *et al.*, 2007).

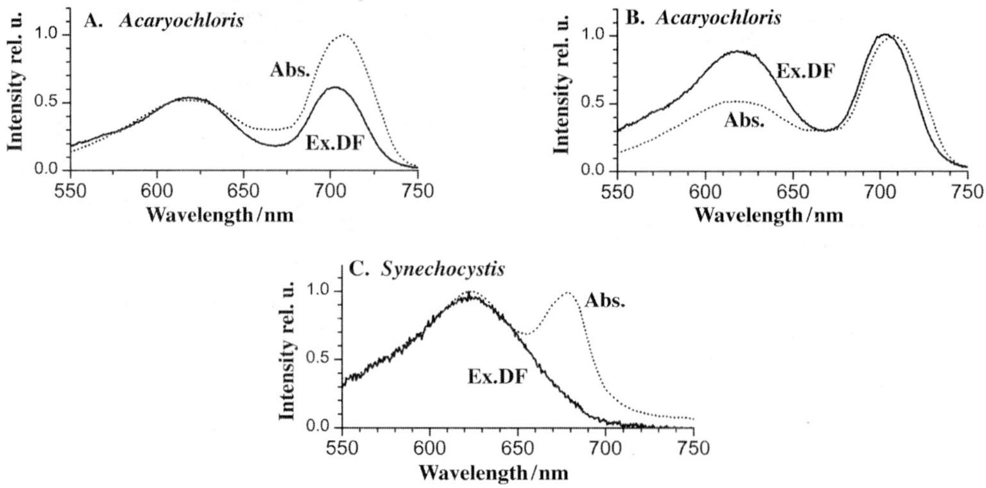

Fig. 6.5 Comparison of the excitation spectra of DF with the absorption spectra in *Acaryochloris* and *Synechocystis* 6803 cells. A and B, absorption spectrum (broken lines) and the excitation spectrum for the 720-nm DF (solid lines) were measured in *Acaryochloris* cells and shown after normalization at the phycocyanin peak (A) and chlorophyll *d* peak (B), respectively. C, absorption spectrum (broken line) and the excitation spectrum for the 680-nm DF measured in *Synechocystis* cells were shown.

Effects of DCMU on Decay Kinetics and Temperature Dependence of DF

We measured the DF decay in the longer time scale at 0.1-10 s after the excitation with a single turnover Xenon flash in the presence and absence of DCMU. DF was detected with the photon counting system. In the presence of DCMU, the decay of DF became slower and the intensity became higher in this time range (Fig. 6.6A). At 300 K, the deconvolution of the decay indicated the half decay times of 1.1 s (with an initial amplitude of 67 %) and 6 s (30%). Most of DF (integrated intensity) in the presence of DCMU, thus, is emitted in these two phases that seem to reflect the recombination of charges on S_2 and Q_A^- (Rappaport *et al.*, 2002, 2005; Rappaport, 2002). At the lower temperatures, the decay became slower and the initial amplitude of the corresponding phases decreased significantly (see a trace at 278K in Fig. 6.6A). Similar analysis was done at each temperature and the sum of the initial amplitudes (I_0) of these two phases was calculated.

The dark decay rate (with an apparent rate constant k_r) of DF is expected to represent the major decay phase of the $S_2Q_A^-$ state in the presence of DCMU. It occurs by the charge recombination reaction through the thermally activated intermediate state X, which seems to be either P^+H^- sate or $P^+Q_A^-$, where P represents P680 and H represents the electron acceptor pheophytin *a* in the Chl *a*-based PS II (Rappaport, 2002; Rappaport *et al.*, 2005). The log (I_0) and log (k_r) values were plotted against inverse of absolute temperature in Fig. 6.6B and C to calculate the activation enthalpy ΔH_I and ΔH_{II}, respectively. For the calculation of k_r, we used the decay time constant that corresponds to the slower phase. Both the plots in *Acaryochloris* showed steeper slopes suggesting the larger ΔH_I and ΔH_{II} values for the reproduction of Chl d^* or X from $S_2Q_A^-$, respectively, compared to those in *Synechocystis* cells (Table 6.1).

Delayed Fluorescence from Chl *d** in *Acaryochloris*

In oxygenic photosynthetic organisms other than *Acaryochloris*, DF has been known to be emitted from Chl a^* in the charge recombination reactions between the negative charges on the reducing side and the

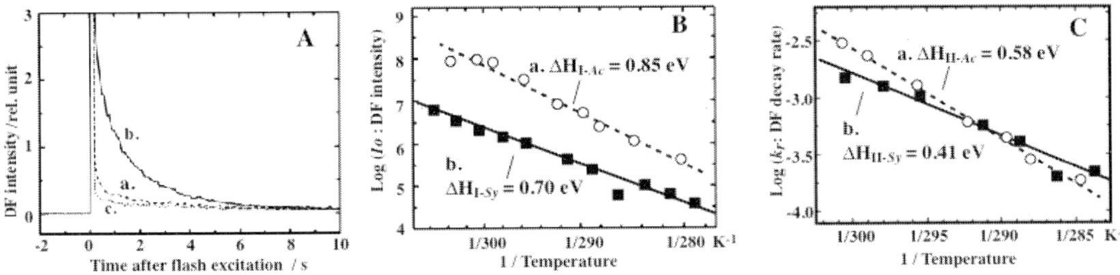

Fig. 6.6 A. Effects of DCMU on the decay kinetics of DF in *Acaryochloris* at different temperatures. a, without DCMU at 300 K. b, with 20 μM DCMU at 300 K. c, with 20 μM DCMU at 278 K. DF was induced by the excitation with a single turnover flash at the saturating intensity. Decay kinetics was simulated as the sum of exponential decays and the intensity of the slow phase extrapolated to time zero (initial intensity I_0) and the decay rate constant k_r values are calculated at each temperature. B. Plots of the initial intensity I_0 of DF in *Acaryochloris* and *Synechocystis* cells. C. Plots of the dark decay rate constant k_r of DF in *Acaryochloris* and *Synechocystis* cells. In B and C, I_0 and k_r values were obtained from the experiments similar to those in A at each temperature by the simulation of the decay curves as described in text and plotted against inverse of absolute temperature.

positive charges on the oxidizing side of PS II (Itoh *et al.*, 1971; Wraight and Crofts, 1971; Grabolle and Dau, 2005; Rappaport *et al.*, 2002, 2005; Rappaport, 2002). It has been assumed that the energy liberated in the charge recombination reaction in PS II excites P680 (or Chl *a* adjacent to P680) to the singlet excited state, and then, the excitation energy is transferred to the antenna Chl *a* molecules to give DF (Rappaport *et al.*, 2002, 2005; Rappaport, 2002; Vass *et al.*, 1981). The other short-wavelength pigments, such as Chls *b, c,* phycocyanin or carotenoids contained in the antenna complexes emit almost no DF because of the low probability of the uphill energy transfer from P680 to these antenna pigments. On the other hand, PS I does not emit DF unless the A_1 (phylloquinone) is reduced under extremely reducing conditions (Kleinherenbrink *et al.*, 1994) or after the extraction of phylloquinone (Iwaki and Itoh, 1988) to stop the fast electron transfer from A_0 to A_1.

The measurements with the phosphoroscope combined with the scanning filter monochromator in the present study indicated that both DF and PF are emitted from Chl *d** giving a peak at 719 nm in *Acaryochloris.* The DF can be interpreted to be produced by the charge recombination reactions in PS II judging from the long time period of the emission, effects of DCMU and the strong temperature dependency. DF from Chl *a* was not detected at 0.1 ms-10 s after the illumination in *Acaryochloris* cells. The reported Chl *a*-type emission detected in the 20-50 ns time range in *Acaryochloris* (Mimuro *et al.*, 1999, 2000, 2004, 2007; Tomo *et al.*, 2007), therefore, might be produced by a mechanism other than the PS II charge recombination reaction. DF from Chl *d** in PS II of *Acaryochloris* is, thus, exceptional in the oxygenic photosynthetic organisms.

Pigment System of *Acaryochloris*

The excitation spectrum of Chl *d*-DF in Fig. 6.5A showed a phycocyanin band at 620 nm and a Chl *d* band at 702 nm, but almost no appreciable contribution of Chl *a* at 680 nm. The positions of two bands almost agreed with the absorption bands of phycocyanin and Chl *d*. The result indicates that the excitation energy absorbed by phycocyanin and Chl *d* is efficiently transferred to the special pair Chl, and produces the Chl *d*-DF at 0.1 ms-10 s.

The Chl *d* excitation peak at 702 nm, which was shorter than the absorption peak at 709 nm of whole Chl *d* band, suggests the shorter absorption peak of Chl *d* band in PS II compared to that in PS I. The result is in line with the reported point-by-point action spectra of PS II and PS I measured by the PS II oxygen evolution and the PS I oxygen absorption activities, respectively, in *Acaryochloris* cells

(Boichenko *et al.*, 2000). The 702 nm excitation peak of Chl *d* band is rather short compared to the 719 nm emission peaks of PF and DF that are assumed to be emitted from the 710-718 nm absorbing forms. The large distinct peaks of Chl *d* detected in the excitation spectra of PF and DF also make clear contrasts to the small peaks of Chl *a* detected in the excitation spectra of PF and DF in *Synechocystis* (Fig. 6.5C). The results agree with the situation that *Acaryochloris* cells have low phycobilin contents and high PS II/PS I ratio of 1.8 (Itoh *et al.*, 2007).

DF Production and PS II Energy Diagram

We attempted to estimate the energy diagram in PS II below. In the presence of DCMU, the regenerated amount of Chl* at 0.1 s to 10 s after the flash excitation can be related to the concentration of flash-formed $S_2Q_A^-$ state, which is expected to be in an equilibrium with Chl *d** with a free energy gap ΔG (= $\Delta H - T\Delta S$) between the two states (Grabolle and Dau, 2005) by assuming the Boltzmann distribution as follows,

$$(\text{Chl*}) = A \exp(-\Delta G/RT) \times (S_2Q_A^-) \tag{1}$$

$$I_{DF} = c\,(\text{Chl*}) \tag{2}$$

where A and c are constants, R is a gas constant, and ΔH and ΔS are enthalpy and entropy changes of the reaction, respectively. (Chl*) in the dark after the flash excitation should be proportional to the intensity I_{DF} of DF by equation 2 (Rappaport *et al.*, 2002, 2005; Rappaport, 2002; Vass *et al.*, 1981). The value of I_{DF} extrapolated to time zero (I_0) (at the time of formation of $S_2Q_A^-$ state, i.e., at a few ms after the flash excitation) can be related to initial extent of ($S_2Q_A^-$) produced after a single laser excitation (Chisholm *et al.*, 1992), which is expected to be almost constant in the dark-adapted cells even at different temperatures. The rate of DF decay, on the other hand, should give information for the decay rate through the major rate-determining step of the charge recombination. This seems to occur through an intermediate level X at the energy level lower than that of P* state. X might be either the P^+H^- state or $P^+Q_A^-$ state (Grabolle and Dau, 2005; Rappaport *et al.*, 2002, 2005; Rappaport, 2002; Vass *et al.*, 1981).

Energy Diagram A: Chl *a*-type PS II of *Synechocystis* and its Modification with Chl *d*

We can start from a typical energy diagram for *Synechocystis* PS II that emits DF of Chl *a* as depicted in Fig. 6.7A. Table 6.1 lists the ΔG and ΔH values required for the regeneration of P680*, antenna Chl *a** or P^+H^- from $S_2Q_A^-$ estimated in the previous studies (Grabolle and Dau, 2005; Rappaport *et al.*, 2002, 2005; Rappaport, 2002).

The temperature dependency of the amplitude I_0 gave a ΔH_I value of 0.70 eV for the regeneration of Chl *a** or P680* from $S_2Q_A^-$ in *Synechocystis* (Table 6.1). This value is almost comparable to the ΔG value of 0.72 eV estimated from the intensity of microsecond DF (Grabolle and Dau, 2005) (after suitable modifications) or 0.78 eV estimated from the Q_A^- decay (Rappaport *et al.*, 2002; Vass *et al.*, 1981), and the thermoluminescence of Chl *a* (Rappaport *et al.*, 2005) P680* then transfers the excitation energy to the antenna Chl *a* molecules that are at the 0.11 eV lower energy level because of the entropy change because of the larger number of antenna Chl molecules (Grabolle and Dau, 2005; Rappaport *et al.*, 2005). DF, thus, is mainly emitted from antenna Chl *a**.

As a simple model for *Acaryochloris* PSII, we tested a case with an antenna Chl *d** level (dotted line) added to diagram A at 0.11 eV below the Chl *a**. This model is almost equivalent to the one that assumes the function of P680 in *Acaryochloris* with Chl *d* antenna (Mimuro *et al.*, 1999, 2000, 2004, 2007). The model also resembles the situation in PS I of ordinary cyanobacteria and plants because the red Chl *a* species in PS I, which emits PF at 720-750 nm, plays a role of low energy antenna as Chl *d* does in this

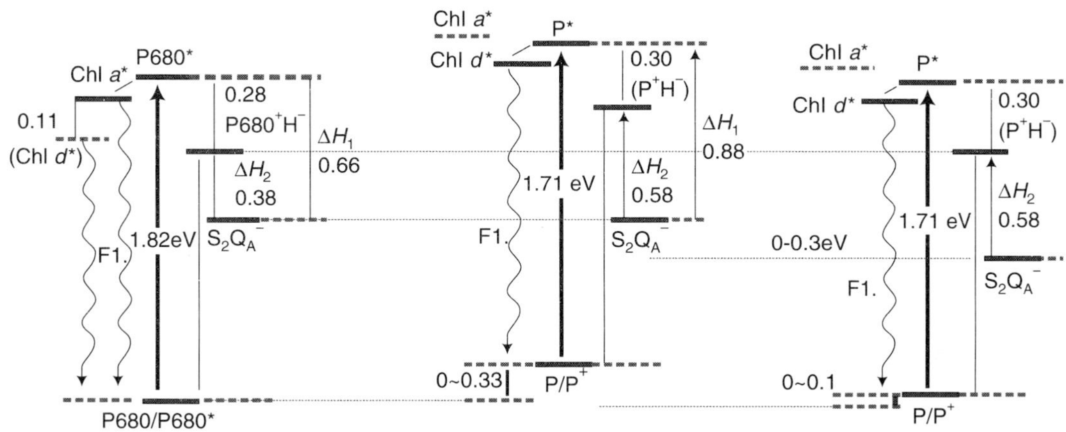

A. *Synecocystic* PSII

B. *Acaryochloris* PSII $E(S_2Q_A^-)$ adjusted to be the as that in *Synecocystis* PSII

C. *Acryochloris* PSII $E(S_2Q_A^-)$ is shifted

Fig. 6.7 Schematic energy diagrams of pigments and cofactors in PS II of *Synechocystis* and *Acaryochloris* based on the results of DF measurements. Energy diagram A, an estimated energy diagram in PS II of *Synechocystis* formulated based on the ΔH_I and ΔH_{II} values measured in the present study. A level of singlet excited state of Chl *d* is also added by assuming the same ground state level as that of Chl *a*. $E_m(P)$ is estimated to be different from $E_m(P680)$. P is assumed to be made of Chl *d*. C, the diagram assumes the levels of P* and P680* to be similar. The scheme also assumes the level of $S_2Q_A^-$ to be lower and $E_m(P)$ to be more positive compared to those in *Synechocystis* PS II.

model (Swingley *et al.*, 2008). The modified diagram A predicts the emission of PF and DF from Chl *d** at the lowest energy level. The model also predicts the ΔH_I value to produce P680* from the $S_2Q_A^-$ state to be the same before and after the addition of Chl *d** level, and suggests the energy to produce Chl *d** to be a little lower.

The ΔH_I value of 0.85 eV obtained in *Acaryochloris* in Table 6.1 is significantly (19%) larger than the ΔH_I value of 0.70 eV obtained in *Synechocystis.* The modified diagram A with Chl *d* antenna cannot interpret this result. The model, therefore, neither interprets the result in the present study, nor the "nanosecond DF" of Chl *a*-type pigment in *Acaryochloris* PS II (Mimuro *et al.*, 1999, 2000, 2004, 2007; Tomo *et al.*, 2007).

Energy Diagrams B and C

The higher ΔH_I value in Table 6.1 suggests either the higher P* level (energy diagram B) or the lower $S_2Q_A^-$ level (diagram C) in *Acaryochloris* compared to those in *Synechocystis*. However, it is rather unlikely to assume the higher P* level because P should oxidize Y_z. It may also be difficult to assume the lower $S_2Q_A^-$ level (C) because the amino acid sequences of D1 and D2 subunit proteins of PS II in *Acaryochloris* are almost homologous to those in *Synechocystis*, except for the phenylalanine to leucine modification nearby H that may affect the redox midpoint potential of H (E_m(H)) (Tomo *et al.* 2007; Swingley *et al.*, 2008). An ESR study suggested the invariant reaction rates between Y_Z and P in *Acaryochloris* compared to that in *Synechocystis* (Razeghifard *et al.*,s 2005). Therefore, both the schemes B and C have some difficulties.

Table 6.1 Values of activation enthalpy ΔH_I and ΔH_{II} for the emission of DF from the $S_2Q_A^-$ state. The ΔH_I and ΔH_{II} values were calculated from the temperature dependencies of the initial intensity I_0 of the slow phase of DF decay and of decay rate constant k_r in the presence of DCMU, respectively, in cells of *Acaryochloris* and *Synechocystis* as shown in the *Arrhenius* plots in Fig. 6.6B and C. See text for the details.

Organism	ΔH_I	ΔH_{II}	*/eV*
Acaryochloris	0.85	0.58	
Synechocystis sp. PCC6803	0.70	0.41	

The values were calculated from Fig. 6.6B and C and are estimated to have error ranges of 5-10 %.

Structure and Function of *Acaryochloris* PSII

Another more plausible case might be that the obtained ΔH_I (and ΔH_{II}) value in Table 6.1 is somewhat deviated from the actual values by the kinetic reason because the dissipation rate of $S_2Q_A^-$ state is determined by the charge recombination rate through P^+H^- (Grabolle and Dau, 2005; Rappaport *et al.*, 2002, 2005; Rappaport, 2002; Vass *et al.*, 1981). If the ΔH_I value obtained in the present study is related to the one between the $S_2Q_A^-$ and the P^+H^- levels, then the scheme expects the larger energy gap (ΔH_I) between $S_2Q_A^-$ and P^+H^- in *Acaryochloris* due to either the positive shift of E_m(P) or the negative shift of E_m(H). The former might be possible if P is made of Chl *d*, but it will be difficult if P is P680. It, however, makes more difficult situation for the function of PS II because the P* level will be further downshifted by the positive shift of E_m(P) in addition to the lower quantum energy available from P*. The latter possibility, on the other hand, might be realized by the protein modification around H as stated above. Another uncertainties for the determination of intermediate energy levels come from the unknown ΔS values, which may compensate the change of ΔH. We expect that the actual situation in *Acaryochloris* PSII occurred with the combined modifications of P and H level just as a mixture of diagrams B and C, although more quantitative studies from different measurements are required to fill these ambiguities.

The structure of PSII reaction center complex of *Thermosynechococcus elongatus* (Loll *et al.*, 2005) is shown with the amino acid residues that are modified uniquely in D1 and D2 subunits of PSII of *Acaryochloris* in Fig. 6.8 according to Itoh *et al.* (2007). The modified residues, which are drawn by space filling models, indicate that the structure of the reaction center is conserved quite well also in this organism, except for a residue nearby pheophytin *a*. This modification might affect the level of H, although it remains to be confirmed. The other cofactors seem to be surrounded by the conserved residues.

CONCLUSION

Acaryochloris emits DF and PF from Chl *d* at 719 nm in PS II RC in contrast to other oxygenic organisms that emit DF and PF from Chl *a* at 680 nm. Although Chl *d*-DF does not directly indicate the nature of the special pair Chl of PS II by itself, the results in the present study suggest the function of Chl *d*-type electron donor (P720), which is different from P680. The energy levels or kinetic parameters in the electron transfer chain of PS II are modified to meet the use of Chl *d*. The features of DF/PF of PS II Chl *d* in *Acaryochloris* were very similar to those of DF/PF of PS II Chl *a* of *Synechocystis* except for their different peak wavelengths. It indicates that the oxygenic photosynthesis in *Acaryochloris* with Chl

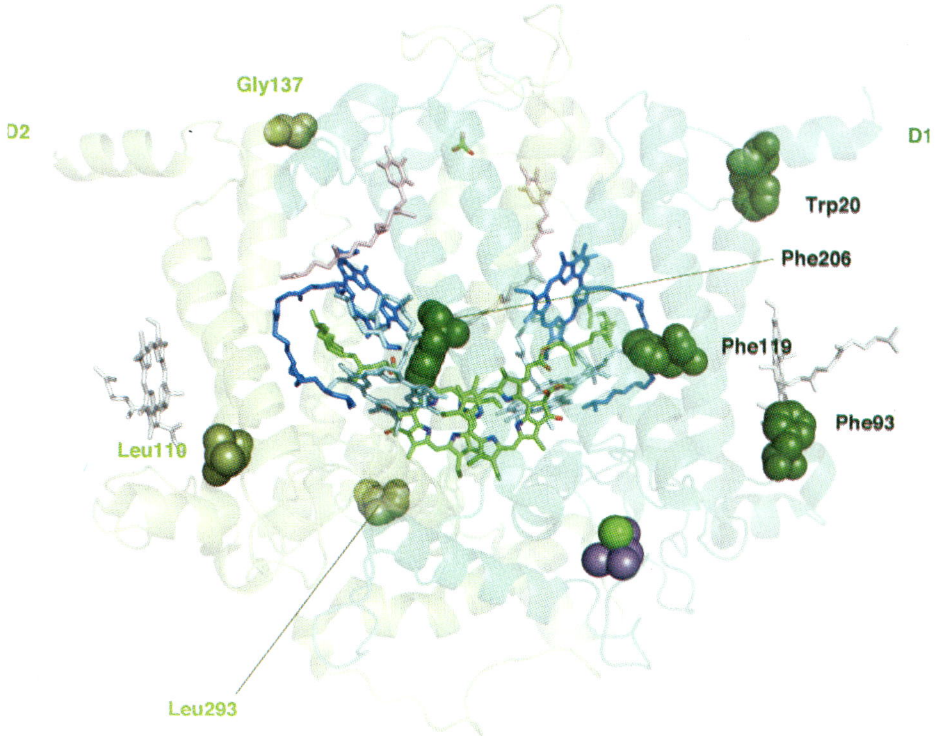

Fig. 6.8 Structure of PS II reaction center complex. Structure of PSII reaction center complex of *Thermosynechococcus elongatus* (Loll *et al.*, 2005) is shown with the amino acid residues (drawn by space filling models) that are modified uniquely in D1 and D2 subunits of PSII of *Acaryochloris* according to Itoh *et al.* (2007).

d-based PS II and PS I is performed efficiently with low charge recombination probabilities, although it had never been imagined before the discovery of *Acaryochloris*.

ACKNOWLEDGEMENTS

The authors thank Mr. M. Maeda for his excellent help in the setup of phosphoroscope, to Mr. K. Itoh, and Mrs. Y. Nakamura for culture of *Acaryochloris*, to Drs. H. Mino, Y. Shibata, T. Uzumaki and T. Tomi for their valuable discussions during the study.

REFERENCES

Boichenko, VA, Klimov, VV, Miyashita, H, Miyachi, S (2000) Functional characteristics of chlorophyll *d*-predominating photosynthetic apparatus in intact cells of *Acaryochloris marina*. Photosynth Res 65: 269-277

Chen, M, Bibby, TS, Nield, J, Larkum, AWD, Barber, J (2005a) Structure of a large photosystem II super complex from *Acaryochloris marina*. FEBS Lett 579: 1306-1310

Chen, M, Telfer, A, Lin, S, Pascal, A, Larkum, AWD, Barber, J, Blankenship, RE (2005b) The nature of the photosystem II reaction center in the chlorophyll *d*-containing prokaryote, *Acaryochloris marina*. Photochem Photobiol Sci 4: 1060-1064

Chisholm, SW, Frankel, SL, Goericke, R, Olson, RJ, Palenik, B, Waterbury, JB, West-Johnsrud, L, Zettler, ER (1992) *Prochlorococcus marinus* nov. gen. nov. sp: An oxyphototrophic marine prokaryote containing divinyl chlorophyll *a* and *b* Arch Microbiol 157: 297-300

Grabolle, M and Dau, H (2005) Energetics of primary and secondary electron transfer in photosystem II membrane particles of spinach revised on basis of recombination-fluorescence measurements. Biochim Biophys Acta 1708: 209-218

Hideg, E, Scott, R, Inaba, H (1991) Spectral resolution of long term (0.5-50 s) delayed fluorescence from spinach chloroplasts. Arch Biochem Biophys 285: 371-372

Hu, Q, Miyashita, H, Iwasaki, I, Kurano, N, Miyachi, S, Iwaki, M, Itoh, S (1998) A photosystem I reaction center driven by chlorophyll *d* in oxygenic photosynthesis. Proc Natl Acad Sci USA 95: 13319-13323

Hu, Q, Marquardt, J, Iwasaki, I, Miyashita, H, Kurano, N, Morschel, E, Miyachi, S (1999) Molecular structure, localization and function of biliproteins in the chlorophyll *a/d* containing oxygenic photosynthetic prokaryote *Acaryochloris marina*. Biochim Biophys Acta 1412: 250-261.

Itoh, S, Murata, N, Takamiya, A (1971) Studies on delayed light emission in spinach chloroplasts. I. Nature of 2 phases in development of millisecond delayed light emission during intermittent illumination. Biochim Biophys Acta 245: 109-120

Itoh, S, Iwaki, M, Noguchi, T, Kawamori, A, Mino, H, Hu, Q, Iwasaki, I, Miyashita, H, Kurano, N, Miyachi, S, Shen, R (1998) Photosystem I and II reaction centers of a new oxygenic organism *Acaryochloris marina* that use chlorophyll *d*. In Proc Internl Congr Photosynthesis (Electronic Edition), S6-028, Sydney, Australia

Itoh, S, Fukushima, Y, Itoh, K, Maeda, M, Mino H, and Kumazaki, S (2002) Photosystem I and II reaction centers in a new type of oxygenic photosynthesis of *Acaryochloris marina* based on chlorophyll *d*: Studies of delayed fluorescence and triplet ESR. J Photosci 9: 70-73

Itoh, S and Sugiura, K (2004) Fluorescence in Photosystem I in "Fluorescence in Photosynthesis" edited by Papageorgiou, CC, Govindjee. Springer, The Netherlands, pp 231-250

Itoh, S, Mino, H, Itoh, K, Shigenaga, T, Uzumaki, T, Iwaki, M (2007) Function of chlorophyll *d* in reaction centers of photosystems I and II of the oxygenic photosynthesis of *Acaryochloris marina*. Biochemistry 46: 12473-12481

Iwaki, M,Itoh, S (1988) Delayed fluorescence in photosystem I enhanced by phylloquinone (vitamin K-1) extraction with ether. Biochim Biophys Acta 934: 32-38

Jordan, P, Fromme, P, Witt, HT, Klukas, O, Saenger, W, Krauss, N (2001) Three-dimensional structure of cyanobacterial photosystem I at 2.5 Å resolution. Nature 411: 909-917

Kamiya N, and Shen JR (2003) Crystal structure of oxygen-evolving photosystem II from *Thermosynechococcus vulcanus* at 3.7 Å resolution. Proc Natl Acad Sci USA 100: 98-103

Kleinherenbrink, FA, Hastings, G, Wittmerhaus, BP, Blankenship, RE (1994) Delayed fluorescence from Fe–S type photosynthetic reaction centers at low redox potential. Biochemistry 33: 3096-3105

Kobayashi, M, Watanabe, T, Nakazato, M, Ikegami, I, Hiyama, T, Matsunaga, T, Murata, N (1988) Chlorophyll *a'*/P-700 and pheophytin *a*/P-680 stoichiometries in higher plants and cyanobacteria determined by HPLC analysis. Biochim Biophys Acta 936: 81-89

Kobayashi, M, Ohashi, S, Iwamoto, K, Shiraiwa, Y, Kato, Y, Watanabe, T. (2007) Redox potential of chlorophyll *d in vitro*. Biochim Biophys Acta 1767: 596-602

Loll, B, Kern, J, Saenger, W, Zouni, A, and Biesiadka, J. (2005) Towards complete cofactor arrangement in the 3.0 Å resolution structure of photosystem II. Nature 438: 1040

Marquardt, J, Morschel, E, Rhiel, E, Westermann, M (2000) Ultra structure of *Acaryochloris marina*, an oxyphotobacterium containing mainly chlorophyll *d*. Arch Microbiol 174: 181-188

Mimuro, M, Akimoto, S, Gotoh, T, Yokono, M, Akiyama, M, Tsuchiya, T, Miyashita, H, Kobayashi, M, Yamazaki, I (2004) Identification of the primary electron donor in PS II of the Chl *d*-dominated cyanobacterium *Acaryochloris marina*. FEBS Lett 556: 95-98

Mimuro, M, Akimoto, S, Tomo, T, Yokono, M, Miyashita, H, Tsuchiya, T (2007) Delayed fluorescence observed in the nanosecond time region at 77 K originates directly from the photosystem II reaction center. Biochim Biophys Acta 1767: 327-34

Mimuro, M, Akimoto, S, Yamazaki, I, Miyashita, H, Miyachi, S (1999) Fluorescence properties of chlorophyll *d*-dominating prokaryotic alga, *Acaryochloris marina*: Studies using time-resolved fluorescence spectroscopy on intact cells. Biochim Biophys Acta 1412: 37-46

Mimuro, M, Hirayama, K, Uezono, K, Miyashita, H, Miyachi, S (2000) Uphill energy transfer in a chlorophyll *d*-dominating oxygenic photosynthetic prokaryote, *Acaryochloris marina*. Biochim Biophys Acta 1456: 27-34

Miyachi, S, Strassdat, K, Miyashita, H, Senger, H (1997) Quantum requirement of photosynthesis in the primarily chlorophyll *d* containing prokaryote *Acaryochloris marina*. Z Naturforsch C 52: 636-638

Miyashita, H, Ikemoto, H, Kurano, N, Adachi, K, Chihara, M, Miyachi, S (1996) Chlorophyll *d* as a major pigment. Nature 383: 402

Miyashita, H, Adachi, K, Kurano, N, Ikemoto, H, Chihara, M, Miyachi, S (1997) Pigment composition of a novel oxygenic photosynthetic prokaryote containing chlorophyll *d* as the major chlorophyll. Plant Cell Physiol 38: 274-281

Miyashita, H, Ikemoto, H, Kurano, N, Miyachi, S (2003) *Acaryochloris marina* gen. et sp. Nov. (Cyanobacteria), an oxygenic photosynthetic prokaryote containing Chl *d* as a major pigment. J Phycol 39: 1247-1253

Nieuwenburg, P, Clarke, RJ, Cai, ZL, Chen, M, Larkum, AWD, Cabral, NM, Ghiggino, KP, Reimers, JR (2003) Examination of the photophysical processes of chlorophyll *d* leading to a clarification of proposed uphill energy transfer processes in cells of *Acaryochloris marina*. Photochem Photobiol 77: 628-637

Rappaport, F (2002) Structure, dynamics, and energetics of the primary photochemistry of photosystem II of oxygenic photosynthesis. Ann Rev Plant Biol 53: 551-580

Rappaport, F, Guergova-Kuras, M, Nixon, P, Diner, B, Lavergne, J (2002) Kinetics and pathways of charge recombination in photosystem II. Biochemistry 41: 8518-8527

Rappaport, F, Cuni, A, Xiong, L, Sayre, R, Lavergne, J (2005) Charge recombination and thermoluminescence in photosystem II. Biophys J 88: 1948-58

Razeghifard, MR, Chen, M, Hughes, JL, Freeman, J, Krausz, E, Wydrzynski, T (2005) Spectroscopic studies of photosystem II in chlorophyll *d*-containing *Acaryochloris marina*. Biochemistry 44: 11178-11187

Schiller, H, Senger, H, Miyashita, H, Miyachi, S, Dau, H (1997) Light harvesting in *Acaryochloris marina*-Spectroscopic characterization of a chlorophyll *d*-dominated photosynthetic antenna system. FEBS Lett. 410: 433-436

Schlodder, E, Cetin, M, Eckert, H.-J, Schmitt, F.-J, Barber, J, Telfer, A (2007) Both chlorophylls *a* and *d* are essential for the photochemistry in photosystem II of the cyanobacteria, *Acaryochloris marina*. Biochim Biophys Acta 1767: 589-595

Strehler, BL, Arnold, W (1951) Light production by green plants. J Gen Physiol 34: 809-20

Swingley, WD, Chen, M, Cheung, PC, Conrad, AL, Dejesa, LC, Hao, J, Honchak, BM, Karbach, LE, Kurdoglu, A, Lahiri, S, Mastrian, SD, Miyashita, H, Page, L, Ramakrishna, P, Satoh, S, Sattley, WM, Shimada, Y, Taylor, HL, Tomo, T, Tsuchiya, T, Wang, ZT, Raymond, J, Mimuro, M, Blankenship, RE, Touchman, JW (2008) Niche adaptation and genome expansion in the chlorophyll *d*-producing cyanobacterium *Acaryochloris marina*. Proc Natl Acad Sci USA 105: 2005-2010

Tomo, T, Okubo, T, Akimoto, S, Yokono, M, Miyashita, H, Tsuchiya, T, Noguchi, T, Mimuro, M (2007) Identification of the special pair of photosystem II in a chlorophyll *d*-dominated cyanobacterium. Proc Natl Acad Sci USA 104: 7283-7288

Vass, I, Horvath, G, Herczeg, T, Demeter, S (1981) Photosynthetic energy conservation investigated by thermoluminescence, activation energies and half-lives of thermoluminescence bands determined by mathematical resolution of glow curves. Biochim Biophys Acta 634: 140-152

Wraight, CA, Crofts, AR (1971) Delayed fluorescence and the high-energy state of chloroplasts. Eur J Biochem 19: 386-97

CHAPTER 7

Delayed Fluorescence in a Millisecond Range: A Probe for the Donor Side-induced Photoinhibition in Photosystem II

Ralphreed A. Gasanov, Samira A. Aliyeva and Fikret Mamedov*

Biophysics Laboratory, Institute of Botany, National Academy of Science, Patamdar shosse 40, Baku AZ 1073, Azerbaijan

* *Molecular Biomimetics, Department of Photochemistry and Molecular Science, Box 523, Ångström Laboratory, Uppsala University, Uppsala SE 75120, Sweden*

ABSTRACT

Photoinhibition of photosystem II (PSII) membranes from spinach was investigated by millisecond delayed fluorescence (ms-DF) at three different pH values (pH 5.5, pH 7.0 and pH 8.5). The photoinhibitory treatment of 4000 μmol photons m^{-2} s^{-1} was used and the fast component of ms-DF, which originates from radiative recombination reactions, and the steady state level of ms-DF were monitored. We have observed strong photoinhibition effect on the components of ms-DF at the elevated pH values. At acidic pH photoinhibition of the fast and steady state components were less pronounced and, in addition, a protective role of an exogenous acceptor PpBQ has been observed. In view of our previous results [Gasanov *et al.*, (2007) *J. Photochem. Photobiol. B* 86, 160-164] we suggest that change in equilibrium between Y_Z $[(P_{680}Q_A^-) \leftrightarrow (Y_Z^{\bullet+})]$ and the $CaMn_4$-cluster $[(P_{680}Q_A^-) \leftrightarrow (S_{i+1}^+)]$ is taking place when pH of medium is changed from low to high values. At high pH this leads to the donor side induced photoinhibition and the destruction of the $CaMn_4$-cluster that can be monitored by ms-DF.

Keywords: Photosystem II, Photoinhibition, Delayed fluorescence.

Abbreviations

Chl chlorophyll;

PSII photosystem II;

ms-DF delayed fluorescence in the millisecond time range;

fs	fast component of ms-DF;
MES	2-(N-morpholino) ethanesulfonic acid;
HEPES	4-(2-hydroxyethyl)-1-piperazineethanesulfonic acid;
PpBQ	phenyl-*p*-benzoquinone;
ss	steady state component of ms-DF.

INTRODUCTION

In nature, photosynthetic organisms are continuously exposed to unfavorable environmental conditions such as wide fluctuations in temperature and light intensity, elevated concentration of salts and metals. Exposure of photosynthetic organisms to strong light leads to inhibition of the photosynthetic activity and photosystem II (PSII)[1] is the main target of this phenomenon, which is also referred to as photoinhibition (Powles, 1984; Prásil *et al.*,1992; Aro *et al.*,1993; Andersson *et al.*, 2001). PSII is a multi-protein-pigment complex where the photochemical reactions involving charge separation and the subsequent transport of electrons from water to plastoquinone take place (Wydrzynski and Satoh, 2005). Since light energy is the driving force for photosynthesis, photoinhibition is unavoidable in photosynthetic organisms (Gombos *et al.*, 1994; Tyystjärvi and Aro, 1996; Andersson and Chow, 2002; Mamedov and Styring, 2003). However, photosynthetic organisms were able to overcome the harmful effects of light through the efficient repair process of inhibited PSII complexes (Andersson and Aro, 2001; Andersson and Chow, 2002; Mamedov and Styring, 2003), although it was suggested that the damaging effect of the reactive oxygen species can slow down the repair process (Nishiyama *et al.*, 2006). Therefore, the extent of photoinhibition depends on the balance between photoinhibition and repair of PSII (Andersson and Aro, 2001; Andersson and Chow, 2002; Mamedov and Styring, 2003).

Fluorescence spectroscopy is a powerful method to study PSII reactions (Krause and Weis, 1991; Mamedov *et al.*, 1998; Grabolle and Dau 2005; Gasanov *et al.*, 2007; Goltsev *et al.*, 2009). Recently, we have shown that the recombinant Chl fluorescence, or delayed fluorescence in the millisecond time range (ms-DF) can be a useful probe to the donor side reactions in PSII (Mamedov and Gasanov, 1993; Gasanov *et al.*, 2007), reflecting an equilibrium between the charge recombination in PSII and the redox state of $Y_Z^{\bullet}$, the intermediate electron carrier between P680, the primary donor and the $CaMn_4$-cluster, the catalytic center of the water oxidation process in PSII (Gasanov *et al.*, 2007). In this article we discuss applications of ms-DF in photoinhibition studies of the PSII reactions at different pH values.

Chloroplast thylakoid membranes were prepared from commercial spinach leaves (*Spinacea oleracea*) in 20 mM Tris-HCl (pH 7.8), 350 mM NaCl and 2 mg/ml Na-ascorbate according to Whatley and Arnon (1963). The final thylakoid pellets were resuspended in 20 mM Tris-HCl (pH 7.2), 35 mM NaCl, 2 mM $MgCl_2$ and 20% glycerol at Chl concentration of 3-4 mg/ml and were stored in liquid nitrogen until used. All steps were performed under dim green light at 4°C.

MATERIALS AND METHODS

Photoinhibition of the thylakoid membranes was performed in a thermostatic cuvette at 20°C with white light (4000 μmol photons m^{-2} s^{-1}) in a buffer containing 400 mM sucrose, 35 mM NaCl and 25 mM MES-NaOH (pH 5.5) or 25 mM HEPES-NaOH (pH 7.0 and 8.5) at a Chl concentration of 5 μg/ml. The oxygen evolving activity was measured with a Clark-type electrode (Hansatech Instruments, UK) at 20°C under saturated illumination with 25 μM PpBQ as an exogenous electron acceptor at a Chl concentration of 20 μg/ml.

The kinetics of ms-DF was measured using a phosphoroscope setup, which was described previously. (Mamedov and Gasanov, 1993; Gasanov *et al.*, 2007). A sample in silica-suprasil cuvette was illuminated with a saturating continuous white light (250 W $m^{-2}s^{-1}$) passing through 2 cm of a $CuSO_4$ (aq) solution. The light was passed through the holes on the rotating wheel of phosphoroscope (three holes, 120° apart) in such a way that 0.3 ms of excitation was followed by 1.25 ms of dark and 0.3 ms of registration of delayed fluorescence (three cycles of excitation/dark/registration per one full turnover of the phosphoroscope). The delayed fluorescence was registered with a photomultiplier via red glass cut off KC19 filter (LOMO Russia, > 680 nm). The signal was passed through the amplifier and was recorded by electrical potential recorder.

RESULTS AND DISCUSSION

After excitation, the charge separated state is known to be stabilized on millisecond time scale through the reduction of the primary quinine acceptor Q_A and formation of the positive charge on the donor side of PSII. Recombination reactions always compete with the forward electron transfer in PSII (Rappaport and Lavergne, 1997; Mamedov *et al.*, 1998; Gasanov *et al.*, 2007) and on the basis of our previous study it was established that recombination partner for Q_A^- on the donor side is determined by the proton availability in the medium, being the $CaMn_4$-cluster at pH above 7.5 and $Y_Z^{\bullet}$ at pH below 7.5 (Gasanov *et al.*, 2007).

In this study, the time course of photoinhibition of PSII in thylakoid membranes isolated from spinach chloroplasts was studied at three different pH values. Two kinetic components of ms-DF — the fast and steady state components showed different pH dependent behavior during photoinhibition (Fig. 7.1). The fast component of ms-DF was more stable at pH 5.5 (Fig. 7.1A), and steady state level of ms-DF increased two times during first minute of photoinhibition treatment and then decreased to 40% of the original value after 9 min (Fig. 7.1B). The ratio of the fast component to the steady-state level of ms-DF was decreasing with photoinhibition time and stayed approximately constant after 1 min of illumination (Fig. 7.3A).

At higher pH values (pH 7.0 and 8.5), the fast component of ms-DF was photoinhibited much faster (Fig. 7.1A), the steady-state level decreased (Fig. 7.1B) and the ratio between these two components rapidly went down within the first minute of photoinhibitory treatment (Fig. 7.3A).

To separate the acceptor side and the donor side related effects during photoinhibition, we performed experiments in the presence of exogenous electron acceptor PpBQ. In this case, the electron pressure from the acceptor side is relieved and the acceptor side effects should be diminished. In the presence of PpBQ, the protection of PSII from photoinhibition has been observed only for the fast component of ms-DF at acidic conditions (Fig. 7.2A and 7.3B).

The pH dependence of the oxygen evolving activity of PSII of chloroplasts has a bell-shaped form with maximal activity at pH 6.3-6.7 and decreased activity at acidic and alkaline sides with apparent pK values of ~5.5 and ~8.0, respectively (Bumann and Oesterhelt 1994; Bernat *et al.*, 2002; Gasanov *et al.*, 2007). On the acidic side, this effect arises from the inhibition of the S state transitions (Bernat *et al.*, 2002; Suzuki *et al.*, 2005) and alteration of the structure of the $CaMn_4$-cluster due to possible protonation of the ligands (most probably carboxylates) or the *oxo*-bridges between the Mn ions (Havelius and Styring 2007; Sjoholm *et al.*, 2009).

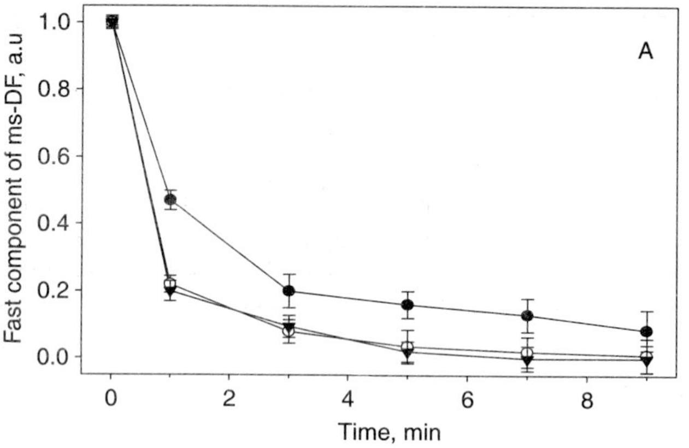

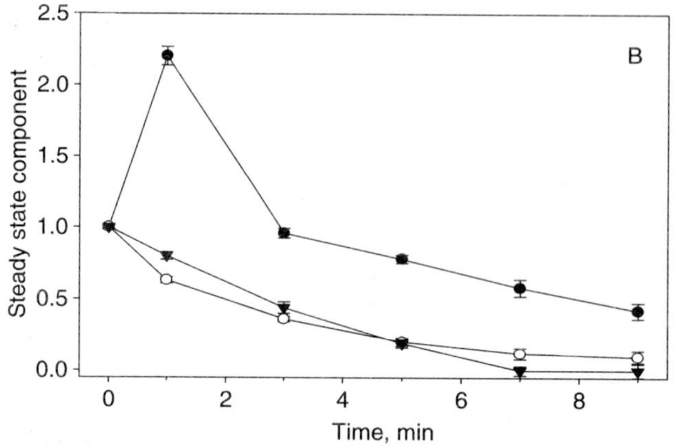

Fig. 7.1 Time course of the ms-delayed fluorescence components during the photoinhibitory illumination of the thylakoid membranes at pH 5.5 (•), pH 7.0 (○) and pH 8.5 (▼). **A**-fast component of ms-DF; **B**-steady-state level of ms-DF.

On the high pH side, this inhibition also reflects the inhibition of the S state transitions and/or destruction of the $CaMn_4$-cluster due to the Mn release (Bernat *et al.*, 2002; Suzuki *et al.*, 2005; Havelius and Styring 2007; Sjoholm *et al.*, 2009). In addition, inhibition of the acceptor side reactions at non-physiological pH values should also be taken into account (Petrouleas and Diner, 1987).We have observed strong photoinhibition effect at the elevated pH values. The origin of this effect can be attributed to the donor side induced photoinhibition since at high pH values, the destruction of the $CaMn_4$-cluster is more pronounced (Schansker and van Rensen 1996; Bernat *et al.*, 2002; Havelius and Styring, 2007; Sjoholm *et al.*, 2009). This is corroborated by our experiments in the presence of PpBQ (Fig. 7.2). At acidic pH we have observed that exogenous acceptor has a protective role during photoinhibition. At low pH values, the $CaMn_4$-cluster is more stable during photoinhibition, and the presence of PpBQ eliminates the acceptor side induced mechanisms for photoinhibition.

It is known that under strong light, changes in the rates of forward electron transfer and recombination can be responsible for photoinhibition mechanism (Vass *et al.*, 1992; Keren *et al.*,1997). In our case, it has been observed by ms-DF as change in equilibrium between Y_Z $[(P_{680}Q_A^-) \leftrightarrow (Y_Z^{\bullet+})]$ and the $CaMn_4$-cluster $[(P_{680}Q_A^-) \leftrightarrow (S_{i+1}^+)]$, indicating the $CaMn_4$-cluster as a site of photodamage at high pH values. It may be concluded that the fast component of DF reflecting the recombination between Q_A^- and $Y_Z^{\bullet}$ at pH

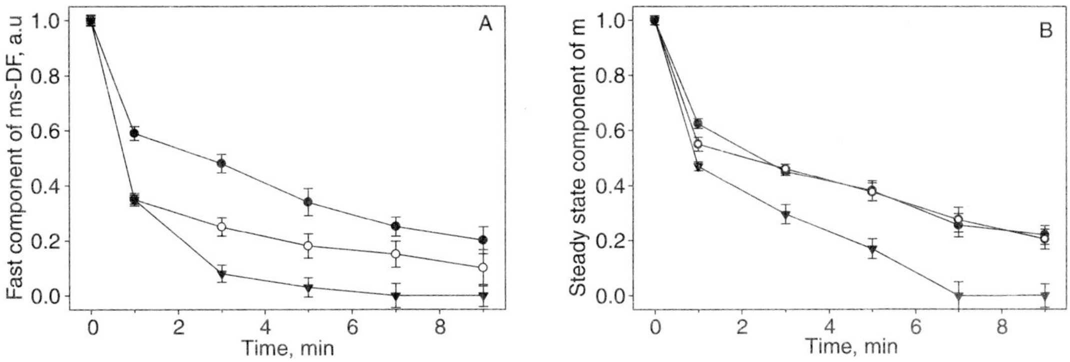

Fig. 7.2 Time course of the ms-delayed fluorescence components during the photoinhibitory illumination of the thylakoid membranes in the presence of 0.6 mM PpBQ at pH 5.5 (•), pH 7.0 (O) and pH8.5 (▼). **A**-fast component of ms-DF; **B**-steady-state level of ms-DF.

< 7.5 is more stable during photoinhibition, whereas the recombination between Q_A^- and $CaMn_4$-cluster (steady state level of ms-DF) at pH > 7.5 is more sensitive to photoinhibition. In addition, our data also show that the protective effect of PpBQ was more visible at pH 5.5 and was relatively weak at pH 7.0.

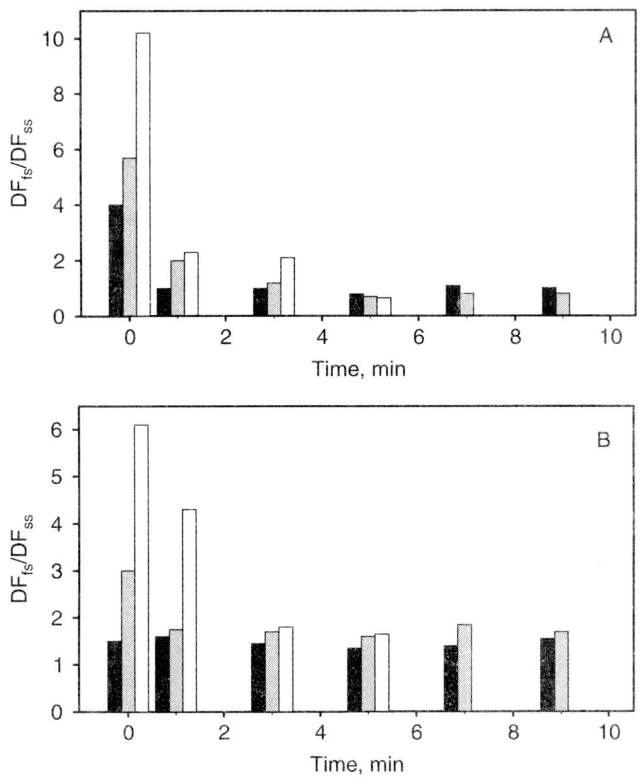

Fig. 7.3 Changes in ratio between the ms-delayed fluorescence fast component and steady-state level under photoinhibitory illumination of the thylakoid membranes at different pH values (pH 5.5 - black, pH 7.0 - grey and pH8.5 - white). **A**-no additions, **B**- in presence of 25 μM PpBQ.

Thus, ms-DF can be a useful tool in monitoring and elucidating photoinhibition in PSII. Detection of the ms-DF is possible on the wide range of the photosynthetic organisms, making this non-invasive method also a useful tool to dissect photoinhibitory reactions *in vivo* in plants, algae and cyanobacteria.

ACKNOWLEDGEMENTS

FM is grateful to the Swedish Research Council and the Knut and Alice Wallenberg Foundation for financial support.

REFERENCES

Andersson B, Aro EM (2001) Photodamage and D1 protein turnover in photosystem II. in: Aro EM, Anderson B (eds) Regulation of Photosynthesis. Kluwer Academic Publishers, Dordrecht, The Netherlands, pp 377-393

Andersson JM, Chow WS (2002) Structural and functional dynamics of plant photosystem II. Philos Trans R Soc Lond B 357: 1421-1430

Aro EM, Virgin I, Andersson B (1993) Photoinhibition of photosystem II. Inactivation, protein damage and turnover. Biochim Biophys Acta 1143: 113-134

Bernat G, Morvaridi F, Feyziyev Y, Styring S (2002) pH dependence of four individual transitions in the catalytic S-cycle during photosynthetic oxygen evolution. Biochemistry 41: 5830-5843

Bumann D, Oesterhelt D (1994) Purification and characterization of oxygen evolving photosystem II core complexes from the green algae *Chlamydomonas reinhardtii*. Biochemistry 33, 10906-10910.

Gasanov R, Aliyeva S, Arao S, Ismailova A, Katsuta N, Kitade H, Yamada S, Kawamori A, Mamedov F (2007) Comparative study of the water oxidizing reactions and the millisecond delayed chlorophyll fluorescence in photosystem II at different pH. J Photochem Photobiol B 86: 160-164

Goltsev V, Chernev P, Zaharieva I, Lambrev P, Strasser RJ (2005) Kinetics of delayed chlorophyll a fluorescence registered in milliseconds time range. Photosynth Res 84: 209-215

Goltsev V, Zaharieva I, Chernev P, Strasser RJ (2009) Delayed fluorescence in photosynthesis. Photosynth Res 101, 217-232

Gombos Z, Wada H, Murata N (1994) The recovery of photosynthesis from low-temperature photoinhibition is accelerated by the unsaturation of membrane lipids: a mechanism of chilling tolerance. Proc Natl Acad Sci USA 91: 8787-8791

Grabolle M, Dau H (2005) Energetics of primary and secondary electron transfer in photosystem II membrane particles of spinach revisited on basis of recombimation fluorescence measurements. Biochim Biophys Acta 1708: 209-218

Hakala M, Tuominen I, Keränen M, Tyystjärvi T Tyystjärvi E (2005) Evidence for the role of oxygen-evolving manganese complex in photoinhibition of photosystem II. Biochim Biophys Acta 1706: 68-80

Havelius KGV, Styring S (2007) pH dependent competition between Y-Z and Y-D in photosystem II probed by illumination at 5 K. Biochemistry 46: 7865-7874

Keren N, Berg A, van Kan PJM, Levanon H, Ohad I (1997) Mechanism of photosystem II photoinactivation and D1 protein degradation at low light: the role of back electron flow. Proc Natl Acad Sci USA 94: 1579-1584

Krause GH, Weis E (1991) Chlorophyll fluorescence and photosynthesis – the basics. *Annu. Rev. Plant Physiol.* Plant Mol Biol 42: 313-349

Mamedov F, Gasanov, RA (1993) Effect of removal of peripheral polypeptides on photoinhibition of oxygen-evolving PSII particles from spinach. Biochemistry (Moscow) 58: 809-814

Mamedov F, Sayre TR, Styring S (1998) Involvement of histidine 190 on the D1 protein in electron/proton transfer reactions on the donor side of photosystem II. Biochemistry 37: 14245-14256

Mamedov F, Styring S (2003) Logistics in the life cycle of photosystem II during maturation – lateral movement in the thylakoid membrane and activation of the electron transfer. Physiol Plantarum 119: 328-336

Nishiyama Y, Alakhverdiyev SI, Murata N (2006) A new paradigm for the action of reactive oxygen species in the photoinhibition of photosystem II. Biochim Biophys Acta 1757: 742-749

Petrouleas V, Diner BA (1987) Light induced oxidation of the acceptor side Fe(II) of photosystem II by exogenous quinones acting through the QB binding site. 1. Quinones, kinetics and pH dependence. Biochim Biophys Acta 893: 126-137

Powles SB (1984) Photoinhibition of photosynthesis induced by visible light. Annu Rev Plant Physiol 35: 15-44

Prásil O, Adir N, Ohad I (1992) Dynamics of photosystem II: mechanisms of photoinhibition and recovery of process. in: Barber J (ed) The photosystems: structure, function and molecular biology. Vol 11, Elsevier Science Publishers, Amsterdam, The Netherlands, pp 295-348

Rappaport F, Lavergne J (1997) Charge recombination and proton transfer in manganese-depleted photosystem II. Biochemistry 36: 15294-15302

Satoh K, Wydrzynski TJ and Govindjee (2005) Introduction to photosystem II. In: Wydrzynski TJ and Satoh K (eds) Photosystem II: The Light-Driven Water. Plastoquinone Oxidoreductase. Advances in Photosynthesis and Respiration. Vol 22 Springer, Dordrecht, pp 11-22

Schansker G, van Rensen JJS (1996) Mechanism of photoinhibition in pea thylakoids: Effects of irradiance level and pH. Biochim Biophys Acta 1276: 239-245

Sjoholm J, Havelius KGV, Mamedov F, Styring S (2009) The S0 state of the water oxidizing complex in photosystem II: pH dependence of the EPR split signal induction and mechanistic implications. Biochemistry 48: 9393-9404

Suzuki H, Sugiura M, Noguchi T (2005) pH dependence of the flash-induced S-state transitions in the oxygen-evolving center of photosystem II from *Thermosynechoccocus elongatus* as revealed by Fourier transform infrared spectroscopy. Biochemistry 44: 1708-1718

Tyystjärvi E, Aro EM (1996) The rate constant of photoinhibition, measured in lincomycin-treated leaves, is directly proportional to light intensity. Proc Natl Acad Sci USA 93: 2213-2218

Vass I, Styring S, Hundall T, Koivuniemi A, Aro E.-M, Andersson B (1992) Reversible and irreversible intermediates during photoinhibition of photosystem II. 2. Stable reduced QA species promote chlorophyll triplet formation. Proc Natl Acad Sci USA 89: 1408-1412

Whatley FR, Arnon DJ (1963) Photosynthetic phosphorylation in plants. Meth Enzymol 6: 308-313

CHAPTER 8

Redox Control of Leaf Variegation and Chloroplast Biogenesis in *Arabidopsis thaliana*

Dominic Rosso[1], Rainer Bode[1], Wenze Li[1], M. Krol[1], Steven R. Rodermel[2], Denis P. Maxwell[1], and Norman P.A. Hüner[1]

[1]Department of Biology and the Biotron Experimental Climate Change Research Centre, University of Western Ontario, London, ON, Canada N6A 5B7

[2]Department of Genetics, Development and Cell Biology, Iowa State University, Ames, Iowa, 50011

ABSTRACT

IMMUTANS is a single-copy, nuclear gene of *Arabidopsis thaliana* encoding a chloroplastic plastid terminal oxidase, which catalyzes the oxidation of plastoquinol in concert with the reduction of O_2 to water. The knockout mutant, *immutans*, normally exhibits a variegated phenotype. However, this can be completely suppressed by growth under low light such that it is indistinguishable from the wild type. Thus, the absence of *IMMUTANS* expression appears to be necessary but not sufficient to account for the variegated phenotype. We developed a nondestructive imaging technique from which we quantified the extent of variegation *in vivo* as a function of developmental time. The extent of variegation was positively correlated (r^2 = 0.7469) with an increase in excitation pressure irrespective of whether high light, low temperature or continuous illumination were used to induce increased excitation pressure. Similar results were obtained for three other variegated mutants of *Arabidopsis* (*spotty*, *var1* and *var2*). Thus, the redox state of the intersystem plastoquinone pool appears to control variegation in these *Arabidopsis* mutants. Chloroplast biogenesis during the light-dependent greening of etiolated WT and *immutans* cotyledons indicated, that although the disassembly of prolamellar bodies occurred normally, structural and functional analyses indicated that the subsequent assembly of thylakoid membranes was inhibited under high irradiance in *immutans* compared to WT. The presence of IMMUTANS differentially lowered excitation pressure significantly during the early stages of greening in WT compared to *immutans*. We conclude that the redox state of plastoquinone pool governs the sectoring patterns associated with the variegated phenotype of *IMMUTANS*, *SPOTTY*, *VAR1*, and *VAR2* through the regulation of chloroplast biogenesis. We suggest that redox control of leaf variegation provides a simple explanation for the variable sectoring patterns observed in variegation mutants of *Arabidopsis thaliana*.

Keywords: *Arabidopsis thaliana*, Immutants, Leaf variegation.

Abbreviations

PTOX: plastid terminal oxidase, *AOX:* alternative oxidase, *HEP:* high excitation pressure phenotype.

VARIEGATION IN *ARABIDOPSIS THALIANA*

Variegation mutants are plants that display distinct green and white sectored leaves (Kirk and Tilney-Bassett, 1978; Rodermel, 2002; Miura *et al.*, 2007). The green sectors contain normal chloroplasts, whereas the white sectors contain plastids devoid of chlorophyll and/or carotenoids (Rodermel, 2002; Miura *et al.*, 2007). The *immutans* (*im*)mutant of *Arabidopsis thaliana* is one of the oldest and most extensively studied variegation mutants and was isolated and characterized nearly 50 years ago (Röbbelen, 1968; Redei, 1963). The *immutans* mutant is the result of a recessive mutation of the nuclear gene, *IMMUTANS* (Wu *et al.*, 1999; Carol *et al.*, 1999). *In vitro* and *in vivo* evidence indicate that IMMUTANS may be the elusive plastid terminal oxidase (PTOX) involved in the chlororespiratory pathway (Cournac *et al.*, 2000b; Josse *et al.*, 2000; Joët *et al.*, 2002; Peltier and Cournac, 2002; Josse *et al.*, 2003; Fu *et al.*, 2005; Shahbazi *et al.*, 2007). Bennoun (1982) proposed that, in the dark, reducing equivalents from a stromal pool of NAD(P)H reduces plastoquinone (PQ), which is mediated by a NAD(P)H dehydrogenase. The oxidation of plastoquinol (PQH_2) is thought to require the action of IMMUTANS.

IMMUTANS exhibits 37% sequence identity to the mitochondrial alternative oxidase (AOX) (Wu *et al.*, 1999), and it was suggested that IMMUTANS, like AOX, acts as a safety valve under conditions whereby excess photosynthetically-generated electrons are produced (Niyogi, 2000) thereby minimizing the formation of ROS (Maxwell *et al.*, 1999). However, recent research conducted on the function of IMMUTANS has shown that modulation of *IMMUTANS* expression and polypeptide accumulation does not alter the flux of intersystem electrons to P_{700}^+ during steady-state photosynthesis in fully expanded leaves and does not appear to provide significant photoprotection in *Arabidopsis thaliana* (Rosso *et al.* 2006). Moreover, through meta-analyses of published *Arabidopsis* microarray data, Rosso *et al.*, (2006) reported that *IMMUTANS* expression was modulated minimally in response to myriad abiotic stresses. These results do not support the model of IMMUTANS as a safety valve to regulate the redox state of the PQ pool during stress and acclimation in fully expanded leaves of *Arabidopsis thaliana*. In contrast, *IMMUTANS* did appear to be strongly regulated by development in *Arabidopsis* (Rosso *et al.*, 2006). This is consistent with the observations of Aluru *et al.* (2001) who concluded that IM is required for the differentiation of chloroplasts, amyloplasts and etioplasts in *Arabidopsis thaliana*.

Wu *et al.* (1999) proposed that IMMUTANS functions as a plastid terminal oxidase capable of oxidizing plastoquinol. In their model, the activity of the phytoene desaturase is regulated by the redox state of the PQ pool during early chloroplast biogenesis (Aluru *et al.*, 2006; Yu *et al.*, 2007). When the PQ pool is oxidized, phytoene desaturase oxidizes phytoene to ξ-carotene by concomitantly reducing the PQ pool through four subsequent desaturation steps. This results in the biosynthesis of a complete complement of carotenoids required to protect the photosynthetic apparatus from photo-damage by reactive oxygen species. Alternatively, if the PQ pool remains reduced due to the lack of IMMUTANS (IM), carotenoid biosynthesis is blocked at the phytoene desaturase step, phytoene accumulates, and photo-bleaching results from the light-dependent generation of reactive oxygen species (Wetzel *et al.*, 1994; Wu *et al.*, 1999). Consistent with the notion that white sectoring is triggered by photo-oxidation is the fact that the variegated phenotype can be suppressed in *im* plants when grown under low light conditions (Rédei, 1963; Wetzel *et al.*, 1994; Aluru and Rodermel, 2004; Rosso *et al.*, 2006). However, recently, it has been reported that variegation in *im* plants can be completely suppressed if and only if *im* plants are exposed initially to low irradiance (5 μmol photons $m^{-2}s^{-1}$) at 25°C (25/5) for 7 days after seeding and then shifted to growth at 50 μmol photons $m^{-2}s^{-1}$ for 35 days. Subsequently, mature *im* plants exhibit an "all green" phenotype indistinguishable from WT plants regardless of the growth irradiance and age of the plant (Rosso *et al.*, 2006). Furthermore, variegation appears to be an irreversible process.

Variegation in the *yellow variegated* (*var2*) mutant of *Arabidopsis thaliana* is the result of loss of FtsH2 from the thylakoid FtsH complex (Miura *et al.*, 2007). Miura *et al.* (2007) provided strong evidence that leaf variegation in the *var2* mutant is the result of an imbalance between the biosynthesis and degradation of the D1 protein, the reaction center polypeptide of PSII. However, these results still do not explain the variable pattern of sectoring typically observed during the development of a variegated leaf. Since *var2*, like *immutans*, is the result of a nuclear mutation, theoretically all cells should exhibit defective chloroplasts. Since this is not the case, Miura *et al.* (2007) conclude that further research is required to elucidate the physiological basis of the observed variegation patterns. Thus, the control of variegation and the leaf sectoring patterns in mutants of *Arabidopsis thaliana* remain an enigma despite recent molecular genetic advances.

LIGHT REGULATION OF PLANT DEVELOPMENT AND PHOTOACCLIMATION

Plants sense light through an array of photoreceptors including phytochromes (Rockwell and Lagarias, 2006; Bae and Choi, 2008), crypotchromes (Li and Yang, 2007) and the more recently discovered phototropins (Christie, 2007) that are critical for plant growth and development. However, in addition to the requirement for photoreceptors sensitive to spectral quality, the oxidation-reduction (redox) state of the photosynthetic intersystem electron transport chain (PET) acts as a sensor of cellular energy imbalances. Such imbalances may occur whenever the absorption and transformation of light by the extremely fast, temperature-insensitive photochemical reactions of photosynthesis either exceeds the capacity to utilize the photosynthetic electrons for reductive C, N and S metabolism or exceeds the capacity of the photosynthetic apparatus to dissipate excess energy non-photochemically as heat (Hüner *et al.*, 1998; Pfannschmidt, 2003; Ensminger *et al.*, 2006; Wilson *et al.*, 2006; Murchie *et al.*, 2009). The redox state of PET has been shown to influence a diversity of phenomena from altering the excitation distribution between photosystems through state transitions (Rochaix, 2004; Kargul and Barber, 2008), to changes in organellar gene expression (Pfannschmidt *et al.*, 1999; Pfannschmidt, 2003) and nuclear gene expression through retrograde regulation (Pfannschmidt *et al.*, 1999; Pfannschmidt, 2003; Ferandez and Strand, 2008; Woodson and Chory, 2008; Brautigam *et al.*, 2009; Pfannschmidt *et al.*, 2009), to changes in plant growth habit and morphology (Gray *et al.*, 1997). Furthermore, tobacco plants deficient in ferredoxin-NADP(H) reductase (FNR), the enzyme that catalyzes the final step in photosynthetic electron transport, exhibit a yellow-green phenotype due to over-reduction of intersystem photosynthetic electron transport chain. The extent of this phenotype is directly dependent upon the irradiance to which the plants are exposed (Palatnik *et al.*, 2003). Consequently, it has been suggested that the chloroplast has a dual role. Not only does it function as a primary energy transducer in all photoautotropns, it also functions as a sensor of environmental change (Hüner *et al.*, 1998; Pfannschmidt, 2003; Wilson *et al.*, 2006; Murchie *et al.*, 2009).

Early research with green algae indicated that a key sensor for photoacclimation was the redox-state of plastoquinone (PQ), a mobile electron carrier that shuttles electrons from photosystem II to the cytochrome *b6f* complex (Escobas *et al.*, 1995; Maxwell *et al.*, 1995). This was based on experiments where the characteristic, yellow-green, high light phenotype brought about by acclimation to high irradiance could be mimicked by chemically modulating the redox status of the intersystem PQ pool with the electron transport inhibitors, DCMU and DBMIB in *Dunalliela tertiolecta* (Escoubas *et al.*, 1995) and *Chlorella vulgaris* (Wilson *et al.*, 2003). Since DBMIB inhibits the oxidation of plastoquinol (PQH_2) by the Cyt b_6/f complex, the PQ pool becomes overly reduced in the light. This induces the high light phenotype which is characterized by low Chl content per cell, high Chl *a/b* ratio (>10), accumulation of the carotenoid-binding protein (cbr) (Hüner *et al.*, 1998) but suppression of both Lhcb2 accumulation and *Lhcb2* expression, the nuclear gene that encodes the major PSII light-harvesting antenna polypeptide, *Lhcb2*. In contrast, since DCMU prevents the exit of electrons from PSII, the PQ pool remains oxidized in the light. Under these conditions, cells exhibit a normal green phenotype, which is associated with high Chl content per cell, low Chl *a/b* ratio (3.0-3.5), high levels of *Lhcb2* expression and *Lhcb2* accumulation. The high light phenotype could also be duplicated by simply lowering the growth temperature of *Chlorella vulgaris* (Maxwell *et al.*, 1995; Wilson *et al.*, 2003). While low temperature does not affect the rate of light absorption, it severely restricts the rate of downstream enzyme-catalyzed reactions thus reducing the demand for the products of PET—NADPH and ATP.

Both high light and low temperature can result in a comparable accumulation of closed PS II reaction centers as measured by the reduction state of Q_A, the first stable quinone electron acceptor in PSII reaction centers (Schreiber *et al.*, 1994; Kramer *et al.*, 2004; Baker, 2008). This has been referred to as a measure of PSII excitation pressure or excessive excitation (1-qP) (Dietz *et al.*, 1985; Maxwell *et al.*, 1995; Hüner *et al.*, 1998; Karpinski *et al.*, 1999; Swiatek *et al.*, 2001; Falkowski and Chen, 2003; Shahbazi *et al.*, 2007), which reflects the over-reduction of PQ pool in the photosynthetic intersystem electron transport chain due to an imbalance between the rapid, temperature-independent photochemical events within PSII reaction centers versus the capacity of the slower, temperature-dependent, metabolic sinks to utilize the photosynthetically-generated electrons (Hüner *et al.*, 1998; Ensminger *et al.*, 2006; Wilson *et al.*, 2006; Ivanov *et al.*, 2006). Photoautotrophs are in photostasis when the photochemistry induced by the absorption of light is balanced either by utilization of photosynthetically-generated electrons through metabolism and growth or through the capacity to dissipate excess energy as heat through nonphotochemical quenching (NPQ) (Hüner *et al.*, 2003). Under such conditions the excitation pressure is low, the PQ pool is in the oxidized state and the organism exhibits a normal, green low excitation pressure phenotype (LEP). However, a myriad of environmental stresses including high irradiance (Escoubas *et al.*, 1995), low temperature (Maxwell *et al.*, 1994; 1995; Wilson and Huner, 2000; Wilson *et al.*, 2003), nitrogen deficiency (Cruz *et al.*, 2003), and salt stress (Huang *et al.*, 2005) increase excitation pressure. In the single cell green algae, *Chlorella vulgaris*, *Dunaliella salina* and *Dunaliella tertiolecta* (Maxwell *et al.*, 1994; 1995; Escoubas *et al.*, 1995) as well as the cyanobacterium, *Plectonema boryanum* (Miskiewicz *et al.*, 2000; 2002), this typically results in a high excitation pressure phenotype (HEP) characterized by a decrease in Chl per cell, an increase in the Chl *a/b* ratio and a decreased photosynthetic efficiency typically associated with photoacclimation. Green algae adjust the structure and composition of LHCII in response to the over-reduction of PQ pool.

In addition to the redox state of PQ pool, recent research in *Arabidopsis thaliana* indicates that redox factors on the acceptor-side of photosystem I (PSI) may also be important in redox regulation of photoacclimation (Stroher and Dietz, 2006; Dietz, 2008). These as well as additional signals including the procursor of chlorophyll synthesis, magnesium protoporphyrin (Strand *et al.*, 2003), and reactive oxygen species generated by the PET (Wagner *et al.*, 2004) may constitute a complex network of signals involved in the retrograde pathway of communication from the chloroplast to the nucleus (Koussevitzky *et al.*, 2007; Fernandez *et al.*, 2008; Woodson and Chory, 2008), which is required to modulate the expression of nuclear-encoded photosynthetic genes. Recent genetic analyses in *Arabidopsis* has identified that STN7 (Bellafiore *et al.*, 2005; Bonardi *et al.*, 2005), a chloroplast protein kinase, may be involved in redox signalling since it has been shown to be required for state transitions and photosynthetic acclimation. Nevertheless, the nature of the mechanisms by which the redox state of the chloroplast is signalled to the nucleus resulting in altered gene expression remains largely unknown.

REDOX REGULATION OF LEAF VARIEGATION

Does excitation pressure influence the extent of leaf variegation? To answer this question, we established an *in vivo*, non-destructive assay to quantify the extent of variegation as a function of developmental time. The extent of leaf variegation in *im* seedlings was estimated from images captured by a CCD camera (Retiga 1300 monochrome 10 bit, Qimaging, Surrey, Canada) attached to a dissecting microscope (Leica Wild M3B, Heerbrugg, Switzerland). The camera was oriented directly over the center of the plant and the magnification selected on the dissecting microscope ensured that an image of the entire plant was captured for each measurement. Subsequently, digital photos were analyzed using Northern Eclipse Image Analysis Software (Empix Imaging, Mississauga, ON, Canada), which was calibrated with an object of known size for each magnification and the number of pixels was divided by the appropriate conversion factor. Images were then converted to grayscale, thereby creating a binary partitioning of the image intensities. An intensity value called threshold value was determined in order to separate green versus white sectors (Pham *et al.*, 2000; Sezgin and Sankur, 2004). Threshold analysis was performed on each image captured to ensure that all green sectors could be clearly resolved from all white sectors irrespective of the magnification.

Using this nondestructive approach, both the rate of development of the variegated phenotype as well as the overall extent of variegation in *im* seedlings were strongly dependent upon growth irradiance with no variegation detected at a growth irradiance of 50 μmol photons $m^{-2}s^{-1}$ at 25°C and an 8 h photoperiod. However, the variegated phenotype could not be explained as a simple irradiance effect, since growth at the same, low irradiance combined with low temperature (12°C) resulted in significant increase in variegation of *im* seedlings. As a consequence, the development of variegation in *im* seedlings appears to be a complex interaction of irradiance and temperature. Figure 8.1 illustrates a strong, positive correlation between excitation pressure and the extent of variegation, regardless of how the excitation pressure was generated, i.e., either by irradiance, low temperature or continuous light. Since both irradiance and temperature both affect excitation pressure, the extent of variegation in *im* seedlings cannot be controlled by irradiance and temperature per se, but rather is governed by the redox state of the PQ pool of the intersystem electron transport chain. To test the generality of this response to excitation pressure, we also examined three additional variegated mutants of *Arabidopsis thaliana*: *spotty*, *var1* and *var2*. A common feature of all the mutants tested is that the mutations affect components of the photosynthetic electron transport chain. First, *immutans* and *spotty* are knockout mutants of IMMUTANS, a plastid terminal oxidase of the thylakoid membrane which uses PQH_2 as an electron donor. Second, *var1* and *var2* are genes encoding *AtFtsH5* and *AtFsH2*, respectively, which appear to be involved in PSII photodamage and repair (Aluru *et al.*, 2006; Miura *et al.*, 2007). Figure 8.2 illustrates that increasing irradiance resulted in increased variegation in all mutants tested. Furthermore, the extent of variegation increased at all light intensities during growth at low temperature in all mutants tested. Thus, the regulation of leaf variegation by excitation pressure, and hence the redox state of the PQ pool, is not specific to *immutans*, but rather it appears to be a general phenomenon in these variegated mutants of *Arabidopsis thaliana*.

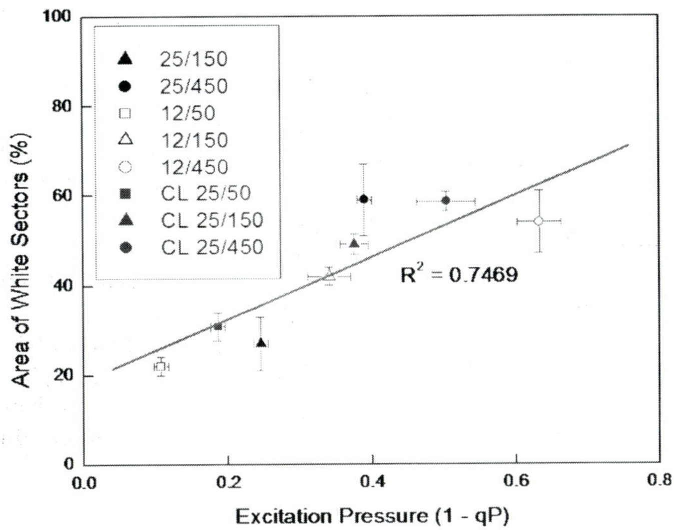

Fig. 8.1 The correlation between variegation measured as % white sector area and excitation pressure measured as 1-qP in *IM* seedlings. Growth conditions of temperature/irradiance are indicated in the legend in the figure. CL, continuous light. Under all other conditions, seedlings were grown under an 8h photoperiod.

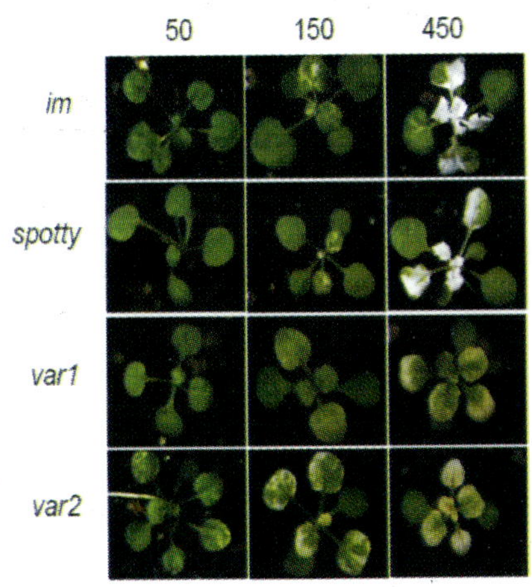

Fig. 8.2 Representative photographs of *im(3157)*, *spotty*, *var1* and *var2* seedlings grown at 25°C under CL at an irradiance of either 50, 150, or 450 μmol photons $m^{-2}s^{-1}$, respectively.

THE ROLE OF IMMUTANS

PTOX has been reported to transfer effectively excess electrons to oxygen, thus playing an important photoprotective role in the high alpine plant species *Ranunculus glacialis* acclimated to low temperature (Streb *et al.*, 2005). Furthermore, Stepien and Johnson (2008) reported that plastid terminal oxidase acts as an alternative electron sink accounting in salt-stressed halophyte *Thellungiella*. These effects were accompanied by a significant increase in the relative abundance of PTOX/IMMUTANS protein in *Ranunculus glacialis* (Streb *et al.*, 2005) and *Thellungiella*, but not in *Arabidopsis thaliana* (Stepien and Johnson, 2008). Furthermore, up-regulation of PTOX and the chloroplast NDH-complex has been reported in plants subjected to heat and high light stresses (Quiles, 2006; Diaz *et al.*, 2007; Tallón and Quiles, 2007). These data support the role of PTOX as an alternative electron sink in alleviating over-reduction of the PQ pool under unfavorable environmental conditions where PSI is limited on the acceptc side. In contrast, Rosso *et al.* (2006) showed that the absence of IM in the *immutans* knockout mutaı exhibited a similar capacity to keep the PQ pool oxidized as WT *Arabidopsis thaliana*. They conclude that IM does not appear to play a major role in photoprotection in fully expanded leaves of *Arabidopsı* (Rosso *et al.*, 2006), Thus, it appears that the role of PTOX in photoprotection remains elusive.

We hypothesized that IM may play an important role in photoprotection during early chloropla: biogenesis and assembly of the photosynthetic apparatus prior to the establishment of full photosyntheti competence. Biochemical and functional data indicate that the biogenesis of thylakoid membranes durin greening of etiolated seedlings is inhibited by increased irradiance in *im* but not in WT cotyledons. Firs although greening at low irradiance indicated minimal differences in the accumulation of *Lhcb2*, th major light harvesting polypeptide associated with PSII, in WT and *im* (Fig. 8.3), *Lhcb2* was detecte only after 24 h and its abundance was reduced during greening at 150 μmol photons m^{-2} s^{-1}. In contras the accumulation of *Lhcb2* in WT during greening at 25/50 was indistinguishable from that observed : 25/150 (Fig. 8.3). Since the disappearance of prolamellar bodies, as indicated by the disappearance (POR, was comparable in *im* and WT cotyledons regardless of the irradiance, we conclude that the ligh dependent inhibition of the etioplast to chloroplast conversion in *im* cotyledons is not due to an inhibitio of the disassembly of prolamellar bodies, but because of an inhibition of the biogenesis and assembly (photosynthetically active thylakoid membranes.

Second, although the rate of development of maximum PSII photochemical efficiency (Fv/Fm) wa stimulated by greening at 150 μmol photons $m^{-2}s^{-1}$ in WT cotyledons (Fig. 8.4A, closed symbols

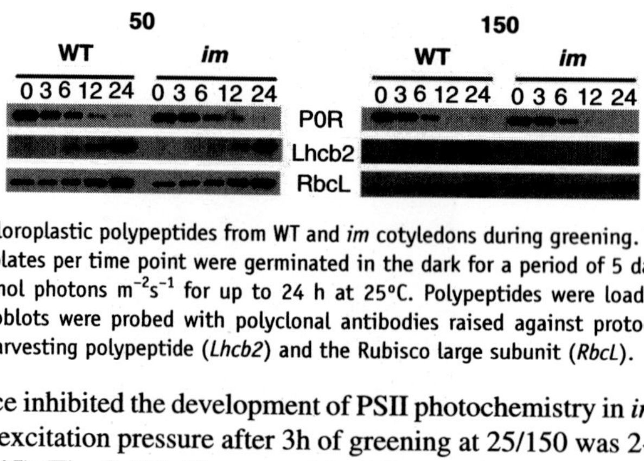

Fig. 8.3 Immunoblots of chloroplastic polypeptides from WT and *im* cotyledons during greening. Twenty-five *IM* seeds per ag plate and a total of 25 agar plates per time point were germinated in the dark for a period of 5 days and subsequently expos to CL of either 50 or 150 μmol photons $m^{-2}s^{-1}$ for up to 24 h at 25°C. Polypeptides were loaded on an equal protein bas (7 μg protein/lane). Immunoblots were probed with polyclonal antibodies raised against protochlorophyllide oxidoreducta (POR), the major PSII light harvesting polypeptide (*Lhcb2*) and the Rubisco large subunit (*RbcL*).

greening at this irradiance inhibited the development of PSII photochemistry in *im* cotyledons (Fig. 8.4/ open symbols). Finally, excitation pressure after 3h of greening at 25/150 was 2-fold higher in *im* (0.8(than WT cotyledons (0.45) (Fig. 8.4B). However, this difference in excitation pressure between *im* ar WT cotyledons gradually decreased as a function of greening time, such that after 24h, the difference i excitation pressure between *im* and WT cotyledons was only 30%. This indicates that *im* seedlings have greater propensity to accumulate closed PSII reaction centers than WT seedlings during the early stag(of greening.

The higher excitation pressure exhibited by *im* cotyledons during greening at 25/150 reflects an ove reduction of the PQ pool. Inhibition of the accumulation of *Lhcb2* during greening at 25/150 in *i* seedlings (Fig. 8.3) is consistent with retrograde regulation between the chloroplast and the nucleu whereby reduction of the PQ pool represses *lhcb2* expression (Taylor, 1989; Aluru *et al.*, 2001; Roderme 2001; Nott *et al.*, 2006; Koussevitzky *et al.*, 2007; Fernandez and Strand, 2008). However, excitatic pressure in *im* cotyledons does decrease as a function of time during greening at 25/150, albeit mo slowly than in WT (Fig. 8.4). Since the only detectable difference in the components of the photosynthet

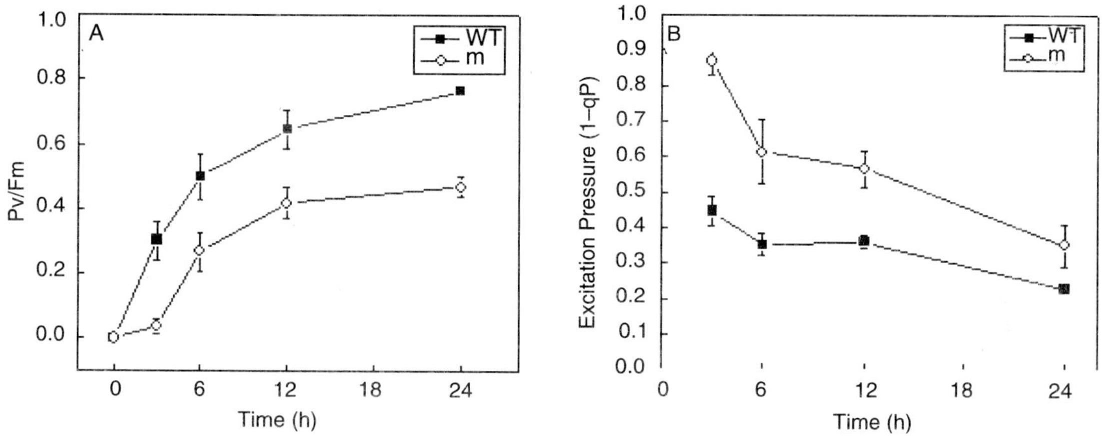

Fig. 8.4 The effects of greening on the development of photosystem II photochemistry. Maximum PSII photochemical efficiency (F_V/F_M) of etiolated WT and *im* cotyledons during greening at a PAR of 150 µmol photons $m^{-2}s^{-1}$ and a temperature of 25°C. Fluorescence of cotyledons was measured at their respective growth irradiance and growth temperature. Data represent the mean ± SE from 5-11 different seedlings per time point. **B.** The effect greening time at a PAR of 150 µmol photons $m^{-2}s^{-1}$ and a temperature of 25°C on the excitation pressure estimated as the relative reduction state of Q_A (1-qP) of etiolated WT and *im* cotyledons. All seedlings were dark adapted for 20 min prior to measurement of Fv/Fm and 1-qP. Fluorescence of cotyledons was measured at their respective growth irradiance and growth.

electron chain between mature WT and *im* seedlings is the absence of IM in the latter (Rosso *et al.*, 2006), we suggest that the difference in the excitation pressure response during greening of *im* and WT cotyledons must be due to the absence of IM. Furthermore, this difference in excitation pressure is about 4 times greater at the early stages (3 h of greening) than the latter stages (24h) of greening (Fig. 8.4). This is consistent with our hypothesis that IM has a putative role in photoprotection by lowering the excitation pressure during the initial, early stages of chloroplast biogenesis in *Arabidopsis*. However, this photoprotective role of IM becomes minimal when the chloroplast becomes fully competent photosynthetically. This is consistent with previous results, which indicate that IM cannot compete with PSI for PSII generated electrons, nor is its presence photoprotective in fully expanded leaves of either WT, *im* seedlings expressing an 'all green phenotype' but lacking IM or two over-expressing lines of *IM* (Rosso *et al.*, 2006).

SUMMARY

Recently, Miura *et al.* (2007) concluded that the development of variegation in *var2* mutant of *Arabidopsis* is the result of a dosage dependency of FtsH. Insufficient levels of FtsH arrested chloroplast development due to photoinhibition of PSII due to the lack of D1 repair cycle (Miura *et al.*, 2007). The absence of IM can also induce variegation in *im* seedlings of *Arabidopsis*. However, we show that IM is necessary but not sufficient to account for this variegated phenotype. In fact, the variegated phenotype can be completely suppressed in *im* seedlings (Rosso *et al.*, 2006). Normal chloroplast development in *im* seedlings as well as *spotty*, *var1* and *var2* appears to be dependent upon excitation pressure. Consequently, we conclude that the control of variegation by the absence of IM is indirect. Rather, we suggest that variegation in *im* seedlings is a direct response to the redox state of the PQ pool which can be modulated either by high irradiance, low temperature or photoperiod which cause an imbalance between PSII photochemistry versus the capacity to utilize the photosynthetically-generated reductant during chloroplast biogenesis and assembly of the photosynthetic apparatus.

We agree with Eckardt (2007) and Miura *et al.* (2007) that variegation in *Arabidopsis* is a consequence of balance. However, the notion of a balance between D1 protein synthesis and degradation, as suggested by Miura *et al.* (2007) fails to account for the heterogeneity in chloroplast development evident by the presence of white and green sectors in the *var2* mutant of *Arabidopsis*. We suggest that the controlling factor for variegation is not protein turnover, but rather, photosynthetic redox energy balance as indicated by modulation of the extent of variegation by excitation pressure. A heterogeneous internal light environment within a developing leaf could account for the presence of green (normal chloroplasts) and white sectors (defective chloroplasts) during leaf initiation and expansion. Thus, we suggest that the control of leaf variegation by cellular energy balance provides a rather simple explanation for the variable sectoring typically observed during leaf variegation in several mutants of *Arabidopsis thaliana*.

ACKNOWLEDGEMENTS

This work was supported by the Natural Sciences and Engineering Research Council of Canada (Discovery Grants to DPM, and to NPAH) and by the US Department of Energy (Energy Biosciences; grant no. DF-FG02-94ER20147 to SRR). DR was the recipient of an NSERC Postgraduate Scholarship (CGS-D).

REFERENCES

Aluru MR, Bae H, Wu D, Rodermel SR (2001) The *Arabidopsis immutans* mutation affects plastid differentiation and the morphogenesis of white and green sectors in variegated plants. Plant Physiol 127: 67-77

Aluru MR, Rodermel SR (2004) Control of chloroplast redox by the IMMUTANS terminal oxidase. Physiol Plant 120: 4-11

Aluru MR, Yu F, Fu A, Rodermel SR (2006) *Arabidopsis* variegation mutants: new insights into chloroplast biogenesis. J Exp Bot 57: 1871-1881

Bae G, Choi G (2008) Decoding of light signals by plant phytochromes and their interacting proteins. Ann Rev Plant Biol 59: 281-311

Baker NR (2008) Chlorophyll fluorescence: a probe of photosynthesis *in vivo*. Ann Rev Plant Biol 59: 89-113

Bellafiore S, Barneche F, Peltier G, Rochaix, J-D (2005) State transitions and light adaptation require chloroplast thylakoid protein kinase STN7. Nature 433: 892-895

Bennoun P (1982) Evidence for a respiratory chain in the chloroplast. Proc Natl Acad Sci USA 79: 4352-4356

Bonardi V, Pesaresi P, Becker T (2005) Photosystem II core phosphorylation and photosynthetic acclimation require two different protein kinases. Nature 437: 1179-1182

Bräutigam, K, Dietzel, L, Kleine, T, Ströter, E, Wormuth, D, Dietz, K-J, Radke, D, Wirtz, M, Hell, R, Dormann, P, Nunes-Nesi, A, Schauer, N, Fernie, AR, Oliver, SN, Geigenberger, P, Leister, D, and Pfannschmidt, T (2009) Dynamic plastid redox signals integrate gene expression and metabolism to induce distinct metabolic states in photosynthetic acclimation in *Arabidopsis*. Plant Cell 21: 2715-2732

Carol P, Stevenson D, Bisanz C (1999) Mutations in the *Arabidopsis* gene *IMMUTANS* cause a variegated phenotype by inactivating a chloroplast terminal oxidase associated with phytoene desaturation. Plant Cell 11: 57-68

Christie JM (2007) Phototropin blue-light receptors. Ann Rev Plant Biol 58: 21-45

Cournac L, Redding K, Ravenel J (2000b) Electron flow between photosystem II and oxygen in chloroplasts of photosystem I-deficient algae is mediated by a quinol oxidase involved in chlororespiration. J Biol Chem 275: 17256-17262

Cruz JL, Mosquim PR, Pelacanil PR, Araujo CR, DaMatta FM (2003) Photosynthesis impairment in cassava leaves in response to nitrogen deficiency. Plant Soil 257: 417-423

Diaz M, De Haro V, Munoz R, Quiles MJ (2007) Chlororespiration is involved in the adaptation of *Brassica* plants to heat and high light intensity. Plant Cell Environ 30: 1578-1585

Dietz KJ (2008) Redox signal integration: from stimulus to networks and genes. Physiol Plant 133: 459-468

Dietz KJ, Schreiber U, Heber U (1985) The relationship between the redox state of Q_A and photosynthesis in leaves at various carbon-dioxide, oxygen and light regimes. Planta 166: 219-226

Eckardt NA (2007) Thylakoid development from biogenesis to senescence, and ruminations on regulation. Plant Cell 19: 1135-1138

Ensminger I, Busch F, Hüner NPA (2006) Photostasis and cold acclimation: sensing low temperature through photosynthesis. Physiol. Plant 126: 28-44

Escoubas JM, Lomas M, LaRoche J, Falkowski PG (1995) Light intensity regulation of cab gene transcription is signaled by the redox state of the plastoquinone pool. Proc Natl Acad Sci USA 92: 10237-10241

Falkowski PG, Chen YB (2003) Photoacclimation of light harvesting systems in eukaryotic algae. In: Green BR, Parson WW (eds) Advances in Photosynthesis and Respration. Light Harvesting Antennas in Photosynthesis, Kluwer Academic Publishers, Dordrecht, The Netherlands, Vol 13, pp 423-447

Fernández AP, Strand Å (2008) Retrograde signalling and plant stress: plastid signals initiate cellular stress responses. Curr Opin Plant Biol 11: 509-513

Fu A, Park S, Rodermel SR (2005) Sequences required for the activity of IM (IMMUTANS), a plastid terminal oxidase. *In vitro* and *in planta* mutagenesis of iron-binding sites and a conserved sequence that corresponds to exon 8. J Biol Chem 280: 42489-42496

Gray GR, Chauvin L-P, Sarhan F, Hüner NPA (1997) Cold acclimation and freezing tolerance. A complex interaction of light and temperature. Plant Physiol 114: 467-474

Huang C, He W, Guo J (2005) Increased sensitivity to salt stress in an ascorbate deficient *Arabidopsis* mutant. J Exp Bot 56: 3041-3049

Hüner NPA, Öquist, G., and Sarhan, F. (1998) Energy balance and acclimation to light and cold. Trends Plant Sci 3: 224-230

Hüner NPA, Öquist G, Melis A (2003) Photostasis in plants, green algae and cyanobacteria: the role of light harvesting antenna complexes. In: Green BR, Parson WW (eds) Advances in Photosynthesis and Respiration. Light Harvesting Antennas in Photosynthesis, Kluwer Academic Publishers, Dordrecht, The Netherlands, Vol 13, pp 401-421

Ivanov AG, Sane PV, Krol M (2006) Acclimation to temperature and irradiance modulates PS II charge recombination. FEBS Lett. 580: 2797-2802

Josse EM, Simkin AJ, Gaffé J (2000) A plastid terminal oxidase associated with carotenoid desaturation during chromoplast differentiation. Plant Physiol 123: 1427-1436

Josse EM, Alcaraz JP, Labouré AM, Kuntz M (2003) *In vitro* characterization of a plastid terminal oxidase (IM). Eur J Biochem 270: 3787-3794

Joët T, Genty B, Josse EM (2002) Involvement of a plastid terminal oxidase in plastoquinone oxidation as enhanced by expression of the *Arabidopsis thaliana* enzyme in tobacco. J Biol Chem 277: 31623-31630

Kargul J, Barber J (2008) Photosynthetic acclimation: Structural reorganisation of light harvesting antenna - role of redox-dependent phosphorylation of major and minor chlorophyll *a/b* binding proteins. FEBS J 275: 1056-1068

Karpinski S, Reynolds H, Karpinska B (1999) Systemic signaling and acclimation in response to excess excitation energy in *Arabidopsis*. Science 284: 654-657

Kirk JTO, Tilney-Basett RAE (1978) The Plastids: Their Chemistry, Structure, Growth, and Inheritance. Elsevier, Amsterdam

Koussevitzky S, Nott A, Mockler TC (2007) Signals from chloroplasts converge to regulate nuclear gene expression. Science 316: 715-719

Kramer DM, Johnson G, Kiirats O, Edwards GE (2004) New fluorescence parameters for the determination of Q_A redox state and excitation energy fluxes. Photosyn Res 79: 209-218

Li Q-H, Yang H-Q (2007) Cryptochrome signalling in plants. Photochem Photobiol 83: 94-101

Maxwell DP, Falk S, Hüner NPA (1994) Growth at low temperature mimics high light acclimation in *Chlorella vulgaris*. Plant Physiol 105: 535-543

Maxwell DP, Laudenbach DE, Hüner NPA (1995) Redox regulation of light-harvesting complex II and *cab* mRNA abundance in *Dunaliella salina*. Plant Physiol 109: 787-795

Maxwell DP, Wang Y, McIntosh L (1999) The alternative oxidase lowers mitochondrial reactive oxygen production in plant cells. Proc Natl Acad Sci USA 96: 8271-8276

Miskiewicz E, Ivanov AG, Williams JP, Khan MU, Falk S, Hüner NPA (2000) Photosynthetic acclimation of the filamentous cyanobacterium, *Plectonema boryanum* UTEX 485, to temperature and light. Plant Cell Physiol 41: 767-775

Miskiewicz E, Ivanov AG, Hüner NPA (2002) Stoichiometry of the photosynthetic apparatus and phycobilisome structure of the cyanobacterium *Plectonema boryanum* UTEX 485 are regulated by both light and temperature. Plant Physiol 130: 1414-1425

Miura E, Kato Y, Matsushima R (2007) The balance between protein synthesis and degradation in chloroplasts determines leaf variegation in *Arabidopsis yellow variegated* mutants. Plant Cell 19: 1313-1328

Murchie EH, Pinto M, Horton P (2009) Agriculture and the new challenges for photosynthesis research. New Phytol 181: 532-552

Niyogi KK (2000) Safety valves for photosynthesis. Curr Opin Plant Biol 3: 455-460

Nott A, Jung H-S, Koussevitzky S, Chory J (2006) Plastid-to-nucleus signalling. Ann Rev Plant Biol 57: 739-759

Palatnik JF, Tognetti VB, Poli HO (2003) Plant J 114: 467-474

Peltier G, Cournac L (2002) Chlororespiration. Ann Rev Plant Biol 53: 523-550

Pfannschmidt T (2003) Chloroplast redox signals: how photosynthesis controls its own genes. Trends Plant Sci 8: 33-41

Pfannschmidt T, Nilsson A, Allen JF (1999) Photosynthetic control of chloroplast gene expression. Nature 397: 625-628

Pfannschmidt, T, Bräutigam, K, Wagner, R, Dietzel, L, Schröher, Y, Steiner, S, and Nykytenko, A (2009). Potential regulation of gene expression in photosynthetic cells by redox and energy state: approaches towards better understanding. Ann Bot (Lond.) 103: 599-607

Pham DL, Xu C, Prince JL (2000) Current methods in medical image segmentation. Annu Rev Biomed Eng 2: 315-337

Quiles MJ (2006) Stimulation of chlororespiration by heat and high light intensity in oat plants. Plant Cell Environ 29: 1463-1470

Rochaix J-D (2004) Genetics of the biogenesis and dynamics of the photosynthetic machinery in eukaryotes. Plant Cell 16: 1650-1660

Rockwell NC, Lagarias JC (2006) The structure of phytochrome: a picture is worth a thousand spectra. Plant Cell 18: 4-14

Rodermel SR (2001) Pathways of plastid to nucleus signaling. Trends Plant Sci 6: 471-478

Rodermel SR (2002) *Arabidopsis* variegation mutants. In: Somerville CR, Myerowitz ER (eds) The *Arabidopsis* Book. American Society of Plant Biologists, Rockville, pp 1-28

Rosso D, Ivanov AG, Fu A (2006) IMMUTANS does not act as a stress-induced safety valve in the protection of the photosynthetic apparatus of *Arabidopsis* during steady-state photosynthesis. Plant Physiol 142: 574-585

Rédei GP (1963) Somatic instability caused by a cysteine-sensitive gene in *Arabidopsis*. Science 139: 767-769

Rédei GP (1975) *Arabidopsis* as a genetic tool. Ann Rev Genet 9: 111-127

Röbbelen G (1968) Genbedingte rotlicht-empfindlichkeit der chloroplastendifferenziering bei *Arabidopsis*. Planta 80: 237-254

Schreiber U, Bilger W, Neubauer C (1994) Chlorophyll fluorescence as a non-intrusive indicator for rapid assessment of *in vivo* photosynthesis. In: Schulze Ed, Caldwell MM (eds) Ecophysiology of Photosynthesis, Springer-Verlag, Berlin, pp 49-70

Sezgin M, Sankur B (2004) Survey over image thresholding techniques and quantitative performance evaluation. J Electronic Imaging 13: 146-165

Shahbazi M, Gilbert M, Labouré A, Kuntz M (2007) The dual role of the plastid terminal oxidase (PTOX) in tomato. Plant Physiol 145: 691-702

Stepien P, Johnson GN (2008) Contrasting responses of photosynthesis to salt stress in the glycophyte *Arabidopsis thaliana* and the halophyte *Tellungiella halophila*. Role of the plastid terminal oxidase as an alternative electron sink. Plant Physiol 149: 1154-1165

Strand A, Asami T, Alonso J, Ecker JR, Chory J (2003) Chloroplast to nucleus communication triggered by accumulation of Mg-protoporphyrin IX. Nature 421: 79-83

Streb P, Josse E-M, Gallouet E, Baptist F, Kuntz M, Cornic G (2005) Evidence for alternative electron sinks to photosynthetic carbon assimilation in the high mountain plant species *Ranunculus glacialis*. Plant, Cell and Environment 28: 1123-1135

Stroher E, Dietz K-J (2006) Concepts and approaches towards understanding the cellular redox proteome. Plant Biol 8: 407-418

Swiatek M, Kuras R, Sokolenko A (2001) The chloroplast gene ycf9 encodes a photosystem II (PS II) core subunit, PsbZ, that participates in PS II supramolecular architecture. Plant Cell 13: 1347-1368

Tallon C, Quiles MJ (2007) Acclmation to heat and high light intensity during the development of oat leaves increases the NADP DH complex and PTOX levels in chloroplasts. Plant Sci 173: 438-445

Taylor WC (1989) Regulatory interactions between nuclear and plastid genomes. Ann Rev Plant Physiol Plant Mol Biol 40: 211-233

Wagner D, Przybyla D, op den Camp R (2004) The genetic Basis of singlet oxygen-induced stress responses of *Arabidopsis thaliana*. Science 306: 1183-1185

Wetzel CM, Jiang CZ, Meehan LJ (1994) Nuclear-organelle interactions: the *immutans* variegation mutant of *Arabidopsis* is plastid autonomous and impaired in carotenoid biosynthesis. Plant J 6: 161-175

Wilson KE, Hüner NPA (2000) The role of growth rate, redox-state of the plastoquinone pool and the trans-thylakoid ΔpH in photoacclimation of *Chlorella vulgaris* to growth, irradiance and temperature. Planta 212: 93-102

Wilson KE, Ivanov AG, Öquist G (2006) Energy balance, organellar redox status, and acclimation to environmental stress. Can J Bot 84: 1355-1370

Wilson KE, Krol M, Hüner NPA (2003) Temperature-induced greening of *Chlorella vulgaris*. The role of the cellular energy balance and zeaxanthin-dependent nonphotochemical quenching. Planta 217: 616-627

Woodson JD, Chory J (2008) Coordination of gene expression between organellar and nuclear genomes. Nature 9: 383-395

Wu D, Wright DA, Wetzel CM (1999) The IMMUTANS variegation locus of *Arabidopsis* defines a mitochondrial alternative oxidase homolog that functions during early chloroplast biogenesis. Plant Cell 11: 43-55

Yu F, Fu A, Aluru M (2007) Variegation mutants and mechanisms of chloroplast biogenesis. Plant Cell Environ. 30: 350-365

CHAPTER 9

Transcriptional Response to Low-O_2 Conditions in *Synechocystis* sp. Strain PCC 6803 Includes Induction of a Gene Cluster Containing *psbA* and *petC2*

Tina C. Summerfield, Jorg Toepel and Louis A. Sherman*

Purdue University, Department of Biological Sciences, Hansen Hall, West Lafayette, IN 47907 USA

ABSTRACT

Microarray analysis indicated that low-O_2 conditions resulted in up-regulation of *psbA1*, the normally low abundance transcript that encodes the D1' protein of photosystem II in *Synechocystis* sp. PCC 6803. Using a Δ*psbA2*:Δ*psbA3* strain, we show that the *psbA1* transcript is translated and the resultant D1' is inserted into functional PSII complexes. Two other cyanobacterial strains have *psbA* genes that were induced by low-O_2. In two of the three strains examined, *psbA* was part of an up-regulated gene cluster including an alternative Rieske iron-sulfur protein. We conclude that this cluster may represent an important adaptation to changing O_2 levels that cyanobacteria experience.

Keywords : Rieske Fe-S protein, *psbA*, low-O_2, *Synechocystis*.

Abbreviations

Chl: Chlorophyll, *PS II:* Photosystem II.

INTRODUCTION

It has become clear that the evolution of oxygenic photosynthesis and cyanobacteria took place primarily in an anaerobic environment. The sequencing of genomes of numerous photosynthetic bacteria has led to the proposal that the first phototrophs were procyanobacteria (anoxygenic ancestors of the extinct cyanobacteria) (Berman-Frank *et al.*, 2003; Mulkidjanian *et al.*, 2006). Since these cyanobacterial ancestors were responsible for adding oxygen to the environment, the external environment was still low in oxygen until around 2 billion years ago (Bya). In addition, the level of O_2

in the atmosphere has varied during this time period to a high of nearly 30% at some times, but often closer to 10%, especially during the past 400 million years (Ward *et al.*, 2006). Thus, it would seem likely that cyanobacteria had developed a regulatory system capable of inducing important genes under low oxygen and/or anaerobic conditions. Despite the importance of understanding this situation, very little information is available for the way in which cyanobacteria respond to low oxygen conditions. We have recently presented the first solid evidence in this area and shown that a series of important gene clusters were induced in a series of cyanobacteria, both diazotrophic and nondiazotrophic (Summerfield *et al.*, 2008).

Before the development of oxygenic photosynthesis in cyanobacteria, the atmosphere was much more reducing in nature. Indeed, there was a glut of reducing equivalents on earth's surface (Falkowski, 2006). During this period, electrons were made available from copious quantities of donors such as H_2, H_2S and CH_4. The electrons could be donated easily, because of a relatively small energy differential, to acceptors such as CO_2 and, to a lesser extent, SO_4. Since relatively high amount of energy was needed to oxidize water, the biggest electron donor pool, H_2O, remained inaccessible until the advent of oxygenic photosynthesis. Although the exact date for the development of oxygenic photosynthesis and the cyanobacteria is somewhat elusive, the dates from different types of evidence are reaching some congruence. It is very likely that the evolution of cyanobacteria occurred prior to 2.4 Bya because their metabolism is required, at least in part, to explain the appearance and the rise of environmental oxygen at that time. The most recent evidence indicates the period for the rise of atmospheric O_2 to be between approximately 2.4 and 2.3 Bya and that this was followed by the onset of an extreme ice age (Rasmussen *et al.*, 2008; Fischer, 2008). Indeed, others have argued the case for a late origin of photosystem II (PSII) evolution and suggested that oxygenic photosynthesis could have evolved close in geological time to the Makganyene Snowball Earth Event and suggest a causal link between these two important events (Kirschvink and Kopp, 2008). Based on this analysis, higher levels of oxygen began to pollute the atmosphere approximately 2.2 Bya. Although our work does not directly impinge on such dates, we are comfortable with the situation in which atmospheric O_2 increased rather substantially soon after the development of water splitting function.

We have recently studied the impact of low-O_2 growth on global transcription in the unicellular, nondiazotrophic cyanobacterium *Synechocystis* sp. PCC 6803 (hereafter *Synechocystis*) (Summerfield *et al.*, 2008). We used our *Synechocystis* microarray system to compare gene expression between aerobically grown and these low-O_2 incubated cells. These experiments led to some surprising findings, not the least of which was the induction of *psbA1*, the third copy of the genes encoding the main Photosystem II reaction center protein. In addition, an alternative Rieske iron-sulfur protein (*petC2*) was located in the same gene cluster, suggesting that changes in the cytochrome b_6f complex might also occur during low-O_2 growth. We will discuss the implications of these results and reflect on the reasons why the cell might need to make changes in key proteins within these photosynthetic complexes.

RESULTS AND DISCUSSION

Low-Oxygen Induction of Cryptic psbA Genes in Cyanobacteria

Cultures were grown in a 6 L bioreactor illuminated with 100 µm photons/m^2/s and bubbled with air for 3 days (Summerfield *et al.*, 2008; Toepel *et al.*, 2008). RNA was harvested from cells grown aerobically ($t = 0$), and after that the cells were bubbled with 99.9% N_2 and 0.1% CO_2 to produce low O_2 conditions for 1, 2 and 6 h. Microarrays were used to compare gene expression between aerobically grown and low O_2 bubbled cells. A major surprise from the microarray data was that a cluster of genes (slr1181-slr1185) containing *psbA1* and *petC2* (genes encoding alternative isoforms of the Photosystem II protein,

D1, and the Cytochrome b_6f Rieske iron-sulfur center protein, PetC, respectively) was up-regulated during growth under low-O_2. For verification of these results, we used semiquantitative RT-PCR as shown in Fig. 9.1E. The *Synechocystis psbA2* and *psbA3* genes encode the same D1 protein with *psbA2* transcribed at high levels constitutively and *psbA3* being stress induced. Due to the similarity of their nucleotide sequences, the *psbA2* and *psbA3* transcripts could not be distinguished in this experiment. The *psbA2/psbA3* transcripts were present at high levels under aerobic and low-O_2 conditions, whereas the *psbA1* transcript was not detected under aerobic conditions, but was strongly induced by 6 h under low O_2 (Fig. 9.1E).

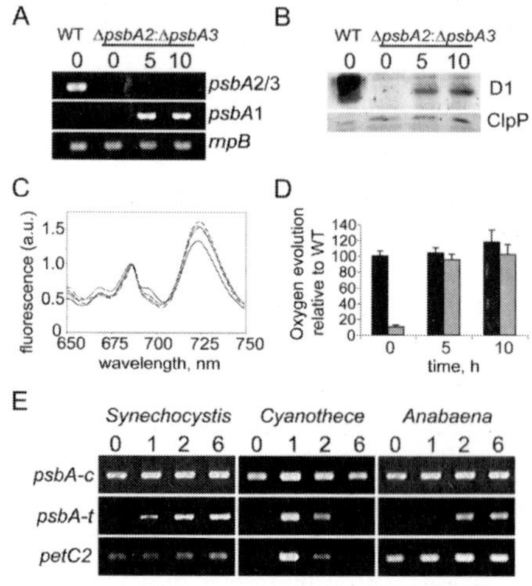

Fig. 9.1 Increased abundance of *psbA* transcript and D1 protein under low O_2 conditions. (A)-(D) *Synechocystis* sp. PCC 6803 wild type and a Δ*psbA2*:Δ*psbA3* mutant. (A) Semi-quantitative RT-PCR using primers to *psbA2/A3*, *psbA1* and *rnpB*; (B) Western blot using antibody to D1 and ClpP. (A), (B) The number above each lane represents time, h under low-O_2 conditions. (C) 77 K fluorescence after excitation at 435 nm WT cells grown in air (____) and the Δ*psbA2*:Δ*psbA3* mutant after growth in air (.....), and after low-O_2 induction for 5h (_._._) and 10 h (- - - -). (D) O_2 evolution for wild-type cells (black bars) and the Δ*psbA2*:Δ*psbA3* mutant (grey bars) under aerobic (0 h) and low-O_2 conditions (5, 10 h). Oxygen evolution was normalized to wild type at 0 h (423 µmol 02 evolved/mg Chl/h). (E) The transcript levels of *psbA* and *petC2* were shown by semi-quantitative RT-PCR for cells grown in air (0 h) and then bubbled with 99.9% N_2 plus 0.1% CO_2 for 1, 2 and 6 h. *psbA*-c represents the consitutively expressed gene for each strain: *Synechocystis* (*psbA2, psbA3*); *Cyanothece* (*psbA1, psbA5*) and *Anabaena* (*psbA1*). *psbA*-t represents the low-O_2 induced genes: *Synechocystis* (*psbA1*), *Cyanothece* (*psbA2*) and *Anabaena* (*psbA0*). *petC* gene in *Anabaena* (*all4511*) (as first described in 4).

We were most interested to determine whether or not these *psbA1* transcripts were translated into protein, and if the protein could be inserted into productive PSII complexes. Since *psbA2/3* transcripts are very abundant in *Synechocystis* wild type, we used the double knockout mutant Δ*psbA2:*Δ*psbA3*. This strain cannot normally grow under photoautotrophic conditions, and any D1 protein or PSII activity would have to come from PSII complexes produced after transition to low-O_2 conditions i.e, after insertion of the *psbA1* gene product D1′. As expected, semiquantitative RT-PCR showed that the wild type transcribed *psbA2/3* under aerobic conditions, whereas no *psbA2/3* transcripts were detected in the Δ*psbA2:*Δ*psbA3* strain (Fig. 9.1A). Western blot analysis demonstrated clearly that there was no D1 protein in the double mutant under aerobic conditions, but significant D1′ protein was detected by 5 h (Fig. 9.1B). The 77K fluorescence spectra of the mutant strain indicated that intact complexes were present under low-O_2 conditions (Fig. 9.1C). In particular, the 695 nm peak, corresponding to excitation of Chl *a* associated with CP 47 and indicative of assembled PSII centers, was at a significant level by 10 h under low-O_2 conditions. All this was accompanied by dramatic increase in PSII O_2 evolving activity by 5 and 10 h after growth under low-O_2 conditions (Fig. 9.1D). We conclude from these experiments that the *psbA1* gene was transcribed and translated under low-O_2 conditions and that D1′ was inserted to produce functional PSII complexes.

We next determined if similar low-O_2 induction occurred in other cyanobacterial strains, including those capable of N_2 fixation. Therefore, we examined the unicellular, N_2-fixing strain *Cyanothece* sp. ATCC 51142 (hereafter *Cyanothece*), a filamentous heterocystous strain *Anabaena* sp. PCC 7120 (hereafter *Anabaena*) and the ancient and divergent strain *Gloeobacter violaceus* sp. PCC 7421 (hereafter *Gloeobacter*). The semiquantitative RT-PCR results for *Cyanothece*, *Anabaena* and *Synechocystis psbA* genes are compared in Fig. 9.1E. For the sake of simplicity, we call the normal constitutive and highly expressed genes *psbA-c* and the genes normally transcribed in only trace amounts under aerobic conditions *psbA-t* (trace). Both *Cyanothece* and *Anabaena* strains showed an increase in *psbA-t* transcript level under low-O_2 growth, but with differing kinetics (Fig. 9.1E). *Cyanothece* demonstrated a rapid induction that peaked around 1 h and then subsequently declined. *Anabaena* kinetics resembled those of *Synechocystis* and increased around 2 h and stayed high through 6 h. *Cyanothece* resembles *Synechocystis* in that both have a *psbA* gene that is clustered near *petC2*. In *Cyanothece*, the levels of both transcripts were strongly increased by 1 h under low-O_2 conditions followed by a fairly rapid decline. In *Anabaena*, there was no *petC* gene close to *psbA-t* and none of the 4 *petC* genes showed induction similar to that of *Synechocystis* or *Cyanothece* (Summerfield *et al.*, 2008). However, the *Anabaena petC2* gene *all4511* (transcribed under aerobic conditions) did show increased transcript levels under low-O_2 conditions (Fig. 9.1E). In low-O_2 incubated *Gloeobacter*, there was no specific gene induction of any of the 5 copies of *psbA*, (including the 2 genes expressed at trace levels under normal aerobic conditions) or the single copy of *petC* (Summerfield *et al.*, 2008). The up-regulation of a *psbA* gene normally transcribed at low levels has recently been reported for *Thermosynechococcus elongatus* BP-1 (hereafter *Thermosynechococcus*) under similar low-O_2 conditions (Sicora *et al.* 2009).

D1 Structure

What would be the purpose of an alternative D1 in PS II under low-O_2 conditions? The potential involvement of this D1′ has been studied extensively for over a decade, but with no satisfactory answer (Sicora *et al.*, 2004; Sicora *et al.*, 2006; Kos *et al.*, 2008; Sicora *et al.*, 2008). When *psbA1* was expressed under the control of the *psbA*2 promoter in cells lacking *psbA*2 and *psbA*3, the modified strain could grow photoautotrophically (Salih and Jansson, 1997). In addition, D1′ centers were found to be functional but less efficient than normal PSII centers (Funk *et al.*, 2001). A number of hypotheses that might explain the need for D1′ were proposed, but could not be verified experimentally. We suggest that one reason for the change can be found in the structure shown in Fig. 9.2. This model represents two of the D1 helices (C and CD) and depicts three amino acids that are changed in *Synechocystis* D1′. As indicated in Fig. 9.2, F 158 L (on helix C) and T 268 A (on helix E, the rest of helix is omitted for clarity) are altered in all of the low-O_2 inducible *psbA* genes identified to date (as was G 80 A which is located on the lumen loop between helices A and B and is not shown in Fig. 9.2); in addition, residue F 186 L (on helix CD) is altered in *Synechocystis*. In a thorough review of the crystal structure of cyanobacterial PSII that included a close look at the antenna system, Muh *et al.* (2008) discussed the relevant interactions of the chlorin molecules with amino acid side chains as portrayed by the 3.0 Å structure. F 158 is an aromatic residue and T 286 is a polar, non-aromatic residue, both of which are within 6.0 Å of the π-system. Thus, it is possible that changes in these amino acids slightly alter the binding of the chlorophylls to D1. Whether or not that is an important aspect for the construction or function of new PSII centers under low-O_2 conditions is something that must await future research. It should be noted that after the induction of *psbA*1, these transcripts represent about 15% of the total *psbA* transcripts that were made during the

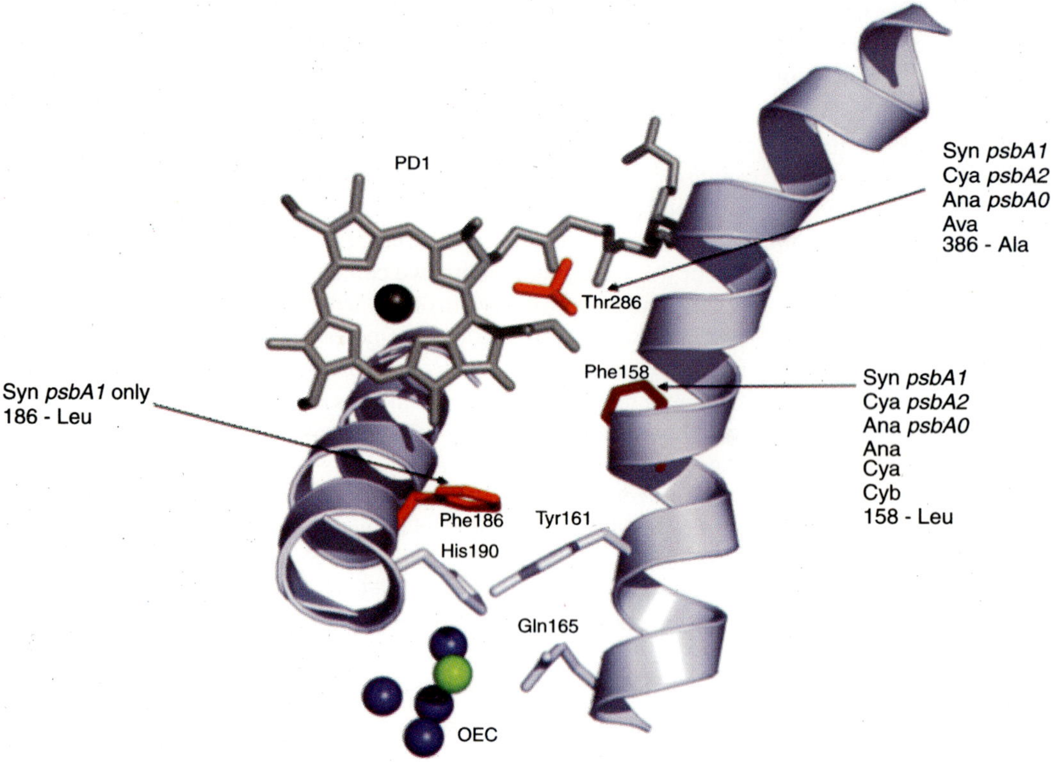

Fig. 9.2 Crystal structure D1 helices C and CD and Thr 286 of helix E. Residues highlighted in red are altered in the low oxygen-induced D1′ proteins of *Synechocystis*, *Cyanothece*, *Anabaena* and *Thermosynechococcus*.

6 h of low-O_2 growth for *Synechocystis*. Thus, this fraction is not large in total, but may represent a larger proportion of those transcripts translated into D1 protein and inserted into the complex.

Alternative Rieske Fe-S Protein of Cytochrome b_6f Complex

The Rieske Fe-S protein is involved in quinol binding and electron transfer, making it an essential component of the cytochrome b_6f complex. *Synechocystis*, *Cyanothece* and *Anabaena* have multiple copies of the *petC* gene that encode alternative copies of this Rieske Fe-S protein of the cytochrome b_6f complex (Table 9.1). PetC1 is the major Rieske protein in *Synechocystis*. However, in *Synechocystis, Cyanothece* and *Anabaena,* the *petC2* gene is up: regulated under low-O_2 conditions. Overlapping functions of PetC1 and PetC2 in *Synechocystis* were demonstrated by photoautotrophic growth of Δ*petC1* and *petC2* strains, but not the strain lacking both *petC1* and *petC2* (Schneider *et al.*, 2004). However, aerobic growth of the Δ*petC1* strain was retarded compared with wild type and the Δ*petC2* strain, indicating that PetC2 did not fully replace PetC1 under these conditions. PetC2 has the conserved cysteines and histidines of a Rieske protein and a high midpoint potential similar to PetC1, but amino acid differences at the "hinge region" may affect quinol oxidation (Fig. 9.3). For example, changes in the hinge region may account for slowing of electrons through cytochrome b_6f complex in the absence of PetC1 under aerobic conditions (Schneider *et al.*, 2004). In contrast, PetC3 has conserved cysteines and histidines of a Rieske protein, but exhibits amino acid variation in these regions (Box 1 and Box 2 in Fig. 9.3) and has shortened N and C termini. The lower midpoint potential of PetC3 makes its function

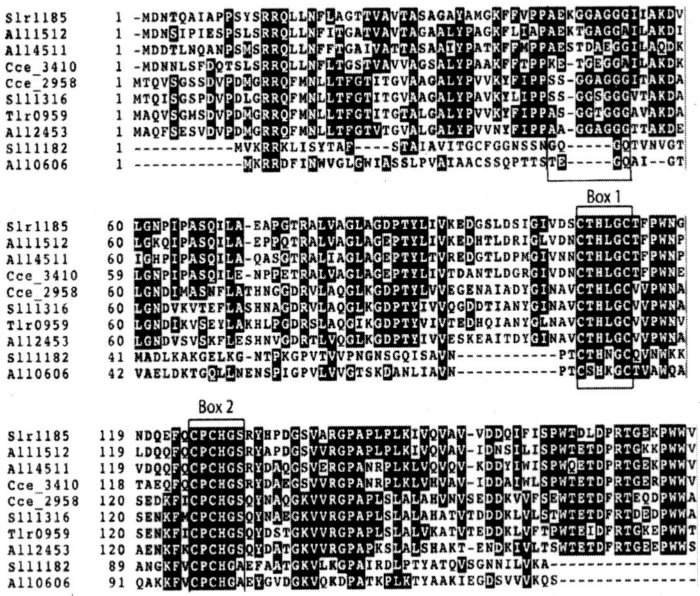

Fig. 9.3 Alignment of PetC proteins of *Synechocystis*, *Cyanothece*, *Anabaena* and *Thermosynechococcus*. PetC1 proteins are sll1316, cce_2958, all2453, and tlr0959, PetC2 proteins are slr1185, cce_3410, all1512, and all4511. PetC3 proteins are sll1182, all0606.

unclear; however, PetC3 is associated with the cytochrome b_6f complex and the Δ*petC1*: Δ*petC3* strain had a more severe phenotype than the Δ*petC1* strain. These data have been interpreted to mean that during aerobic growth there is a heterologous population of cytochrome b_6f complexes. Under low oxygen conditions, we suggest that this population includes more PetC2-containing complexes as the *petC2* gene is upregulated but the *petC3* gene is not. The PetC2 containing cytochrome b_6f complexes may play a specific role under these conditions. The up-regulated *petC2* genes of *Synechocystis*, *Cyanothece* and *Anabaena* encode PetC2 proteins that have similar protein sequences (as determined by per cent identity). The exception is *Thermosynechococcus* that has only one copy of *petC*. In *Synechocystis* and *Cyanothece,* but not *Anabaena,* the induced *petC* copy is located in a cluster with the low oxygen induced *psbA* gene (Fig. 9.3 a-c).

Coordinately Regulated Gene Clusters

A clue for this low-O_2 induction of *psbA* and *petC* genes may be found in the genome organization of these clusters (Fig. 9.4). Figures 9.4a and b highlight the clusters in *Synechocystis* and *Cyanothece* that we have studied by microarray and RT-PCR analysis. It is evident that both clusters contain a series of genes that are annotated as hypothetical or unknown and the entire cluster is up-regulated under low-O_2 conditions. Figures 9.4C represents the two gene clusters on separate parts of the *Anabaena* genome with the inducible *psbA* gene (alr3742) and *petC2* (all4511). In addition, we provide similar information for *Thermosynechococcus* in Fig. 9.4d. Note that *Cyanothece* and *Anabaena* have an *sqr* gene located close to a low oxygen induced *petC2* gene. *Synechocystis* and *Thermosynechococcus* also contain *sqr* genes. In *Synechocystis*, the *sqr* gene is located on its largest plasmid, pSYSM. Neighbouring the *sqr* genes of *Synechocystis*, *Cyanothece* and *Thermosynechococcus* are hypothetical genes that share more than 50% identity. Similarity between other genes in these clusters is shown by the use of shading in Fig. 9.4. Could the fact that all these strains contain *sqr* genes be significant? It should be noted that the

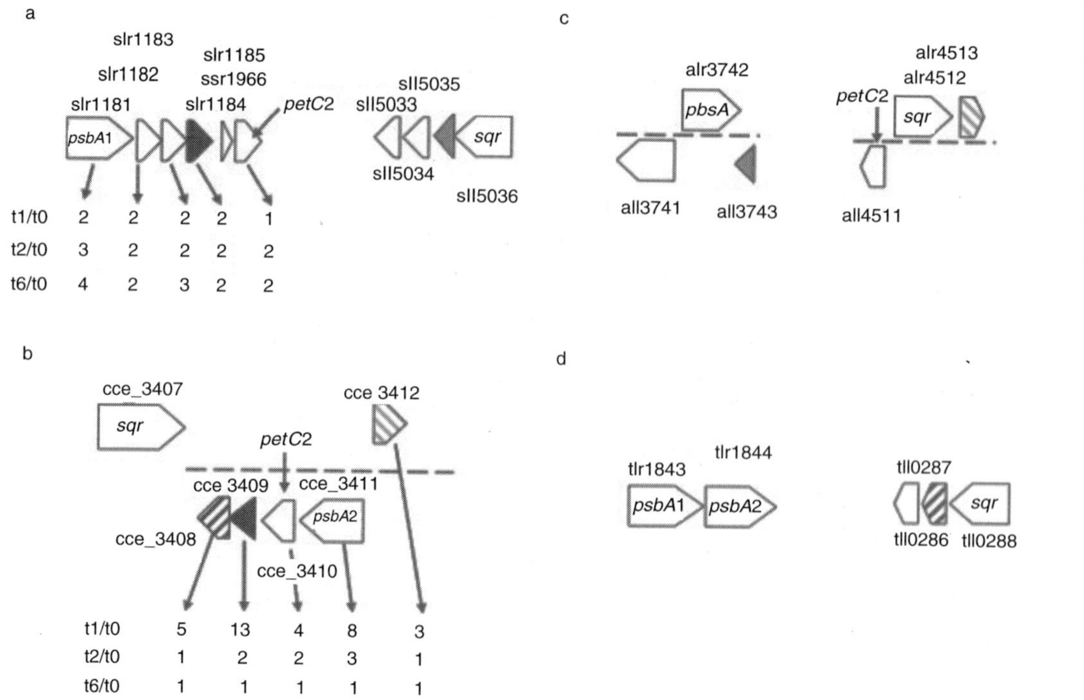

Fig. 9.4 Gene clusters co-ordinately regulated under low O_2 conditions containing the *psbA* gene that encodes the alternative D1′ polypeptide: (a) *Synechocystis* sp. PCC 6803 and (b) *Cyanothece* sp. ATCC 51142. Numbers represent fold change of transcript abundance on transition from aerobic to low-O_2 conditions. Genes neighboring low O_2 induced *psbA* genes and *petC* gene in (c) *Anabaena* sp. PCC 7120 and (d) *Thermosynechococcus elongatus* BP-1. Genes with sequence similarity in two or more strains have the same shading. Both slr1184 and cce_3409 (shaded black) encode proteins containing rhodanese-like domains. The predicted proteins encoded by cce_3412 and alr4513 share 60% identity. The sulfide quinine reductase (*sqr*) proteins of *Cyanothece* sp. ATCC 51142 and *Anabaena* sp. PCC 7120 share 61% identity. There is only one copy of *petC* in *Thermosynechococcus elongatus* BP-1 (not shown).

up-regulation of *psbA* under low-O_2 conditions may help maintain internal aerobic cellular environment: the importance of this aspect will vary between strains as cyanobacteria differ in their sensitivity to O_2 levels. Furthermore, low-O_2 conditions are often associated with an increased level of H_2S that can damage PSII. Importantly, O_2 can inactivate H_2S. Increasing D1′ levels could accelerate the replacement of damaged D1, both increasing the rate of oxygenic photosynthesis, but also limiting further damage to PSII. The SQR protein, encoded by *sqr*, transfers electrons from H_2S to the quinone pool and is present in many anoxygenic phototrophs, as well as *Aphanothece halophytica*, where it may have a role in detoxification of sulfide. Alternatively, low-O_2 conditions on the earth, when oxygenic photosynthesis arose, raises the possibility that these low-O_2 induced *psbA* genes, as well as PetC and SQR, represent ancient genes that were important metabolically prior to the huge increase of O_2 in the atmosphere, approximately, 2.5 bya.

Cyanobacteria: *psbA* and *petC* Gene Families

What are the implications of these findings in cyanobacterial physiology? Many cyanobacteria contain *psbA* gene families (Table 9.1). These frequently encode two D1 isoforms, D1:1 and D1:2, with the D1:2 isoform being more abundant under stress conditions such as high light. This is not always the case; in

Synechocystis, there is no D1:2 isoform and stress conditions increase the transcription of *psbA3* (which encodes the same D1:1 as the constitutively expressed *psbA2*). In all four strains with low-O_2-induced *psbA* genes, these encoded a distinct D1 isoform called D1′, and phylogenetic analyses indicated that these genes were divergent from the other copies of *psbA* within three of the four strains. In addition, all four of these low-O_2-induced *psbA* genes were expressed at only trace amounts under normal growth conditions. Do many cyanobacteria have low-O_2-induced *psbA* genes? Table 9.1 presents an analysis of most of the cyanobacteria for which complete genome sequences are now available. This indicates that all strains, except *Prochlorococcus* strains, *Microcystis aeruginosa* NIES-843, and *Trichodesmium erythraeum* IMS101 contain multiple *psbA* genes that encode for two or more versions of D1. *Microcystis aeruginosa* NIES-843, *Trichodesmium erythraeum* IMS101a and several *Prochlorococcus* strains have multiple copies of *psbA,* but these encode the same D1 isoform. In these *Prochlorococcus* strains, multiple *psbA* genes are predicted to have arisen from recent gene duplications, and these are found in both high (e.g., *Prochlorococcus marinus* str. MIT 9312) and low light ecotypes (e.g., *Prochlorococcus* marinus str. MIT 9313) (Kettler *et al.,* 2007). The marine *Synechococcus* strains contained 3 or 4 copies of *psbA* that encode two D1 isoforms for each strain. Overall these data are consistent with the reduced genome sizes observed in the marine cyanobacteria with *Prochlorococcus* strains containing fewer *psbA* genes than *Synechococcus* strains (Table 9.1). The nitrogen-fixing strains *Nostoc punctiforme* and *Anabaena variabilis* ATCC 29413 contain multiple copies of *psbA* encoding 3 and 4 D1 isoforms, respectively. There is no data to indicate whether increased transcription of any of these genes forms part of the response to low-O_2 conditions. Based on the presence of the three conserved amino acid substitutions, we could predict the *Anabaena variabilis* ATCC 29413 gene Ava_1583, but not the *Nostoc punctiforme psbA* genes, to be up-regulated under low oxygen conditions. The extent of low-O_2-induction of *psbA* genes in cyanobacteria remains to be determined and these data would provide insight into the role of these alternative D1 proteins.

There are fewer *petC* genes than *psbA* genes in each strain with the majority of strains having only one *petC* gene. In fact only 8 of the 35 strains had more than one *petC* copy. However, having only one *petC* copy does not rule out having a low-O_2 induced *psbA* gene as has already been reported in the case of *Thermosynechococcus* (Sicora *et al.,* 2009).

CONCLUSION

We have identified a strategy used by many freshwater, as well as marine diazatrophic, cyanobacteria to respond to low ambient O_2 levels. The key feature to this response is the induction of a *psbA* gene that is normally transcribed at very low levels. This *psbA* gene is transcribed and a modified D1 protein, called D1′ in *Synechocystis,* is translated and inserted into functional PSII complexes. This response is often paralleled by the induction of a gene encoding an alternative Rieske iron-sulfur protein that also can be functionally inserted into complexes. Although the exact reason for these changes will require additional studies, it is clear that low-oxygen represents an important condition for many cyanobacteria. Ultimately, the main question is whether or not these environmental responses help the organism to grow under these low-O_2conditions. In the case of *Synechocystis*, the cells do continue to grow and divide for about 18 h. We have also identified a mutant that grows even better under these conditions. The main gene clusters that are up-regulated under these low-O_2 conditions included sll0789 and sll0790, a response regulator and histidine kinase, *hik31*, respectively. Experiments using an insertion mutant in *hik31* suggests that Hik31 is involved in regulation under low oxygen conditions. This gene and its involvement in regulation under low-O_2 conditions is currently being studied in our lab. Without question, this response is of sufficient evolutionary significance that it elicits a series of structural and regulatory changes that we are just beginning to identify and understand.

Table 9.1 Number of *psbA* genes, D1 isoforms and *petC* genes identified in different cyanobacteria.

Cyanobacterial strain	*psbA genes*	*D_1 isoforms*	*petC genes*	*genome size (Mb)*
Acaryochloris marina MBIC11017	3	2	2	8.36
Anabaena sp. PCC 7120	5	3	4	7.21
Anabaena variabilis ATCC 29413	6	4	3[c]	7.07
Arthrospira maxima CS-328[a]	2	2	1	~6.00
Crocosphaera watsonii WH8501[a]	2	2	1	4.92
Cyanothece sp. ATCC 51142	5	4	2	5.46
Gloeobacter violaceus PCC 7421	5	3	1	4.66
Microcystis aeruginosa NIES-843	5	1	2	5.84
Nostoc punctiforme ATCC 29133	5	3	3[c]	9.01
Prochlorococcus marinus MED4	1	1	1	1.66
Prochlorococcus marinus MIT9313	2	1	1	2.41
Prochlorococcus marinus SS120	1	1	1	1.75
Prochlorococcus marinus str. AS9601	2[b]	1	1	1.67
Prochlorococcus marinus str. MIT 9211	1	1	1	1.69
Prochlorococcus marinus str. MIT 9215	1	1	1	1.74
Prochlorococcus marinus str. MIT 9301	1	1	1	1.64
Prochlorococcus marinus str. MIT 9303	2[b]	1	1	2.68
Prochlorococcus marinus str. MIT 9312	2	1	1[d]	1.71
Prochlorococcus marinus str. MIT 9515	2[b]	1	1	1.70
Prochlorococcus marinus str. NATL1A	3	1	1	1.86
Prochlorococcus marinus str. NATL2A	3	1	1	1.84
Synechococcus elongatus PCC 6301	3	2	1	2.75
Synechococcus elongatus PCC 7942	3	2	1	2.74
Synechococcus sp. CC9311	4	2	1	2.61
Synechococcus sp. CC9902	4	2	1[d]	2.23
Synechococcus sp. JA-2-3B'a (2-13)	3	2	1	3.05
Synechococcus sp. JA-3-3Ab	4	2	1	2.93
Synechococcus sp. PCC 7002	3	2	3	3.41
Synechococcus sp. RCC307	4	2	1	2.22
Synechococcus sp. CC9605	3	2	1[d]	2.51
Synechococcus sp. WH7803	4	2	1	2.37
Synechococcus sp. WH8102	4	2	1	2.43
Synechococcus sp. PCC6803	3	2	3	3.96
Thermosynechococcus elongatus BP-1	3	3	1	2.59
Trichodesmium erythraeum IMS101[a]	3	1	1	7.75

ACKNOWLEDGEMENTS

We thank Bronwyn Carlisle of the Department of Biochemistry, University of Otago for the production of Fig. 9.2. This research was funded by a grant from the US Department of Energy. Current addresses: TC Summerfield, Dept. of Botany, University of Otago, Dunedin, New Zealand; J Toepel, Center for Biotechnology, University of Bielefeld, Germany.

REFERENCES

Berman-Frank I, Lundgren P, Falkowski P (2003) Nitrogen fixation and photosynthetic oxygen evolution in cyanobacteria. Res Microbiol 154: 157-164

Falkowski, PG (2006) Tracing oxygen's imprint on earth's metabolic evolution. Science 311: 1724-1726

Fischer, WW (2008) Life before the rise of oxygen. Nature 455:1051-1052

Funk, C, Wiklund, R, Schroder, WP, Jansson C (2001) D1 centers are less efficient than normal photosystem II centers. FEBS Lett 505: 113-117

Kettler, GC, Martiny, AC, Huang, K, Zucker, J, Coleman, ML, Rodrigue, S, Chen, F, Lapidus, A, Ferriera, S, Johnson, J, Steglich, C, Church, G. M, Richardson, P. and Chisholm, P. W. (2007) Patterns and implications of gene gain and loss in the evolution of Prochlorococcus. PLoS Genet 3: e231

Kirschvink, JL, Kopp, RE (2008) Palaeoproterozoic ice houses and the evolution of oxygen-mediating enzymes: the case for a late origin of photosystem II Phil Trans R Soc. B 363: 2755-2765

Kós, PB, Deak, Z, Cheregi, O. and Vass, I. (2008) Differential regulation of psbA and psbD gene expression, and the role of the different D1 protein copies in the cyanobacterium Thermosynechococcus elongatus BP-1. Biochim Biophys Acta 1777: 74-83

Müh, F, Renger, T, Zouni, A. (2008) Crystal structure of cyanobacterial photosystem II at 3.0 Å resolution: A closer look at the antenna system and the small membrane-intrinsic subunits. Plant Physiol Biochim 46: 238-264

Mulkidjanian, AY, Koonin, EV, Makarova, KS, Mekhedov, SL, Sorokin, A, Wolf, YI, Dufresne, A, Partensky, F, Burd, H, Kaznadzey, D, Haselkorn, R. and Galperin, MY (2006) The cyanobacterial genome core and the origin of photosynthesis. PNAS 35: 13126-13131

Rasmussen, B, Fletcher, IR, Brocks, JJ, Kilburn, MR (2008) Reassessing the first appearance of eukaryotes and cyanobacteria. Nature 455: 1101

Salih, GF, Jansson, C. (1997). Activation of the silent psbA1 gene in the cyanobacterium *Synechocystis* sp. strain 6803 produces a novel and functional D1 protein. Plant Cell 9: 869-878

Schneider, D, Berry, S, Volkmer, T, Seidler, A, Rogner, M (2004) PetC1 is the major Rieske iron-sulfur protein in the cytochrome b6f complex of *Synechocystis* sp. PCC 6803. JBC 279: 39383-39388

Sicora, C, Wiklund, R, Jansson, C, Vass, I (2004) Charge stabilization and recombination in Photosystem II containing the D1 protein product of the psbA1 gene in *Synechocystis* 6803. Phys Chem Chem Phys 6: 4832-4837

Sicora, CI, Appleton, SE, Brown, CM, Chung, J, Chandler, J, Cockshutt, AM, Vass, I, Campbell, DA (2006) Cyanobacterial psbA families in *Anabaena* and *Synechocystis* encode trace, constitutive and UVB-induced D1 isoforms. Biochim Biophys Acta 1757: 47-56

Sicora, CI, Brown, CM, Cheregi, O, Vass, I. and Campbell, DA (2008). The psbA gene family responds differentially to light and UVB stress in *Gloeobacter violaceus* PCC 7421, a deeply divergent cyanobacterium. Biochem Biophys Acta 1777: 130-139

Sicora, CI, Ho, FM, Salminen, T. Styring, S, Aro, EM (2009). Transcription of a "silent" cyanobacterial psbA gene is induced by microaerobic conditions. Biochim Biophys Acta 1787: 105-112

Summerfield, TC, Toepel, J, Sherman, LA 2008. Low-oxygen induction of normally cryptic psbA genes in cyanobacteria. Biochemistry 47: 12939-12941

Toepel, J, Welsh, E, Summerfield, TC, Pakrasi, HB, Sherman, LA (2008). Differential transcriptional analysis of the cyanobacterium *Cyanothece* sp. strain ATCC 51142 during light-dark and continuous-light growth. J *Bacteriol* 190: 3904-3913

Ward, PD (2006). Out of Thin Air: Dinosaurs, Birds and Earth's Ancient Atmosphere. Joseph Henry Press, Washington, DC

CHAPTER 10

Mutants of Chlorophyllide Oxygenase

Sujata R. Mishra, Young-Jae Eu, Krishna Nath, Altanzaya Tovu, Ismayil S. Zulfugarov, and Choon-Hwan Lee

Department of Molecular Biology, Pusan National University, Jangjeon–dong, Keumjung-ku Busan 609-735, Republic of Korea

ABSTRACT

Chlorophyllide oxygenase is a single-gene encoded enzyme in the inner envelope/thylakoid membrane of chloroplasts and catalyzes the synthesis of Chl *b* from Chl *a*. It consists of three domains *A*, *B* and *C*. *AB* domain regulates the binding of pigments to apoproteins and domain *C* catalyzes the biosynthesis of Chl *b*. The mutants of CAO are naturally occurring like cyanobacteria or can be induced by physical or chemical mutagens or be generated genetically. CAO mutants are either deficient in Chl *b* or totally lack it. As Chl *b* is responsible for absorption and transfer of light energy, assembly of light harvesting complexes, regulation of antenna size and stable insertion of light-harvesting complexes into the thylakoid membranes, the structure and function of chloroplasts are affected in its absence. In the present review, we will be discussing the different means of CAO mutant generation in different species, the effect of Chl *b* loss on the structure and function of the plants under normal and adverse conditions, i.e. high light, high temperature, senescence and drought, the regulation of expression and function of CAO gene and finally, the importance of CAO mutants in practical application and basic science research.

Keywords: Chlorophyllide oxygenase, ROS, Photochemical quenching, Quantum yield, Apoprotein.

Abbreviations

Chl: Chlorophyll, *LHC:* Light harvesting complex, *PSI:* Photosystem I, *PSII:* Photosystem II.

INTRODUCTION

Chlorophyll *a* oxygenase (CAO) is the chloroplastic enzyme located in the inner envelope and thylakoid membrane, responsible for the synthesis of Chl *b* from Chl *a*. It is present in land plants and algae including prochlorophytes, except cyanobacteria (Table 10.1). It is encoded by a single gene in the nuclear genome and its translation product is imported into the chloroplasts (Hoober *et al.*, 2007). In chloroplasts, it catalyzes two-step oxygenation reactions and converts chlorophyllide *a* to chlorophyllide *b* via 7-hydroxymethylchlorophyllide (Oster *et al.*, 2000)(Fig. 10.1 A and B).

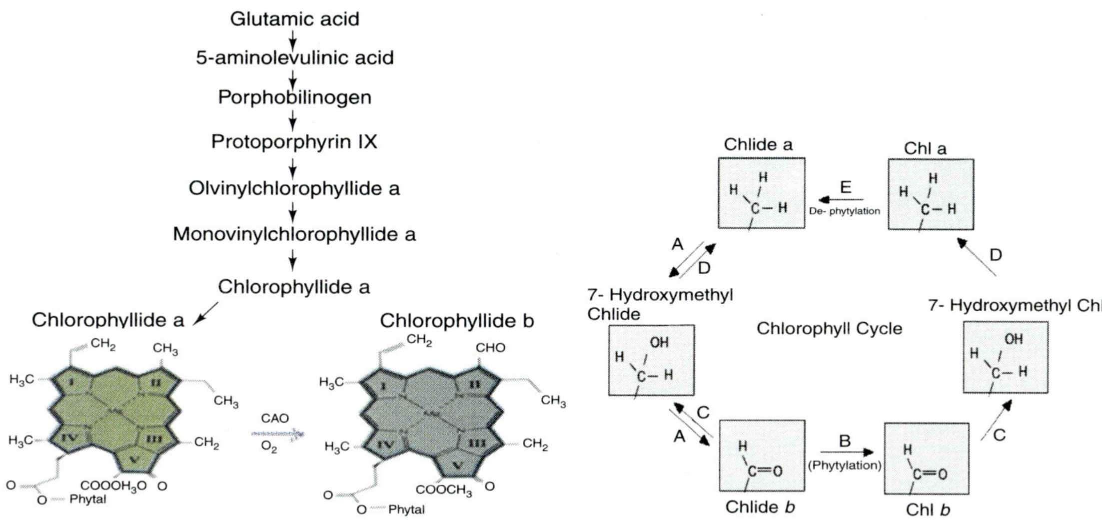

Fig. 10.1 A. Schematic presentation of Chl *b* biosynthesis, B. Chlorophyll cycle: Different intermediates in the cycle has been represented by their functional groups. This model is proposed by Ohtsuka *et al* .,1997.

As CAO is a single nuclear gene, the plants that lack this enzyme are either devoid of chlorophyll *b* (chlorophyll *b* less mutants) or deficient in it (chlorophyll deficient mutants). Chlorophyll *b* mutants can be formed spontaneously or are induced by physical or chemical mutagens or genetically (Table 10.1)

Table 10.1 Chlorophyll *b* mutants.

Cyanobacteria	*Natural mutant*	*Natural*	*Satoh et al., 2001*
Chlamydomonas	*pg 113* cbs1, cbs2, cbs3, cbs4, cbs5, cbs6	Physical mutagenesis (Ultraviolet radiation) Insertional mutagenesis	Michel *et al.,* 1983
Barley	*Chlorina-f_2*	Spontaneous	Highkin and Frenkel. 1962, Bossmann *et al.,*, 1997
Arabidopsis	*ch*-1-1, *ch*1-2, ch1-3 YK1-3, YK1-10, YK2-3 (Over expression mutant)	Physical mutagenesis	Muray and Kohoron, 1991 Tanaka and Tanaka 2005
Tobacco Rice	*Overexpression mutant* Chlorina1-16 VG28-1, VG30 OsCAO1, OsCAO2 *cao1, cao-2, cao-3,* *cao-4, cao-5*	Promotor (Camv35s) induced Ionization radiation, chemical mutagenesis T-DNA insertion γ-irrradiation	Pattnayak *et al.,* 2003 Terao and Yamashita, 1982 Wang *et al.,* 2000 Jeon *et al.,* 2000 Amano 1989
Maize	OY-YG	Chemical mutagenesis (EMS)	Greene *et al.,* 1988
Brassica	Cr3529	Chemical mutagenesis	Zhao *et al.,* 2000
Sweet clover	ch_4, ch_5 ch_6, ch_7, and ch_v.	Chemical mutagenesis (EMS)	Gengenbach, and Gorz, 1970
Soyabean	Clark Y_9Y_9, Clark Y11	Spontaneous	Ghiradi and Melis, 1988
Wheat	CD3	Induced mutation	Duysen *et al.,* 1984
Pea	Greenfeast mutant	Spontaneous	Highkin *et al.,*1969
Tomato	Xantha Xantha-2	Spontaneous β-radiation	Butler and Chang, 1958 Grober, 1963

CONSEQUENCE OF CAO DELETION, OVEREXPRESSION AND INTRODUCTION

Consequence of Deficiency in CAO

The main functions of chlorophyll *b* are the absorption and transfer of light energy, assembly of light-harvesting complexes, regulation of antenna size and stable insertion of light harvesting complexes into the thylakoid membranes. As no redundant genes or biochemical pathway is available for CAO, mutations in the CAO gene result in the loss of chlorophyll *b* (Tanaka *et al.,* 1998), causes structural changes, and affects the plant function.

Structural changes

The lack of Chl *b* can affect the abundance of certain proteins and pigment-protein complexes (Falbel *et al.,* 1996). Mutants without Chl *b* fail to accumulate adequate amount of light harvesting chlorophyll *a*/*b* binding proteins and that lead to the reduction in photosynthetic unit size and development of abnormal thylakoid membrane system. In the mutant, the ultrastructure of chloroplast has shown decreased grana lamellae and slightly swelling thylakoid (Chen *et al.,* 2007). Nakatani and Baliga (1985) have described a chlorophyll *b*-lacking ch5 mutant (U374) in sweet clover, where grana are nearly as normal as those in the wild type.

Pigment composition

The CAO mutants either have no chlorophyll (Chl) *b* (Type –I :Chl *b*-less mutant, Chl *a*/*b* >20) or are deficient in Chl *b* (Type II: Chl *b*-deficient mutant, Chl *a*/*b* 6-10) (Terao *et al.,* 1985, Falbel *et al.,* 1996). Chl *b*-less mutants of barley (*chlorina f*2) and *Arabidopsis* (*ch-1*) have almost normal amounts of Chl *a* (Murray and Kohorn, 1991; Falbel and Staehelin, 1996). However, lower Chl *a* content in comparison to the wild type (Wt) has been reported in barley chlorina *f2* mutant and Chl *b* less mutant of rice (Lin *et al.,* 2003) as evident from the absorption spectra and HPLC analysis of pigments. The pool size of xanthophyll pigments (Violaxanthin+Antheraxanthin+Zeaxanthin) has been found to be enhanced in Chl *b* mutants (Lin *et al.,* 2003, Stroch *et al.,* 2004, Havaux *et al.,* 2007).

Pigment-protein complexes and LHC proteins

The CAO1 mutant of rice lacks LHC II trimer (unpublished data, Fig. 10.2A) and *ch*-1 of *Arabidopsis* is also devoid of LHC II trimer and monomer (Havaux *et al.,* 2007, Fig. 10.2A, B). The green gel analysis of CAO mutants of rice by Morita *et al.* (2005) shows the loss of LHC trimer, dimer and monomer. The *ch*1-3 mutant of *Arabidopsis* also has been shown to have great decrease in all six LHC proteins, but in the double CAO mutant *ch*1-3*lhcb*5, there has been further decrease in LHCS with complete absence of Lhcb5 (Kim *et al*., 2009).

Chlorina f2 mutant of barley fails to accumulate three out of 10 LHC proteins (Bossman *et al.,* 1997). These are antenna proteins of PSII, Lhcb1 and Lhcb6, and of PSI, Lhca4. In addition, it has reduced amounts of LHCII proteins Lhcb2, Lhcb3 and Lhcb4. Similarly, *ch*-1 of *Arabidopsis* is also reduced in all LHCs except Lhcb5 and Lhcb3, 4 and 6 are not detected at all (Havaux *et al.,* 2007).

The biosynthesis of these proteins is not affected by the lack of Chl *b*. They are synthesized and imported into the chloroplasts and assembled in the thylakoid membranes, but degraded after they are inserted into the membranes because they need Chl *b* for their stabilization (Bellemare *et al.,* 1982, Preiss and Thornber, 1995). The sizes of antenna are also decreased by 30 and 80% of PSI and PS II respectively in *chlorina f2* mutant of barley (Harrisson *et al.,* 1993). The reduction in antenna size is also evident in CAO mutant of barley (Ghiradi *et al.,* 1986), rice (Hsu and Lee, 1995, Table 10.2) and *Arabidopsis* (Havaux *et al.,* 2007; Kim *et al.*, 2009).

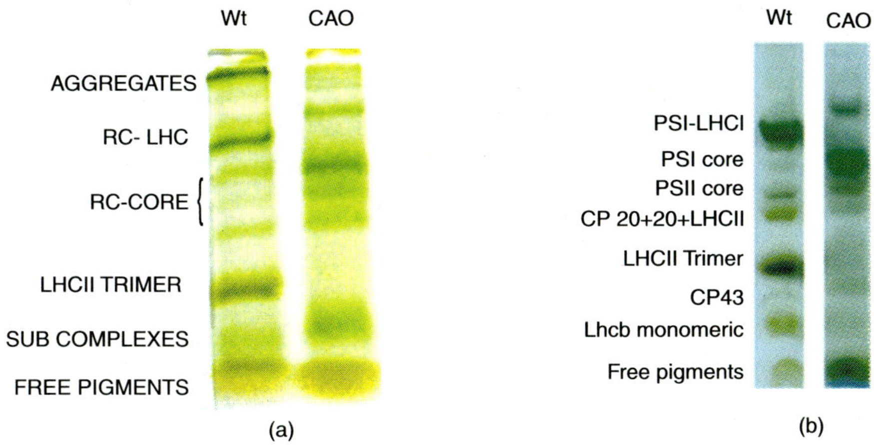

Fig. 10.2 Pigment protein complexes of Wt and CAO mutant of rice (A) (unpublished result) and *Arabidopsis* (B) (Havaux *et al*; 2007).

Table 10.2 The antenna sizes of PSII of various geno types of rice.

Rice type	*Chl a:b*	*Antenna size*
Wild type	2.9	250
Chlorina-12	10	55
Chlorina-14	15	50
Chlorina-7	∞	35

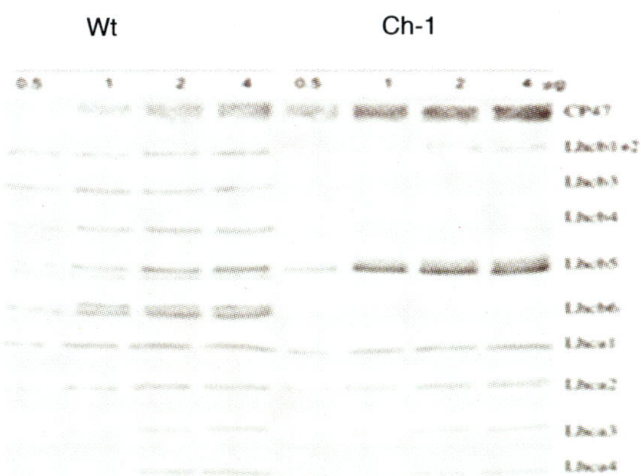

Fig. 10.3 Light-harvesting antenna proteins of PSII (Lhcbs) and PSI (Lhcas) in Wt and ch-1 (CAO mutant) *Arabidopsis* (Western blot image from Havaux *et al.*, 2007). Different amount of Chl used were 0.5, 1, 2 and 4. CP 47 analysis was for normalization purposes.

Functional Changes

The loss of Chl *b* affects the plant function in various ways. It affects the phenotype of plants and leaf becomes pale green in colour (Fig. 10.4 A and C). There is retardation of growth (Fig 10.4B), delay of flowering, reduction in leaf area (Lin *et al.*, 2003) and lower yield. However, it has been reported that Chl *b* loss does not affect plant productivity in the rice CAO (OsCAO) mutants (Zhou *et al.*, 2006).

Fig. 10.4 Growth and morphological changes in *Arabidopsis* (A) and rice (B) in response to Chl *b* loss. (C) Wild type rice leaf and pale green colour of rice CAO mutant.

Chl a fluorescence characteristics

Chl *b* loss does not affect photochemical efficiency Fv/Fm, (Lin *et al.,* 2003) of CAO mutant. However, it affects energy dissipation processes like energy dependent quenching and state transition. Non-photochemical quenching is decreased (Gilmore *et al.,* 1996) and state transition is reduced (Fig. 10.5 A and B).

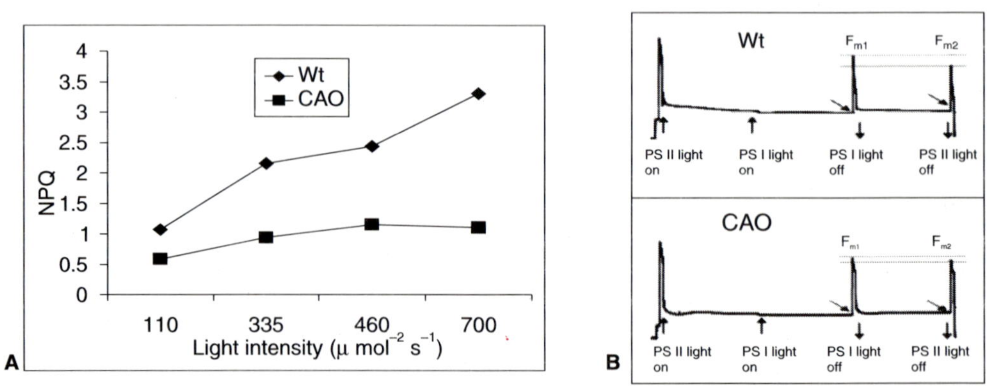

Fig. 10.5 Non-photochemical quenching (A) and state transitions (B) in Wt and CAO mutant of rice (unpublished results). Non-photochemical quenching measured at both low and high light intensities. B. Pulse-modulated chlorophyll fluorescence measurement of the state transitions.

There is reduction in net photosynthetic rate (P_N), total electron transport rate (JF), photochemical quenching (qP), quantum yield of PS II dependent non-cyclic electron transport ΦPS2 (Lin *et al.,* 2003; Kim *et al.,* 2009). However, higher net photosynthetic rate (P_N), lower non-photochemical quenching (qN), but higher maximal photochemical efficiency (F_v/F_m), higher excitation energy capture efficiency of PS II reaction centers (F_v'/F_m'), higher photochemical quenching (qP), higher effective PS II quantum yield (ΦPSII), and higher non-cyclic electron transport rate (ETR) have also been reported in Chl *b* mutant of rice (Zhou *et al.,* 2006).

77 K fluorescence

Chl *a* fluorescence emission at 77K shows the origin of emission in the long wavelength region, which is believed to be emanated from photosystem I. Wild type shows emission peak at 740-742nm. As a result of mutation in CAO, emission peak of PS I has been shown to be shifted 12-18 nm lower (Bossmann *et al.,* 1997; Lee *et al.,* 2005) and there is reduction in emission peak indicating the loss of Lhca protein (Fig. 10.6).

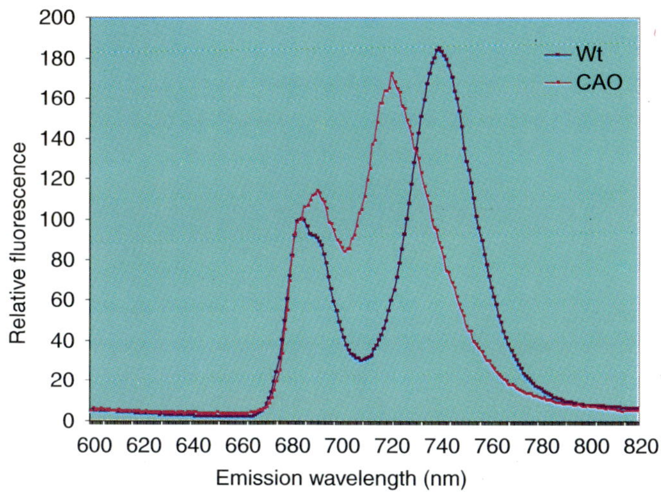

Fig. 10.6 Fluorescence emission spectra at 77 K measured in leaf segments from wild-type (Wt) and CAO mutant of rice. Spectra were normalized to their 685 nm emission.

Stress Response of CAO Mutants

High temperature

CAO mutant has been shown to be affected more as compared to wild type plants in response to high temperature (55°C). It has been reported that most of the Chl-deficient mutants of sweetclover (*Melilotus alba*) grown under 17°C and 26°C are thermo-sensitive (Yang *et al.,* 1990). Photochemical efficiency (Fv/Fm), NPQ and qP of rice CAO mutant is reduced to large extent as compared to wild type (Chen *et al.,* 2007, Fig. 10.7 B). Fo-temperature curve shows that the critical temperature for F_0 shifted to 3°C lower due to CAO mutation, implying the thermosensitivity of CAO mutant as compared to wild type. It has been found that oxygen evolution in CAO mutant is more susceptible to high temperature than that in wild type (Georgieva *et al.,* 2003). Although the structure of grana and stroma thylakoids in chloroplasts of both wild type and CAO mutant have been found to be swollen and distorted, the thylakoids of CAO mutant become more contorted and dilated and large amount of osmophilic granules accumulated (Lin *et al.,* 2005, Fig. 10.7 A).

It has been found that generally oxygen evolution is susceptible to high temperature, and as a response to heat, there is the release of oxygen evolving proteins (Takeuchi and Thornber, 1994). It has been

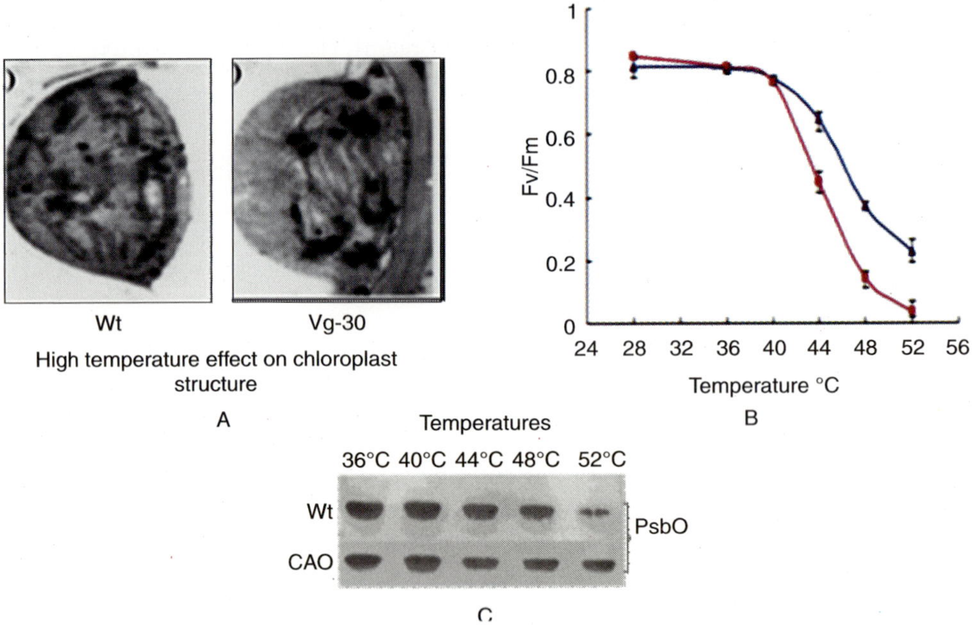

Fig. 10.7 Effect of high temperature on Wt and CAO mutant (Vg-30) of rice (A) thylakoids structure (B) photochemical efficiency (Fv/Fm) circled line Wt, triangled line CAO mutant and (C) oxygen evolving complex protein PsbO in PS II.

reported that although the extent of decrease in photochemical efficiency in rice CAO mutant is more, the heat induced release of PsbO (oxygen-evolving complex protein) is less in CAO mutant indicating more thermostability of PS II in later (Fig. 10.7 C). As evident from the proteomic analysis, this stabilizing capacity may be correlated to the increase in PSII assembly/stability factor HCF136 along with an increase in other proteins like H+-transporting ATP synthase, fructose-bisphosphate aldolase in CAO mutant (Chen *et al.,* 2007).

High light

In a Chl *b*-deficient barley CAO mutant, both the light harvesting efficiency and endurance to strong irradiance was decreased (Havaux and Tardy, 1997). On the contrary, decreased photoinhibition under strong irradiance was observed in Chl *b*-deficient mutants of *Brassica chinensis* (Guo *et al.*, 2003) and rice (Dai *et al.*, 2003), due to the lower photon absorption rate and the higher xanthophylls cycle capacity in the mutant (Dai *et al.,* 2003). High sensitivity of rice has also been observed in OsCAO1 mutant. Although long term high light exposure in the field condition has been shown not to affect the photochemical efficiency of OsCAO1, short term high light stress decreases the photochemical efficiency of CAO mutant more than the wild type (unpublished data) and it leads to enhanced generation of reactive oxygen species (ROS), photoinhibition and D1

Fig. 10.8 High light sensitivity of *Arabidopsis* Wt and CAO mutant (Image rearranged from Havaux *et al.*, 2007)

protein degradation. Acclimation to high light results in the reduction of functional antenna size of both PSII and PSI, but this reduction was even higher for CAO mutant (Stroch *et al.*, 2004). It has been observed that CAO mutants are more sensitive to high light induced photooxidation (Havaux *et al.*, 2007, Fig 10.8).

Senescence and Drought

Senescence is an oxidative process that involves the degradation of cellular structure and redistribution of the degraded products in different compartments of the cell. During leaf senescence, there dismantling of chloroplasts and other cellular organelles, degradation of macromolecules such as chlorophyll (Chl), proteins, nucleic acids and lipids, enhancement in lipid peroxidation, membrane leakiness and redistribution of nitrogen and other nutrients (Gan and Amasino, 1997). Although most of the studies on Chl *b* mutant involve the Chl biosynthetic system and photosynthetic characteristic, little, if any, information is available on the senescence course of the Chl *b*-deficient mutant. It has been found in cr3529 mutant of *Brassica* is slower as compared to wild type (Qin *et al.*, 2006). It has been observed that the expression of CAO is not affected by water deficit (Munne-Bosch *et al.*, 2009).

Consequence of Overexpression and Introduction of CAO

Photosynthetic organisms adapt to their light environment by adjusting the antenna size of their photosystems. In the shade habitat, plants have larger antenna size and lower Chl *a*/*b*, whereas in the sun habitat, the antenna size is reduced and Chl *a*/*b* increases. As Chl *b* is synthesized by CAO, it has been speculated that the expression of CAO determines Chl *a*/*b*. Correlation of Chl *a*/*b* and steady state m-RNA level has been reported in *Dunaliella salina* under low and high light conditions (Masuda *et al.* 2003). In *Arabidopsis thaliana*, the change in Chl *a*/*b* during the shift from moderate light to shaded conditions is also correlated with proteins and mRNA levels (Harper *et al.*, 2004). When CAO is overexpressed in transgenic plants of *Arabidopsis*, Chl *a*/*b* has been reported to be decreased and there is predominant effect of steady state CAO mRNA level on Chl *a*/*b* and accumulation of LHC proteins. CAO m-RNA level increased upto 3.5 at moderate light intensity and decreased to 3-3.3 under high light in overexpresssed plants, while they increased to 4.2 in wild plants (Tanaka and Tanaka, 2005). However, it has been reported that there is no correlation between change in Chl *a*/*b* and LHC accumulation in transgenic tobacco plants overexpressed for CAO (Pattanayak *et al.*, 2005).

Chlorophyllide a oxygenase (CAO) is associated with LHC II proteins in chloroplast envelope. Chl *b* synthesized by CAO is directly transferred to LHC II apoproteins (Hoober and Eggnik, 2001). Due to the direct transfer to the apoproteins, there is preferential incorporation of Chl *b* into LHC II. So, any defect in CAO interferes with the normal distribution of Chl *b*. It has been found that when defective CAO gene in *Arabidopsis ch1-1* mutant is replaced by the introduction of normal PhCAO gene, the normal phenotype is being rescued (Hirashima *et al.*, 2006). Although Chl *b* is synthesized, it accounts for only 10 % of the total chlorophyll; but when CAO is introduced into the *Synechocystis* PCC 6803 in the transient presence of LHC II, Chl *b* becomes the major pigment that replaces Chl *a* in PS II complex (Xu *et al.*, 2001).

Cyanobacteria unlike higher plants and green algae do not have Chl *b* because of the absence of CAO. When higher plant CAO was introduced into the cyanobacterium *Synechocystis* PCC 6803, the Chl *b* synthesis occurred (Satoh *et al.*, 2001). The Chl *b* can bind to CP 43', which is expressed in cells grown in low level of iron and normally binds to Chl *a*. Binding of CP 43 to Chl *b* converts it to Chl *a*/Chl *b* binding protein that resembles the Chl *a*/Chl *b*-binding Pcb protein of prochlorophytes (Duncan *et al.*, 2003).

Structure and function of Chlorophyllide oxygenase

The sequence of CAO is available for *Arabidopsis*, rice and prochlorophytes; only partial sequence is available for *Zea mays* and *Chlamydomonas reinhardtii*. Mature peptides of CAO are composed of 537, 542 and 367, amino acids in *Arabidopsis* (At CAO), rice (OsCAO) and prochlorophytes (PhCAO) respectively, and organized into 3 domains i.e., A, B and C and a transit peptide at N-terminal end (Fig. 10.9A). A and B domains are present as N-terminal extension and composed of 134 and 30 amino acids respectively. This N-terminal extension is the unique structure of CAO, as sequence similar to A domain has not been found in other similar enzymes like CAO, but sequence similar to B domain has been found in those enzymes. Both A and B domains are absent in prochlorophytes, instead, they have C-terminal extension consisting of 32 amino acids. C domain is constituted of 337 amino acids in AtCAO and OsCAO and 335 amino acids in prochlorophytes. Transit peptide is composed of 36 amino acids and it is not found in prochlorophyte CAO.There is a high sequence similarity in A domain between AtCAO and OsCAO, and the sequence of this domain does not significantly match with any other protein in the database. The homologus protein of CAO i.e., leaf-spot1 homolog (Lls1), lethal leafspot1-like gene, putative Rieske iron-sulfur protein of TIC55 homolog and choline monooxgenase do not sequence corresponding to the CAO A domain (Nagata *et al.,* 2004). There is a suggestion that a relationship between LHC II and A domain exists. The C domain has a Rieske-type [2Fe-2S] cluster and a mononuclear iron-binding site, indicating their role in Chl *b* biosynthesis. This domain is conserved for all the CAO sequence from land plants, green algae and cyanobacteria. B domain is a linker between A and C domains and is found to have no sequence similarities between AtCAO and OsCAO. While Cdomain has catalytic function in Chl *b* biosynthesis, A and B play a regulatory role in assembly of pigments with apoproteins.

Isolation of CAO genes and cDNAs (Tomitani *et al.,* 1999) from *Prochlorothrix hollandica*, *Prochloron didemni* (Prochlorophyta), *Dunaliella salina* (Chlorophyceae), *Marchantia polymorpha* (Bryophyta), and *Oryza sativa* (Angiospermae) and their phylogenetic analyses have indicated that these species share a common evolutionary origin and that progenitors of oxygenic photosynthetic bacteria, including chloroplast ancestors, had both Chl *b* and phycobilins (Fig. 10.9B).

Regulation of CAO gene expression and function

Chlorophyllide oxygenase catalyzed chlorophyll *a* to chlorophyll *b* conversion involves the sensing of chlorophyll *b* level. Chlorophyll *b* contributes to the antenna size of the photosystem II, as in response to low light intensity, the Chl *a*/*b* ratio decreases and in higher light intensity it increases. The size of the antenna is also determined by the light harvesting complexes. There is a correlation between the level of light harvesting complex and the chlorophyll *b* synthesis. Therefore, chlorophyll *b* has a role in the regulation of antenna size. As the excess light energy absorbed by the photosystems causes photoinhibition and inadequate light energy capture affects plant growth, therefore, there should be some dynamic mechanism to regulate the Chl *b* level. It has been observed that in the plants like *Arabidopsis*, tobacco, *Dunaliella* in response to low light intensity, the CAO mRNA level increases along with the decrease in Chl *a*/*b* ratio indicating that CAO is regulated at transcription level. The post-transcriptional regulation of CAO has also been suggested based on the observation that the overexpression mutant has limited effect on CAO mRNA level in *Arabidopsis* (Tanaka *et al.,* 2001). The function of CAO is also determined by its stability (Tanaka and Tanaka, 2005). The regulation of CAO occurs through the destabilization of N-terminal domain (AB domain) and the conjugated protein C in response to Chl b accumulation. It has been shown that Clp protease takes part in the destabilization and degradation of CAO protein. There are three suggestions regarding the degradation of AB domain containing protein.

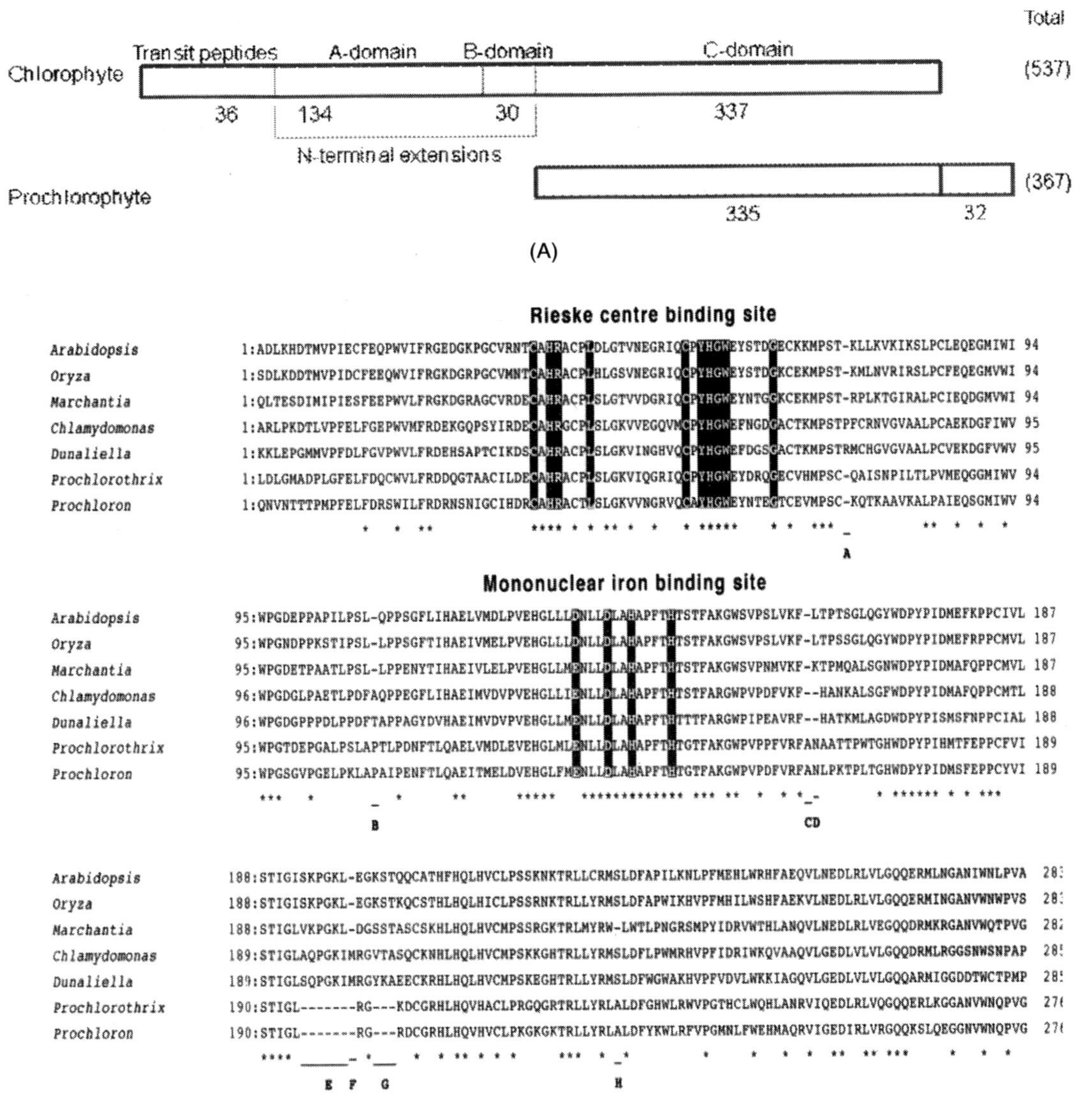

(B)

Fig. 10.9 A. Schematic presentation of domain structure of CAOP, B. Alignment of CAO aminoacid sequences from chlorophytes *A. thaliana*, *O. sativa*, *M. polymorpha*, *C. reinhardtii* and *D. salina*, and prochlophytes *P. hollandica* and *P. didemni*. Asterisks indicate conserved identical amino acid residues. The residues of the two binding sites for a [2Fe-2S] Rieske centre and a mononuclear iron are highlighted in black. The eight sites where amino acid gaps exist in the alignment are underlined and labelled A to H (Tomitami *et al.,* 1999)

Firstly, the Chl *b* that cannot be incorporated into the photosystem directly binds to the AB domain, and as a result, the conformation of AB domain changes that makes it accessible to degradation by Clp protease. Secondly, AB domain may sense the physiological status of the cell i.e., redox state or energy level and change their structure. Thirdly, there are some unknown factors expressed in response to Chl *b* accumulation and facilitates the interaction between recognition sequence in the AB domain and the

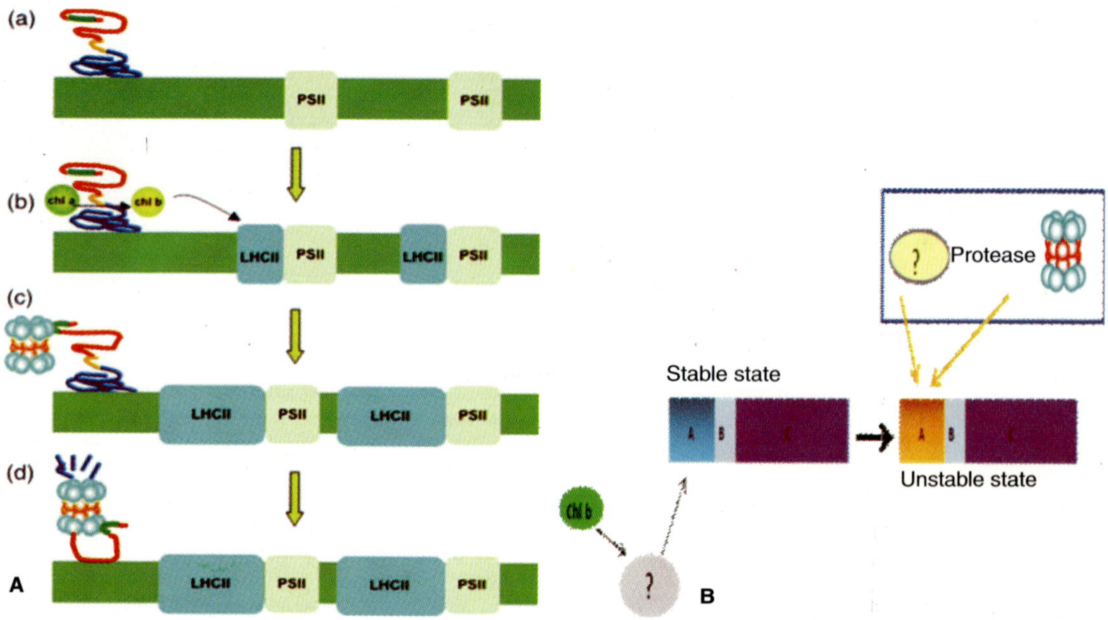

Fig. 10.10 **A.** ***A tentative model for the regulation of chlorophyll b synthesis by Clp protease (Nakagawara* et al., *2007).*** The green line designates the signal peptide with CAO; red, yellow and blue lines represent the A, B and C domains of CAO. Blue and orange balls represent the ClpC1 and ClpP subunits of ClpCP protease, respectively. **B.** Proposed model for degradation mechanism of CAO protein involving recognition factors (Kanematsu et al., 2008).

Clp protease. The model proposed (Nakagawara *et al.*, 2007, Kanematsu *et al.*, 2008) for the degradation mechanism of CAO protein is given in Fig 10.10 A and B.

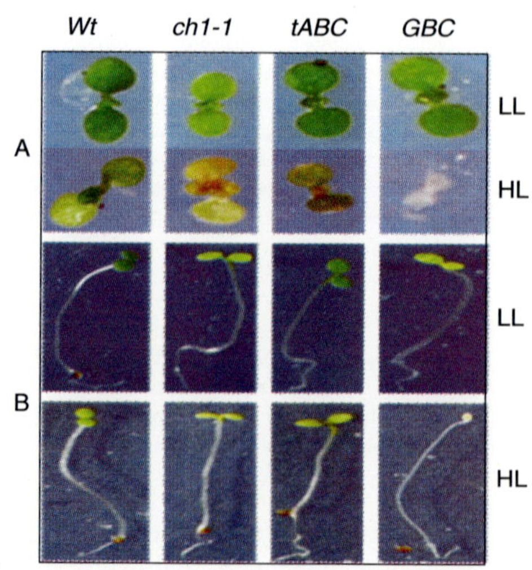

Fig. 10.11 Light sensitivity of arabidopsis seedlings. **A.** Arabidopsis seedlings (unetiolated) **B.** Etiolated seedlings tGABC and tGABC are transgenic plants with A-domain and without A-domain respectively. Arabidopsis seedlings were germinated and grown continuously under low (70 μE m^{-2} s^{-1}) or high light (300 μE m^{-2} s^{-1}) conditions for 7 days, and etiolated seedlings were grown for 4 days and transferred to low or high light conditions for an additional 3 days.(Figures rearranged from Yamasato et al. (2009).

The signal sequence for CAO degradation resides in the A domain. In the absence of Chl *b*, the signal sequence is protected against the Clp (Fig 10.10A a) protease and CAO can accumulate and synthesize Chl*b* (Fig 10.10A b). When sufficient amount of Chl *b* is synthesized, some factors influence the structure of A domain and signal sequence is exposed (Fig. 10.10A c). Interaction of ClpC1 of ClP protease with the CAO makes the later to be dragged into the proteolytic chamber of ClP protease (Fig. 10.10A d).

Physiological function of A domain has been found in *Arabidopsis ch1-1* mutant (Yamasato *et al.*, 2009). It has been reported that loss of A domain impairs one of the step in chlorophyll biosynthesis i.e., the conversion of divinylprotochlorophyllide *a* to monovinyl protochlorophyllide *a* (see Fig. 11A) under dark conditions and induces photodamage during greening of etiolated seedlings (Fig. 10.11).

Perspectives of CAO research

The research on Chl *b* has been studied for more than 50 years. The study on CAO started when it was found that this enzyme is responsible for the Chl *a* to Chl *b* formation. The research on CAO has both practical implications and is valuable for basic sciences.

It has been found that reduced antenna size facilitates the better solar energy penetration, enhances solar energy conversion by 3-4 times and improves productivity that can be achieved by pigmented cells (Melis *et al.,* 1999). Reduced antenna size facilitates the probability of individual cells at surface of the culture getting light energy from sun is less and thus, allowing the more penetration of irradiance dipper into the culture and enabling the more number of cells to contribute to photosynthesis and H_2 production. As CAO mutants of microalgae reduced in their antenna size, are proved to be efficient solar energy converter and potential source of biological hydrogen production.

Overexpression of CAO may have the agricultural benefits as it increases the antenna size, the plants can grow in the low light intensity thereby reducing the cost of energy for growing in growth chamber.

Although CAO is found to be the enzyme for conversion of Chl *a* to Chl *b*, but still it is a challenge to find out and study the gene responsible for the reverse process i.e., Chl *b* to Chl *a* conversion. This would test the hypothesis that Chl degradation is the trigger for plant senescence (Vincentini *et al.*, 1995; Matile *et al.,* 1996). If this hypothesis tests positive, the senescence of plants can be reduced by overexpressing the antisense oriented genes for Chl *b* reducing enzymes.

The introduction of CAO gene into the cyanobacterium *Synechocystis* PCC 6803 which is a natural CAO mutant, can enable it to produce Chl *b* (Satoh *et al.*, 2001) explaining that how the photosynthetic bacterium acquires the pigments in the course of evolution and suggests that the evolution of pigments started earlier than the pigment-binding proteins. These pigment-binding proteins are flexible in binding with any pigments indicating an important feature of the evolution of photosynthesis.

REFERENCES

Amano E (1989) Mutant induction method using leaf colour mutant. Technical News 34:1-2

Bellemare G, Bartlett SG, Chua N-H (1982) Biosynthesis of chlorophyll *a*/*b*-binding polypeptides in wild type and the chlorina f2 mutant of barley. J Biol Chem 257: 7762-7767

Bossmann1 B, Knoetzel J, Jansson S (1997) Screening of *chlorina* mutants of barley (*Hordeum vulgare* L.) with antibodies against light-harvesting proteins of PS I and PS II: Absence of specific antenna proteins. Photosynth Res 52: 127-136

Butler L, Chang LO (1958) Genetics and physiology of the xanthophyllous mutant of the tomato. Can J Bot 36: 251-267

Chen Xi, Zhang W, Xie Y, Lu W, Zhang R (2007) Comparative proteomics of thylakoid membrane from a chlorophyll *b*-less rice mutant and its wild type. Plant Sci 173: 397-407

Dai XB, Xu XM, Lu W, Kuang TY (2003) Photoinhibition characteristics of a low chlorophyll *b* mutant of high yield rice. Photosynthetica 41: 57-60

Duncan J, Bibby T, Tanaka A, Barber J (2003) Exploring the ability of chlorophyll *b* to bind to the CP43' protein induced under iron deprivation in a mutant of *Synechocystis* PCC 6803 containing the cao gene. FEBS Lett 541: 171-175

Duysen ME, Freeman TP, Williams ND, Olson LL (1984) Regulation of excitation energy in a wheat mutant deficient in light-harvesting pigment protein complex. Plant Physiol 76: 561-566

Falbel TG, Staehelin LA (1996) Partial block in the early steps of the chlorophyll synthesis pathway: A common feature of chlorophyll *b*-deficient mutants. Physiol Plant 97: 311-320

Falbel TG, Meehl JB, Staehelin, LA (1996) Severity of mutant phenotype in a series of chlorophyll-deficient wheat mutants depends on light intensity and the severity of the block in chlorophyll synthesis. Plant Physiol 12: 821-832

Gan S and Amasino RM (1997) Making sense of senescence. Plant Physiol 113: 313-319

Gengenbach BG, Gorz HJ, Haskins FA (1970) Genetic studies of induced mutants in *Melilotus alba*. II. Inheritance and complementation of chlorophyll-deficient mutants. Crop Sci 10: 154-156

Georgieva K, Fedina, I, Maslenkova L, Peeva V (2003) Response of chlorina barley mutants to heat stress under low and high light. Funct Plant Biol 30: 515-524

Ghirardi ML, McCauley SW, Melis A (1986) Photochemical apparatus organization in the thylakoid membrane of *Hordeum vulgare* wild type and chlorophyll less *chlorina* f2 mutant. Biochim Biophys Acta 851: 331-339

Ghirardi ML, Melis A ((1988) Chlorophyll *b* deficiency in soybean mutants: I. Effects on photosystem stoichiometry and chlorophyll antenna size. Biochim Biophys Acta 932: 130-137

Gilmore A, Hazlett T, Debrunner P, Govindjee (1996) PS II Chl a fluorescence lifetimes and intensity are independent of the antenna size difference between barley wild type and chlorina mutants: photochemical quenching and xanthophylls cycle dependent non-photochemical quenching of fluorescence. Photosynth Res 48: 171-187

Greene BA, Allred DR, Morishige DT, Staehelin LA (1988) Hierarchical response of light harvesting chlorophyll-proteins in a light-sensitive chlorophyll *b*-deficient mutant of maize. Plant Physiol 87: 357-364.

Greene BA, Staehelin LA, Melis A (1988) Compensatory alterations in the photochemical apparatus of a photoregulatory, chlorophyll *b*-deficient mutant of maize. Plant Physiol 87: 365-370

Grober K (1963) Genelical, physiological behavior of a dominant chlorophyll mutant of tomato. Tom Genel Coop 13: 48-51

Guo SW, Zhang YH, Jin YQ, Shi SY, Tan XY, Liu AM (2003) Characterization of chlorophyll fluorescence in a mutant of *Brassica chinensis* L. with *xantha* seedling leaves. Acta Agron Sin 29: 958-960

Harper AL, Gesjen SE von, Linford AS, Peterson MP, Faircloth RS, Thissen MM, Brusslan JA (2004) Chlorophyllide *a* oxygenase mRNA and protein levels correlate with the chlorophyll *a/b* ratio in *Arabidopsis thaliana.* Photosynth Res 79: 149-159

Harrison MA, Nemson JA, Melis A (1993) Assembly and composition of the chlorophyll *a/b* light-harvesting complex of barley (*Hordeum vulgare* L.): Immunochemical analysis of chlorophyll *b* less and chlorophyll *b*-deficient mutants. Photosynth Res 38 : 141-151

Havaux M, Tardy F (1997) Themostability and photostability of photosystem II in leaves of the chlorina-f2 barley mutant deficient in light-harvesting chlorophyll *a/b* protein complexes. Plant Physiol 113: 913-923

Havaux M, Dall'Osto L, Bassi R (2007) Zeaxanthin has enhanced antioxidant capacity with respect to all other xanthophylls in *Arabidopsis* leaves and functions independent of binding to PSII antennae. Plant Physiol 145: 1506-1520

Highkin HR, Frenkel A (1962) Studies on growth and metabolism of a barley mutant lacking chlorophyll *b*. Plant Physiol 37: 814-820

Highkin HR, Boardman NK, Goodchild DJ (1969) Photosynthetic studies on a pea-mutant deficient in chlorophyll. Plant Physiol 44: 1310-1320

Hirashima M, Satoh S, Tanaka R, Tanaka A (2006) Pigment shuffling in antenna systems achieved by expressing prokaryotic chlorophyllide *a* oxygenase in *Arabidopsis*. J Biol Chem 281: 15385-15393

Hoober JK, Eggink, LL (2001) A potential role of chlorophylls *b* and *c* in assembly of light-harvesting complexes. FEBS Lett. 489: 1-3

Hoober JK, Eggink LL, Chen M (2007) Chlorophylls, ligands and assembly of light-harvesting complexes in chloroplasts Photosynth Res 94: 387-400

Hsu B-D, Lee J-Y (1995) The photosystem I1 heterogeneity of chlorophyll *b*-deficient mutants of rice: a fluorescence induction study. Aust J Plant Physiol 22: 195-200

Jeon, JS, Lee S, Jung KH, Jun SH, Jeong, D-H, Lee J, Kim, C, Jang S, Yang K, Nam J, An K, Han M-J, Sung, RJ, Choi HS, Yu JH, Choi JH, Cho SY, Cha SS, Kim SI and An G (2000) T-DNA insertional mutagenesis for functional genomics in rice. Plant J 22: 561-570

Kanematsu S, Sakuraba Y, Tanaka A, Tanaka R (2008) Characterization of *Arabidopsis* mutants defective in the regulation of chlorophyllide *a* oxygenase Photochem. Photobiol Sci 7: 1196-1205

Kim E-H, Li X-P, Razeghifard R, Anderson JM, Niyogi KK, Pogson BJ, Chow WS (2009) The multiple roles of light-harvesting chlorophyll *a/b*-protein complexes define structure and optimize function of *Arabidopsis* chloroplasts: A study using two chlorophyll *b*-less mutants. Biochim Biophys Acta 1787: 973-984

Lee Sichul, Kim J-H, Yoo E-S, Lee C-H, Hirochika H, An G (2005) Differential regulation of chlorophyll a oxygenase genes in rice. Plant Mol Biol 57: 805-818

Lin Z-F, Peng C-L, Lin G-Z, Ou Z-Y, Yang C-W, Zhang J-L (2003) Photosynthetic characteristics of two new chlorophyll *b*-less rice mutants. Photosynthetica 41: 61-67

Lin Z-F, Peng C-L, Lin G-Z, Ou Z-Y, Yang C-W, Zhang J-L (2005) Thermostability of photosynthesis in two new chlorophyll *b*-less rice mutants *Science in China* Ser. C Life Sciences 48: 139-147

Masuda T, Tanaka A, Melis A (2003) Chlorophyll antenna size adjustments by irradiance in *Dunaliella salina* involve coordinate regulation of chlorophyll *a* oxygenase (CAO) and Lhcb gene expression. Plant Mol Biol 51: 757-771

Matile P, Hortensteiner S, Thomas H, Krautler B (1996) Chlorophyll breakdown in senescent leaves. Plant Physiol 112: 1403-1409

Melis A, Neidhardt J, Benemann JR (1999) *Dunaliella salina* (Chlorophyta) with small chlorophyll antenna sizes exhibit higher photosynthetic productivities and photon use efficiencies than normally pigmented cells. J Appl Phycol 10: 515-52

Michel H, Tellenbach M, Boschetti A (1983) A chlorophyll b-less mutant of *Chlamydomonas reinhardii* lacking in the light-harvesting chlorophyll *a/b*-protein complex, but not in its apoproteins. Biochim Biophys Acta 725: 417-424

Morita R, Kusaba M, Yamaguchi H, Amano E, Miyao A, Hirochika H, Nishimura M (2005) Characterization of Chlorophyllide oxygenase (CAO) in rice. Breeding Science 55: 361-364

Munne´-Bosch S, Falara V, Pateraki Irene, Lo´pez-Carbonell M, Cela J, Kanellis AK (2009) Physiological and molecular responses of the isoprenoid biosynthetic pathway in a drought-resistant Mediterranean shrub, *Cistus creticus* exposed to water deficit. J Plant Physiol 166: 136-145

Murray DL, Kohorn BD (1991) Chloroplasts of *Arabidopsis thaliana* homozygous for the ch-1 locus lack chlorophyll *b*, lack stable LHCPII and have stacked thylakoids. Plant Mol Biol 16: 71-79

Nagata N, Satoh S, Tanaka R, Tanaka A (2004) Domain structures of chlorophyllide a oxygenase of green plants and *Prochlorothrix hollandica* in relation to catalytic functions. Planta 218: 1019-1025

Nakagawara E, Sakuraba Y, Yamasato A, Tanaka R, Tanaka A (2007) Clp protease controls chlorophyll *b* synthesis by regulating the level of chlorophyllide a oxygenase. The Plant J 49: 800-809

Nakatani HY, Baliga V (1985) A clover mutant lacking chlorophyll *a*- and *b*-containing protein antenna complexes. Biochim Biophys Res Commun 131: 182-189

Ohtsuka T, Ito H, Tanaka A (1997) Conversion of chlorophyll *b* to chlorophyll *a* and the assembly of chlorophyll with apoproteins by isolated chloroplasts. Plant Physiol 113: 137-147

Oster U, Tanaka R, Tanaka A, Rüdiger W (2000) Cloning and functional expression of the gene encoding the key enzyme for chlorophyll *b* biosynthesis (CAO) from *Arabidopsis thaliana*. Plant J 21: 305-310

Pattanayak GK, Biswal AK, Reddy VS, Tripathy BC (2005) Light-dependent regulation of chlorophyll *b* biosynthesis in chlorophyllide *a* oxygenase overexpressing tobacco plants. Biochem Biophys Res Commun 326: 466-471

Preiss S, Thornber JP (1995) Stability of the apoproteins of light harvesting complex I and II during biogenesis of thylakoids in the chlorophyll *b*-less barley mutant *chlorina* f2. Plant Physiol 107: 709-717

Qin T, Fu J, Zhang N, Du L (2006) Comparative studies of senescence-related enzymes in the cotyledon of chlorophyll *b*-deficient mutant and its wild type oilseed rape during senescence. Plant Sc 171: 293-299

Satoh S, Ikeuchi M, Mimuro M, Tanaka A (2001) Chlorophyll *b* expressed in cyanobacteria Functions as a light-harvesting antenna in photosystem I through flexibility of the proteins. J Biol Chem 276: 4293-4297

Stroch M, Cajanek M, Kalina J, Spunda V (2004) Regulation of the excitation energy utilization in the photosynthetic apparatus of chlorina f2 barley mutant grown under different irradiances. J Photochem Photobiol B: Biology 75: 41-50

Takeuchi T, Thornber J (1994) Heat-induced alterations in thylakoid membrane composition of barley. Aust J Plant Physiol 21: 759-770

Tanaka A, Ito H, Tanaka R, Tanaka NK., Yoshida K, Okada K (1998) Chlorophyll *a* oxygenase (CAO) is involved in chlorophyll *b* formation from chlorophyll *a*. Proc Natl Acad Sci USA 95: 12719-12723

Tanaka R, Koshino Y, Sawa S, Ishiguro S, Okada K, Tanaka A (2001) Overexpression of chlorophyllide a oxygenase, (CAO) enlarges the antenna size of photosystem II in *Arabidopsis thaliana*. Plant J 26: 365-373

Tanaka R, Tanaka A (2005) Effects of chlorophyllide a oxygenase overexpression on light acclimation in *Arabidopsis thaliana*. Photosynth Res 85: 327-340

Terao T and Yamashita (1982) A *Chlorina* mutants of *Oryza sativa* L. (cultivar Norin 8) induced by ionizing radiation and chemicals. I. Two types of chlorina mutant different in chlorophyll *a/b* ratio and phenotype of F1. Jap J Breed Suppl I: 190-191

Terao T, Yamashita A, Katoh, S (1985) Chlorophyll *b*-deficient mutants of rice. I. Absorption and fluorescence spectra and chlorophyll *a/b* ratios. Plant Cell Physiol. 26: 1361-1367

Tomitani A, Okada K, Miyashita H, Matthijsk HCP, Ohno T, Tanaka A (1999) Chlorophyll *b* and phycobilins in the common ancestor of cyanobacteria and chloroplasts. Nature 400: 159-162

Vincentini F, Hortensteiner S, Schellenberg M, Thomas H, Matile P (1995) Chlorophyll breakdown in senescent leaves: Identification of the biochemical lesion in a stay-green genotype of *Festuca pratensis* Huds. New Phytol 129: 247-252

Wang J, Lin L, Wan XS (2000) Generation and molecular analysis of a population of transgenic rice plants carrying Ds element, Acta Phytophysiol. Sin, 26: 501-506

Xu H, Vavilin D, Vermaas W (2001) Chlorophyll *b* can serve as the major pigment in functional photosystem II complexes of cyanobacteria Proc Natl Acad Sci 98: 14168-14173

Yamasato A, Nagata N, Tanaka, R, Tanaka A (2009) The N-terminal domain of chlorophyllide a oxygenase confers protein instability in response to chlorophyll *b* accumulation. The Plant Cell 17: 1585-1597

Yang CM, Osterman JC, Markwell J (1990) Temperature sensitivity as a general phenomenon in a collection of chlorophyll-deficient mutants of sweetclover (*Melilotus alba*). Biochem Gen 28: 31-40

Zhao Y, Wang ML, Zhang YZ, Du LF (2000) Inheritance and agronomic performance of a chlorophyll-reduced seedling mutant in oilseed rape (*Brassica napus* L.) and its utilization in F1 hybrid production. Plant Breeding 154 :131-135

Zhou X-S, Wu D-X, Shen S-Q, Sun J-W, Shu Q-Y (2006) High photosynthetic efficiency of a rice (*Oryza sativa* L.) *xantha* mutant. Photosynthetica 44 : 316-319

CHAPTER 11

A Comprehensive Study of the Effects of Nitrite Anion on Photosynthetic Electron Transport Chain

Anjana Jajoo* and Sudhakar Bharti
School of Life Sciences, Devi Ahilya University, Indore, India
E-mail: anjanajajoo@hotmail.com

ABSTRACT

Nitrite has been found to have significant effects on the electron transport chain in thylakoid membranes. This mini-review illustrates the effects of nitrite on the rates of electron transport mediated by Photosystem II (PS II) and Photosystem I (PS I). A role of nitrite in regulating distribution of absorbed excitation energy more in favour of PS I has also been reported. The mode of binding of nitrite to PS II plays an important role where migration of PS II antenna is less in response to the tight binding of nitrite and more when both loose and tight binding sites of nitrite are occupied.

Keywords: Photosystem II, Photosystem I, Energy distribution, Nitrite.

Abbreviations

PS I: Photosystem I, *PS II:* Photosystem II, *EPR:* Electron paramagnetic resonance.

INTRODUCTION

The role of cations and anions in the photosynthetic electron transport has been of recurrent interest. The anions have been known to affect various reactions in the PS II. It is a well known fact that chloride and other anions stimulate the electron flow (Hind *et al.,* 1969; Jajoo *et al.,* 2001). Chloride is an essential inorganic cofactor for the normal function of the oxygen evolving complex (OEC). Oxygen evolution is inhibited by the depletion of chloride and is restored by replenishing it (Homann, 1987; Debus, 1992; Renger, 2001). Anions like bromide, nitrate, iodide etc. have been reported to substitute chloride functionally to varying extents (Kelly and Izawa, 1978; Homann, 1988; Wincencjusz *et al.,* 1999; *Hasegawa et al.,* 2002, Jajoo *et al.,* 2005a).

Some of the monovalent anions like formate, nitrate, nitrite and fluoride inhibit oxygen evolution (Jursinic and Stemler, 1988). Anions like formate have been shown to inhibit the reactions on the donor side of PS II (Feyziev *et al.*, 2000) as well as to compete with chloride (Jajoo *et al.*, 2005b). It has also been reported earlier that the inhibition induced by formate or nitrite can be overcome by bicarbonate more effectively in the presence of chloride (Jajoo and Bharti, 1993). Multiple sites of anion binding have been suggested (Jursinic and Stemler, 1988) and hence each anion requires individual characterization.

Nitrite (NO_2^-) is an inorganic monovalent anion which affects the process of photosynthesis quite significantly. It is known that nitrite inhibits the rates of electron transfer between the two photosystems. The rate of oxygen evolution from the broken chloroplasts isolated from *Bumillariopsis filiformis* was reduced in the presence of 10 mM nitrite, and a nitrite sensitive site close to PS II was suggested (Spiller and Boger, 1977). Nitrite treatment resulted in a decrease in PS II partial reaction rate and an increase in PS I partial reaction rate in spinach thylakoid membranes (Sinclair, 1987). Oxygen evolution from maize chloroplasts was inhibited in the presence of 20 mM nitrite (Stemler and Murphy, 1985). In wheat chloroplasts, 5 mM nitrite inhibited noncyclic phosphorylation but stimulated cyclic phosphorylation (Makarova *et al.*, 1998). It was hypothesized that anions increase the density of negative charge that creates a repulsive force between the chlorophyll protein complexes, which in turn redistributes within the plane of chloroplasts membranes. Nitrite may be causing more of the absorbed light energy to reach PS I by causing conformational change in the thylakoid membranes (Sinclair, 1987; Jajoo and Bharti, 1995). However, the mechanism responsible for this phenomenon was not suggested. This review presents a comprehensive study of the effects of nitrite on PS II and PS I.

EFFECTS OF NITRITE ON THE RATES OF ELECTRON TRANSPORT

Rates of Electron Transfer Through PS II and PS I

The rates of electron transfer through PS II decreased by the addition of nitrite in a concentration dependent manner (Table 11.1), where 20 mM nitrite caused 32 % inhibition. The activity of PS I was measured through $DCPIPH_2 \rightarrow MV$ reaction, and the PS I mediated electron transfer rate was 23% higher in the presence of 20 mM nitrite. Further stimulation was not observed on increasing the concentration of nitrite. The optimum concentration of nitrite was found to be 20 mM, which was used for further experiments.

Table 11.1 Change in the PS II electron transport rates ($H_2O \rightarrow$ DCPIP reaction) and PS I electron transport rates ($DCPIPH_2 \rightarrow$ MV reaction) in the presence of nitrite in spinach thylakoids. The reaction mixture for PS II measurements contained 0.33 M sucrose, 50 mM HEPES-NaOH buffer (pH 7.6), 1 mM NaCl, 1 mM $MgCl_2$, 33 μM DCPIP and thylakoid suspension equivalent to 20 μg of Chl/ml. The reaction mixture for PS I measurement contained 0.1 M Sucrose, 20 mM HEPES-NaOH buffer (pH 7.6), 3 mM ascorbate, 100 μM DCPIP, 5 μM DCMU, 0.1 mM methyl viologen, 5 μM sodium azide and thylakoid suspension equivalent to 20 μg Chl/ml. The thylakoids were incubated for 5 min in the dark at room temperature before measurement. Values in parenthesis show normalized values.

Treatment	*PS II activity (μ mol DCPIP reduced mg Chl^{-1} h^{-1})*	*PS I activity (μ mol oxygen consumed mg Chl^{-1} h^{-1})*
Control	141 ± 2 (100)	162 ± 3 (100)
5 mM $NaNO_2$	114 ± 3 (81)	174 ± 2 (107)
10 mM $NaNO_2$	105 ± 5 (75)	182 ± 2 (111)
20 mM $NaNO_2$	96 ± 4 (68)	200 ± 6 (123)
40 mM $NaNO_2$	93 ± 3 (66)	195 ± 4 (119)

To ascertain that all these effects were specific to the added nitrite and not due to the general ion effect, proper controls were put with equimolar concentrations of NaCl. The effect of 20 mM chloride and 20 mM nitrite were found to be altogether different (data not shown), suggesting that the inhibition observed was because of the presence of nitrite and not due to difference in the ionic strength of the samples.

Role of Nitrite in Energy Redistribution

Most of the studies on nitrite were performed in saturating light intensities, and involvement of nitrite in redistribution of absorbed excitation energy could not be explored. State transition alters the distribution of light energy between the two photosystems. State transition is relevant only at limiting light intensities, since it is difficult to observe changes in energy distribution when both photosystems receive saturating excitation light. The effect on antenna size in the presence of nitrite can be seen only at low light intensities. Therefore, the rate of electron transfer through PS II and PS I was measured at limiting light intensity.

The possibility that nitrite may cause state transition was further confirmed by the measurement of Chl *a* fluorescence emission spectra at 77 K in nitrite treated thylakoid membranes. The change in the F_{735}/F_{685} ratio is regarded as an indicator of energy distribution between PS II and PS I, because of the origin of F_{735} and F_{685} from the PS I and PS II antennae respectively (Samson and Bruce, 1995). The ratio of intensity of Chl *a* fluorescence emitted at 735 nm (F_{735}) to that emitted at 685 nm (F_{685}) was increased by the addition of nitrite (Fig. 11.1). Similar results were obtained even in the presence of DCMU (data not shown). The extent of increase in the F_{735}/F_{685} defines the extent of attainment of state II, the state usually caused by exposure to light absorbed in pigment system II (Allen *et al.*, 1981, Cardol *et al.*, 2003). Therefore, these results indicate that nitrite diverts more energy to PS I by state transition.

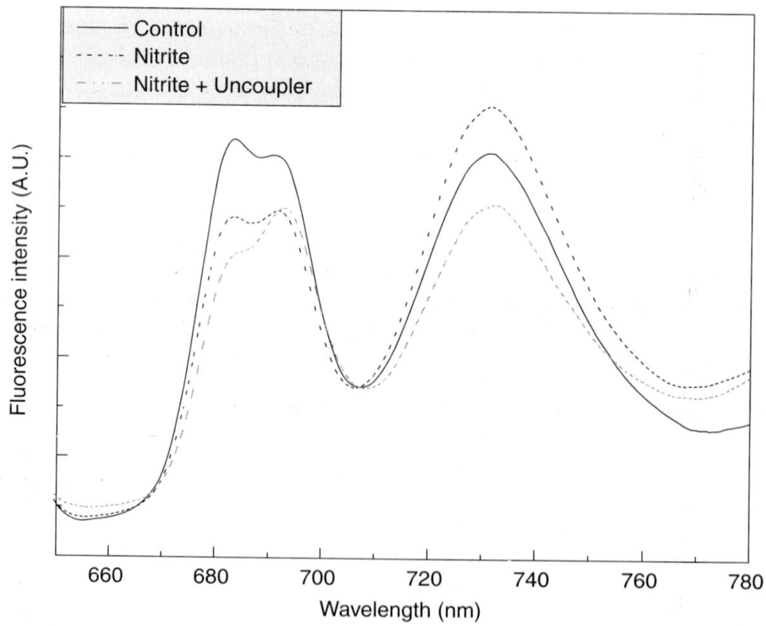

Fig. 11.1 Chl *a* fluorescence emission spectra measured at 77 K in spinach thylakoids membranes. Emission spectra were recorded at the excitation wavelength of 435 nm (preferential excitation of Chl *a*). The reaction mixture contained 0.33 M sucrose, 50 mM HEPES-NaOH buffer (pH 7.6), 1 mM NaCl, 1 mM $MgCl_2$, 50% v/v glycerol and thylakoid suspension equivalent to 10 µg of Chl/ml. Control (——), 20 mM nitrite (----).

State transition involves complementary changes in the absorption cross-section of the two photosystems (Allen, 1992). Fluorescence excitation spectra recordings help in investigating the absorption cross-section of PS I, which is indicative of the changes in the antenna size of PS I. Moreover, antenna size of PS I has been found to increase significantly upon transition to state II (Allen and Forsberg, 2001). Enhancement in the F_{735}/F_{685} ratio may reflect an increase in the absorption cross-section of PS I antenna. Excitation spectra of F_{735} were measured after nitrite treatment, which was compared with the spectrum of control thylakoid membranes (Fig. 11.2A). It is evident from the figure that nitrite treatment increased the shoulder at 650 nm, which belongs to chlorophyll *b* and shifts the peak at 680 nm to 676 nm.

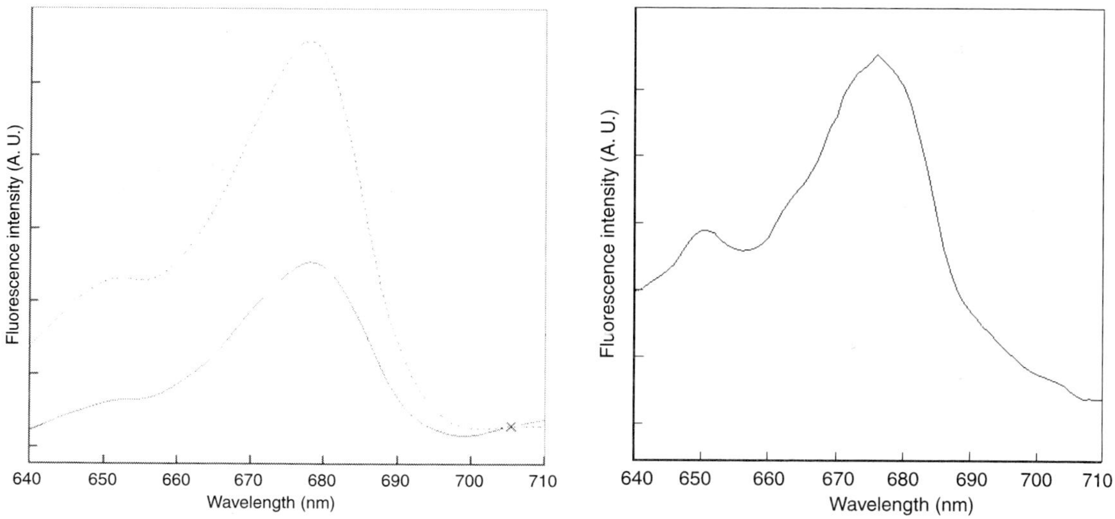

Fig. 11.2 (A) Low temperature 77 K fluorescence excitation spectra of thylakoid membranes, Control (——), 20 mM nitrite (----).'x' shows the wavelength (705 nm) at which spectra have been normalized. The reaction mixture was same as described in Fig. 1. (B) Difference excitation spectra (nitrite treated – control) for F735 of thylakoid membranes. The difference spectra were calculated by treatment excitation spectra – control excitation spectra.

In order to confirm the role of nitrite in causing state transition and to estimate the absorption characteristic of chlorophyll species of PS II responsible for developing the additional cross-section of PS I, difference excitation spectra (difference absorption cross-section spectra) was measured (Fig. 11.2B). In case of nitrite treatment, PS I antenna cross-section increased as indicated by prominent peaks in the difference spectra (nitrite minus control) at 650 nm and 676 nm. The peak at 650 nm belongs to Chl *b*, while short wavelength Chl *a* species absorbed at 674-676 nm, which has been attributed to LHC II (Ruban 1991).

In state transitions, the two photosystems alter their relative extent of light absorption. In state II, the PS II does not divert the absorption energy but simply receives less light because of decreased antenna size. Earlier work (Jajoo *et al.,* 1998) had suggested the role of chloride in causing energy redistribution between the two photosystems. Some anions like chloride, bicarbonate etc. are known to increase membrane fluidity of thylakoid membranes (Jajoo *et al.,* 1994). Nitrite might also have caused unstacking of the thylakoid membranes, which might have helped in attaining state II.

Reversibility of Nitrite Effects

It is well known that the light harvesting complex (LHC) II plays a vital role in the energy distribution (Bennett, 1983; Allen, 1992; Cardol *et al.*, 2003). LHC is not a static structure but is able to interact with both the systems to maximize photosynthetic efficiency. Reversibility of the effects of nitrite was studied in order to identify tight binding (non-dissociable) and loose binding (dissociable) of nitrite in PS II. To study the reversibility of nitrite effects, three types of samples were prepared (Singh *et al.*, 2007): first control (without nitrite treatment), second in which nitrite treatment was given and later thylakoids were washed with nitrite-free (designated $-NO_2$ or nitrite washed), and third in which nitrite treatment was given and thylakoids were suspended in suspension medium containing nitrite (designated $+NO_2$ or nitrite treated).

Nitrite was found to inhibit PS II electron transfer rates. The rates of PS II electron transfer ($H_2O \rightarrow$ DCPIP reaction) have been given in Table 11.2. The nitrite treated samples showed a decrease in the rates of Hill activity by 35%, while the activity in the case of washed samples was found to be inhibited by only 20%. The rates of oxygen consumption were measured to elucidate the effects of nitrite on the electron transfer through PS I. It was found that the rates of PS I mediated electron transport in nitrite treated samples increased by 70%, while for nitrite washed samples an increase of only 4% was observed (Table 11.2). To investigate the effects of nitrite on the donor side of the PS II, the S_2 multiline spectra were monitored for all these three types of samples. As shown in Fig. 11.3, the signal intensity of multiline EPR signal of the three samples was found to be different. The amplitude of multiline EPR signal decreased in nitrite treated samples by 45% and 15% in the nitrite washed samples when compared to that of the control. No difference in the line shape or position was observed in all the three samples. Washing the nitrite-treated samples in nitrite-free medium largely restored the multiline signal. The restoration was found to be 85% of the control.

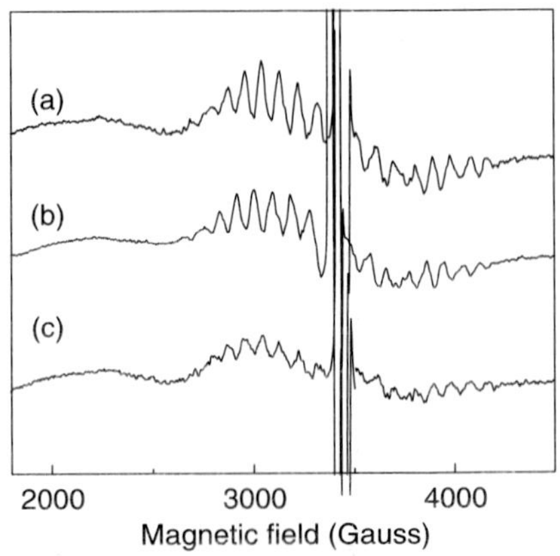

Fig. 11.3 S_2 state EPR spectra of control (a), nitrite washed (b) and nitrite treated (c) thylakoid membranes. All the spectra are (illuminated - minus-dark) difference spectrum and have been normalized by Y_D^+ radical signal intensity observed at a microwave power of 0.8 □W and field modulation amplitude of 3.2 G. Experimental conditions: microwave frequency 9.42 GHz, microwave power 3.2 mW, modulation amplitude 10 G, temperature 6 K. Each of the spectra is a mean of 4 scans.

Low temperature emission spectra of Chl *a* at 77 K was monitored for all the three samples (Fig. 11.4). A distinct decrease in the fluorescence intensities at 680 nm and 695 nm was accompanied by a consistent rise in the 730 nm peak for the control, nitrite washed and nitrite treated samples. The ratio F_{735}/F_{680}, which shows the state of excitation energy distribution, was found to be 0.97 for the control, while it was 0.982 and 1.24 for the nitrite washed and nitrite treated samples respectively.

Changes in the F_{735}/F_{680} ratio have been mostly associated with the rearrangement of CP complexes in thylakoid membranes (Garstka *et al.*, 2005). To understand more about the distribution of the excitation energy, the absorption cross section of PS I at 77K was studied in the three samples. An increase in the membrane cross section of PS I was seen in the case of nitrite washed as well as nitrite treated samples when compared to that of control (Fig. 11.5a). The difference absorption cross section spectra

Table 11.2 Changes in control, nitrite washed ($-NO_2$) and nitrite treated ($+NO_2$) thylakoid membranes as studied by Hill activity, oxygen consumption rates and steady-state fluorescence measurements. Normalized values are given in parenthesis.

Parameters studied	*Control*	*$-NO_2$*	*$+NO_2$*
Hill activity (μmoles of dye reduced (mg chl)$^{-1}$ hr^{-1})	135.53 ± 1.12 (100)	108.07 ± 1.10 (79.73)	87.82 ± 0.90 (64.79)
Oxygen consumption (μmoles of oxygen consumed (mg chl)$^{-1}$ hr^{-1})	299 ± 4.76 (100)	313.12 ± 0.10 (104)	510 ± 8.21 (170)
Steady state fluorescence (Intensity)	70.37 ± 1.4 (100)	65.98 ± 0.68 (93.76)	57.59 ± 1.01 (81.83)

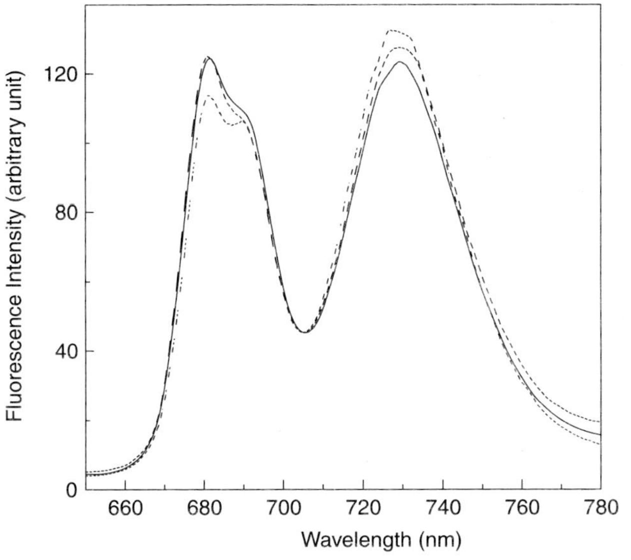

Fig. 11.4 Chl *a* fluorescence emission spectra measured at 77 K. Emission spectra were recorded at excitation wavelength of 435 nm. The spectra were normalized at 705 nm. Control (—), nitrite washed (- - -) and nitrite treated (-···· ···).

(Fig. 11.5b) showed a significant difference in the absorption cross sections of PS I for each of these samples and a clear increase in the antenna cross section of PS I region of 674-676 nm was observed for the nitrite treated samples. The variable cross section spectra, which show the percentage change in the excitation spectra, showed new bands at 668 nm and 687 nm (data not shown).

Thus, the rate of PS II electron transfer was reduced in the case of nitrite treated samples. The PS II activity for nitrite washed samples was found to be higher than that of nitrite treated samples, indicating that a part of nitrite was removed from the treated samples by washing the thylakoids in a nitrite-free medium (up to 15%). This suggested a possibility of nitrite being bound to these pockets tightly and irreversibly, which could not be removed by washing. The other type of binding was observed in case of nitrite treated samples; here the inhibition could be an additive effect of binding of nitrite to both the tight as well as weak binding sites. These types of binding were termed dissociable (loose or low affinity) and non-dissociable (tight or high affinity) binding, which could also be discussed in terms of high and low

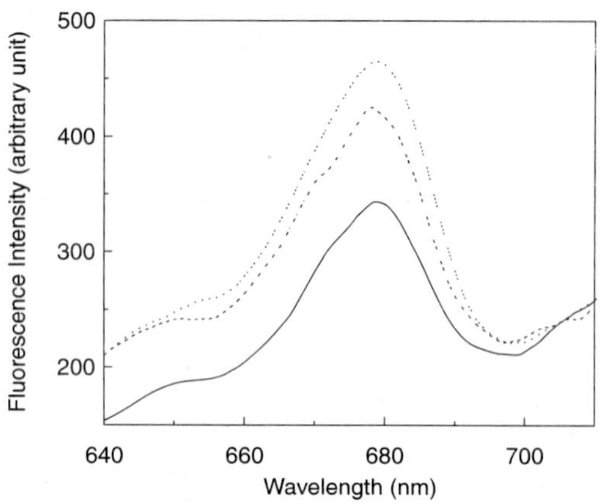

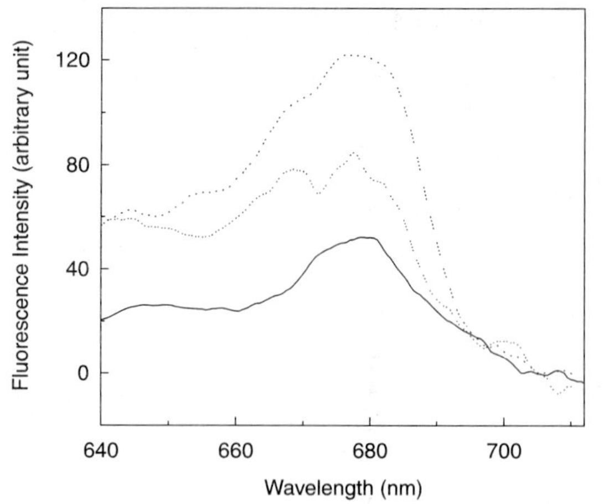

Fig. 11.5 (a)Low temperature F735 fluorescence excitation spectra of thylakoid membranes. Spectra have been normalized at 705 nm. Control (—), nitrite washed (– – –) and nitrite treated (·····). (b) Difference excitation spectra (difference absorption cross section spectra) for F735 of thylakoid membranes. The difference spectra were calculated for different conditions by Treatment excitation spectra – control excitation spectra. Control-control (—), nitrite washed-control (–··–··) and nitrite treated-control (·····).

affinity binding respectively on the PS II. These results also indicated that inhibitory effects of nitrite on the PS II are largely reversible and non-destructive.

An increase in F_{735}/F_{680} ratio, decrease in the peak intensity at 680 nm observed in the steady state fluorescence, and increase in the PS I mediated rates of oxygen consumption for the nitrite treated samples supported the possibility of more energy transfer towards PS I (Samson and Bruce, 1995). A clear increase in the antenna cross section of PS I in the region 674-676 nm (Fig. 5b), which has been attributed to LHC II (Ruban, 1991), suggested that the increase in the PS I antenna cross section is at the expense of Chl *b* (650 nm) and Chl *a* (680 nm).

All these results strongly substantiate that the mode of binding of nitrite anion on the PS II is of two ways: dissociable (loose) and non-dissociable (tight) binding. The inhibition caused by the dissociable binding may be reversible i.e., transitory inhibition as long as nitrite anions are present in the medium, while the inhibition caused by the non-dissociable binding is irreversible i.e., irreversible inhibition that causes a permanent damage. Of the total nitrite that binds to the PS II in the nitrite treated samples, ~60% of the nitrite effects are observed in the nitrite washed samples. Nitrite leads to an increase in the rates of PS I mediated electron transport probably by transfering more energy towards PS I. The transfer of energy towards PS I may be correlated to migration of antenna molecules from PS II to PS I (Ruban, 1991; Allen, 1992; Samson and Bruce, 1995; Cardol *et al.*, 2003). This is evident from an increase in the absorption cross section of PS I in the nitrite treated samples. The extent of the antenna migration and hence the energy transfer depends on the type of binding of nitrite to PS II complex. Antenna migration is less in response to the tight binding of nitrite and more when both loose and tight binding sites of nitrite are occupied. The removal of the loosely bound nitrite leads to reduction in the migration of the antenna molecules and subsequently causes reduction in the energy transferred to PS I (as evident from the fluorescence emission and excitation spectra). Thus it was concluded that nitrite plays a regulatory role in the migration of LHC II towards PS I causing changes in the redistribution of the excitation energy. Singh *et al.* (2007), thus showed for the first time the dependence of antenna migration and thereby the energy transfer between the two photosystems on the dissociable and non-dissociable modes of binding of nitrite anions.

Interaction of Nitrite and Chloride

The concentration response (Fig. 11.6) of chloride and nitrite was studied and 10mM Cl and 20 mM NO_2 were chosen as optimum concentrations. To study the interaction better, concentration of both nitrite and chloride were varied (5, 10, 20 mM). Nitrite effects have been reported on the donor side of PS II (Sahay *et al.*, 2006 a,b). The electron transport rates through PS II were studied for various combinations of nitrite and chloride (Table 11.3). An increase of 27% was observed with 10 mM chloride, while 10 mM nitrite showed a decline of 43% when compared to control. The study was carried out for two sets. In the first set, 10 mM Cl^- was kept constant and to it nitrite was added in varying concentrations (5, 10, 20 mM). There was an increase of 27%, 23% and 32% respectively, when compared to the control. This shows that no effect of nitrite was observed when the samples already had 10 mM Cl^-. For the second set, 10 mM nitrite was kept constant for all the samples and chloride was added to this in varying concentration (5, 10, 20 mM). There was a decrease of 27% when 5 mM in presence of 10 mM Cl^- and 20 mM Cl^-, the inhibition due to nitrite was curbed and Cl^- effects were prominent, an increase of 21% and 34% respectively was noted.

Table 11.3 Changes in PS II electron transport rates ($H_2O \rightarrow$ DCPIP reaction) in saturating light intensities (150 Wm^{-2}) in spinach thylakoid membranes at varying chloride and nitrite concentrations. Normalized values are given in parenthesis.

Treatment	*Hill Activity*
Control	89.42 ± 2.3 (100)
10 mM Chloride	114.65 ± 6.6 (128)
10 mM Chloride + 5 mM Nitrite	113.57 ± 3.2 (127)
10 mM Chloride + 10 mM nitrite	109.95 ± 1.2 (123)
10 mM chloride + 20 mM nitrite	118.39 ± 2.6 (132)
10 mM nitrite	50.61 ± 2.8 (57)
10 mM nitrite + 5 mM chloride	65.72 ± 0.9 (73)
10 mM nitrite + 10 mM Chloride	107.84 ± 1.9 (121)
10 mM nitrite + 20 mM chloride	119.49 ± 3 (134)

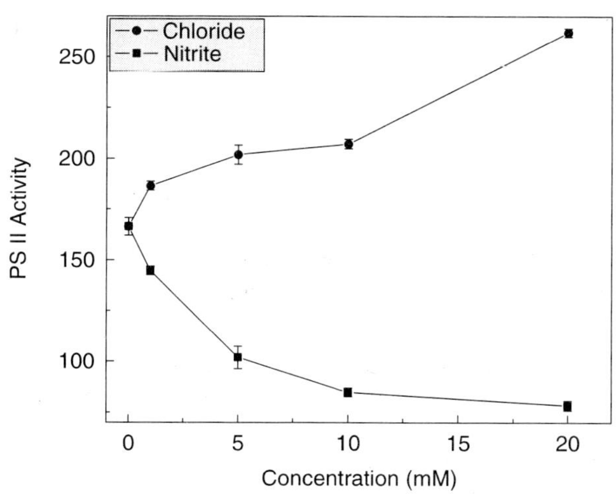

Fig. 11.6 PS II electron transport rates measured as the photoreduction of DCPIP in spinach thylakoid membranes after addition of chloride and nitrite.

Further, low temperature (77 K) Chl *a* fluorescence emission spectra was recorded (Fig. 11.7). An increase in PS I/PS II ratio from 1.055 for control to 1.389 for 10 mM nitrite treated samples were observed. With the addition of 5 mM chloride to these 10 mM nitrite treated samples, the PS I/PS II ratio declined to 1.313 and further decreased to 1.210 and 1.042 with the addition of 10 mM chloride and 20 mM chloride respectively. These results led to the conclusion that nitrite effects decreased in the presence of chloride. The nitrite binding sites reported on the donor side thus have higher affinity for chloride. Thus, chloride can replace nitrite from its binding site.

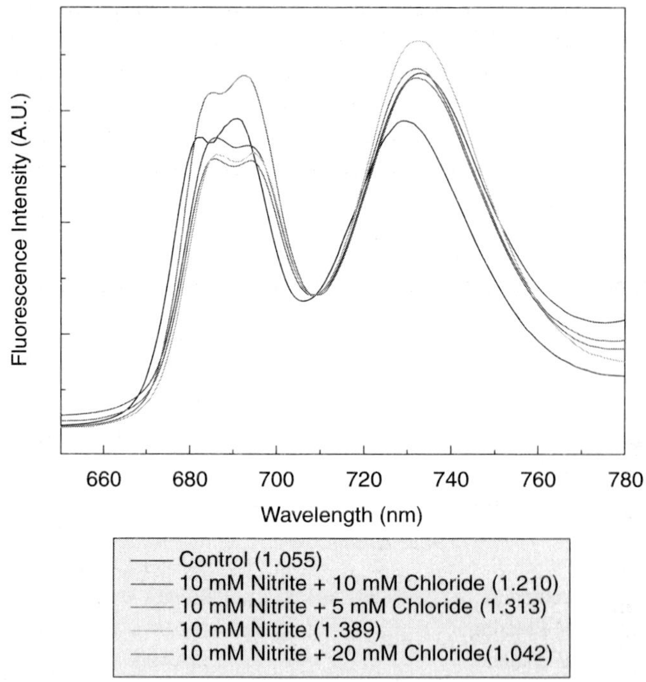

Fig. 11.7 Chl *a* fluorescence emission spectra measured at 77 K. Emission spectra were recorded at excitation wavelength of 435 nm. Spectra have been normalized at 705 nm. [Values in the brackets show the ratio F_{735}/F_{685}]

EFFECT OF NITRITE ON THE STRUCTURAL PROPERTIES OF THYLAKOID MEMBRANES

The effects of nitrite have been studied on the structural integrity of the thylakoid membranes (Jajoo and Kawamori, 2006). The effects of nitrite on the fluidity properties of the thylakoid membranes have been investigated. Doxyl stearic acid (5-SAL) was used as a spin label, which serves as a physical probe to report structural and environmental aspects at a localized site in the thylakoid membrane.

The ESR spectra of 5-SAL in buffer shows an isotropic hyperfine structure consisting of 3 equidistant lines indicating that the paramagnetic molecule is tumbling freely in the solution. Incorporation of the spin label in the thylakoid membranes shows an anisotropic spectrum indicating immobilization of the paramagnetic probe in the surroundings. The spectra recorded in the presence of $NiCl_2$ confirms the incorporation of SAL into the thylakoid membranes.

Figure 11.8 shows the EPR spectra of thylakoid membranes after giving various treatments. As reported earlier (Jajoo *et al.,* 1994), addition of 5 mM Mg^{2+} makes the thylakoid membranes rigid

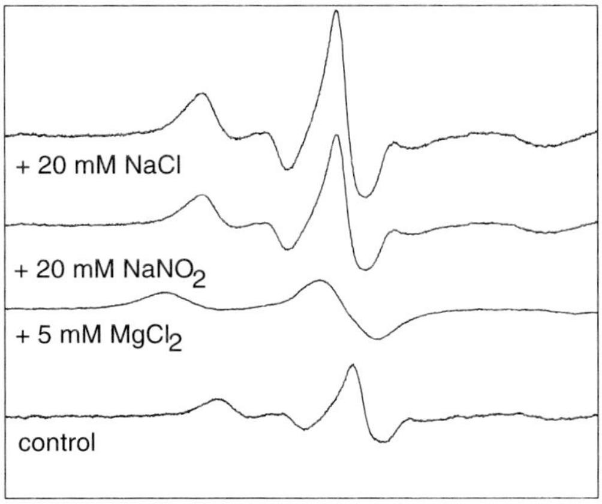

Fig. 11.8 ESR spectra showing effects of 20 mM nitrite on 5-SAL treated thylakoid membranes. Experimental conditions: microwave field: 3386 G, microwave power 2 mW, modulation amplitude 5G.

as evident from the highly anisotropic spectra. Addition of 20 mM of chloride and nitrite resulted in free motion of the probe (Fig. 11.8). This was further confirmed by calculating rotational correlation time (Tc) computed from the ESR spectra presented in Fig. 11.8. Tc is a measure of relative fluidity of the membranes. Addition of anions caused an increase in the value of Tc (Table 11.4), suggesting relatively free motion of the probe in the thylakoid membranes and indicating an increased fluidity of the thylakoid membranes. A comparatively low Tc value obtained in thylakoid membranes denotes a rigid lipid matrix. An order parameter (S) calculated from the ESR spectra is a measure of the mean angular deviation of the rotating hydrocarbon chain from its time-averaged orientation in the bilayer. Values close to 1 correspond to highly anisotropic motion with only a small angular deviation, while a value of zero would correspond to the other extreme of completely isotropic motion. S is used as a parameter describing the fluid nature of the membranes. The S values also showed differences in control and nitrite-treated thylakoid membranes. Nitrite caused decrease in S indicating that the amplitude of motion, which averages out anisotropy, increased. It is suggested that nitrite might be increasing the fluidity of the thylakoid membranes possibly by redistributing the surface charges.

Table 11.4. Rotational correlation time (Tc) and order parameter (S) as calculated in SAL-labeled thylakoid membranes. Means □ SE, $n = 9$. In all samples means differ significantly from control at *P* □ 0.01.

Treatment	*Rotational correlation time (Γc)*	*Order parameter (S)*
Control	4.77 ± 0.04	0.74 ± 0.02
5 mM $MgCl_2$	2.52 ± 0.02	0.86 ± 0.03
20 mM NaCl	5.92 ± 0.03	0.59 ± 0.02
20 mM $NaNO_2$	5.62 ± 0.02	0.62± 0.02

CONCLUSION

1. Nitrite enhanced PS I activity and inhibited PS II activity in low light intensities in thylakoid membranes. An increase in the F_{735}/F_{680} ratio at 77 K in the presence of nitrite suggested that nitrite transferred excitation energy more towards PS I by causing state transition. Quenching of Chl *a* fluorescence and an increase in the antenna size of PS I further confirmed that nitrite

diverts the absorbed excitation energy more towards PS I by the mechanism of state transition (Sahay *et al.,* 2006a).

2. Study of different types of nitrite-treated thylakoid membranes strongly suggest that the mode of binding of nitrite anion on the PS II is in two ways: dissociable (loose) and non-dissociable (tight) binding. The extent of the antenna migration and hence the energy transfer depends on the type of binding of nitrite to PS II complex. Antenna migration is less in response to the tight binding of nitrite and more when both loose and tight binding sites of nitrite are occupied.
3. An interaction between chloride and nitrite was observed where chloride seems to have more affinity and is able to replace nitrite from its binding sites quite effectively.
4. Nitrite anion caused a change in the structural organization of the thylakoid membranes by making the membranes more fluid.

REFERENCES

Allen JF (1992) Protein phosphorylation in regulation of photosynthesis. Biochim Biophys Acta 1098: 275-335

Allen JF, Bennett J, Steinbeck KE, Arntzen CJ (1981) Chloroplast protein phosphorylation couples plastoquinone redox state to distribution of excitation energy between photosystems. Nature 291: 25-29

Allen JF, Forsberg J (2001) Molecular recognition in thylakoid structure and function. Trends Plant Sci 6: 317-326

Bennett J (1983) Regulation of photosynthesis by reversible phosphorylation of the light-harvesting chlorophyll *a/b* protein. Biochem J 212: 1-13

Cardol P, Gloire G, Havaux M, Remacle C, Matagne R, Franck F (2003) Photosynthesis and state transitions in mitochondrial mutants of *Chlamydomonas reinhardtii* affected in respiration. Plant Physiol 133: 1-11

Debus RJ (1992) The manganese and calcium ions in photosynthetic oxygen evolution. Biochim Biophys Acta 1102: 269-352

Feyziev YM, Yoneda D, Yoshii T, Katsuta N, Kawamori A , Watanabe Y (2000) Formate – induced inhibition of the water-oxidizing complex of photosystem II studied by EPR. Biochemistry 39: 3848-3855

Garstka M, Drozak A, Rosiak M, Venema JH, Kierdaszuk B, Simeonova E, van Hasselt PR, Dobrucki J, Mostowska A (2005) Light-dependent reversal of dark-chilling induced changes in chloroplast structure and arrangement of chlorophyll-protein complexes in bean thylakoid membranes. Biochim Biophys Acta 1710: 13-23

Hasegawa K, Kimura Y, Ono T (2002) Chloride cofactor in the photosynthetic oxygen-evolving complex studied by Fourier transform infrared spectroscopy. Biochemistry 41: 13839-13850

Hind G, Nakatani HY, Izawa S (1969) The role of chloride ion in photosynthesis. I. The chloride requirement of electron transport. Biochim Biophys Acta 172: 277-289

Homann PH (1987) The relations between the chloride, calcium and polypeptide requirements of photosynthetic water oxidation. J Bioenerg Biomemb 19: 105-123

Homann PH (1988) Structural effects of chloride and other anions on the water oxidizing complex of chloroplast photosystem II. Plant Physiol 88: 194-199

Jajoo A and Bharti S (1993) On the participation of chloride in bicarbonate-induced reversal of anion inhibition of photosystem II electron transport in spinach thylakoids. Physiol Plant 88: 78-84

Jajoo A, Dube A, Bharti S (1994) Mg^{2+} induced lipid phase transition in thylakoid membranes is reversed by anions, Biochem Biophys Res Commun 202: 1724-1730

Jajoo A, Bharti S (1995) Possible role of anions in the regulation of energy distribution between the two photosystems. Ind J Exp Biol 33: 67-72

Jajoo A, Bharti S, Govindjee (1998) Inorganic anions induce state changes in spinach thylakoid membranes. FEBS Lett 434: 193-196

Jajoo A, Bharti S, Mohanty P (2001) Evaluation of the specific roles of anions in electron transport and energy transfer reactions in photosynthesis. Photosynthetica 39(3): 321-337

Jajoo A, Bharti S, Kawamori A (2005a) Interactions of chloride and formate at the donor and the acceptor side of photosystem II. J Bioenerg Biomemb 37(1): 49-54

Jajoo A, Bharti S, Kawamori A (2005b) EPR characteristics of chloride-depleted photosystem II membranes in the presence of other anions. Photochem Photobiol Sci 4: 459-462

Jajoo A, Kawamori A (2006) Anions affect structural organisation of thylakoid membranes. Biologia Plantarum 50(3): 444-446

Jursinic P, Stemler A (1988) Multiple anion effects on photosystem II in chloroplast membranes. Photosynth Res 15: 41-56

Kelly PM, Izawa S (1978) The role of chloride in the photosystem II. I. Effects of chloride ion on photosystem II electron transport and on hydroxylamine inhibition. Biochim Biophys Acta 502: 198-210

Makarova VV, Kukarskikh GP, Nizovskaya NV, Lavrukhina OG, Krendeleva TE (1998) Effect of nitrite on primary photosynthetic processes in isolated chloroplasts of wheat grown under various conditions of nitrogen supply. Biochemistry (Moscow) 63: 1414-1418

Porra RJ (1990) A simple method for extracting chlorophylls from the recalcitrant alga *Nannochloris atomus* without formation of spectroscopically different magnesium-rhodochlorins derivatives. Biochim Biophys Acta 1019: 137-141

Renger G (2001) Photosynthetic water oxidation to molecular oxygen: apparatus and mechanishm. Biochim Biophys Acta 1503: 210-228

Ruban AV (1991) The use of excitation fluorescence spectroscopy in the study of short term chloroplast membrane reorganization. Biochim Biophys Acta 1058: 411-415

Sahay A, Jajoo A, Bharti S (2006a) Nitrite regulates distribution of excitation energy between the two photosystems by causing state transition. Plant Physiol and Biochem 44(1): 7-12

Sahay A, Jajoo A, Bharti S (2006b) Thermoluminescence study of the effects of nitrite on the photosystem II in spinach thylakoids. Luminescence 21(3): 143-147

Samson G, Bruce D (1995) Complementary changes in absorption cross-sections of photosystems I and II due to phosphorylation and Mg^{2+} depletion in spinach thylakoids. Biochim Biophys Acta 1232: 21-26

Sinclair J (1987) Changes in spinach thylakoid activity due to nitrite ions. Photosynth Res 12: 255-263

Singh P, Jajoo A, Sahay A, Bharti S (2007) Relation between the mode of binding of nitrite and energy distribution between the two photosystems. Physiol Plant 29: 447-454

Spiller H, Boger P (1977) Photosynthetic nitrite reduction by dithio-erythritol and the effect of nitrite on electron transport in isolated chloroplasts. Photochem Photobiol 26: 397-402

Stemler A, Murphy J (1985) Bicarbonate reversible and irreversible inhibition of PS II by monovalent anions. Plant Physiol 77: 974-977

Wincencjusz H, Yocum CF and van Gorkom HJ (1999) Activating anions that replace Cl^- in the O_2^- evolving complex of photosystem II slow the kinetics of the terminal step in water oxidation and destabilize the S_2 and S_3 states. Biochemistry 38: 3719-3725

CHAPTER 12

Mitochondrial Oxidative Metabolism Optimizes Photosynthetic Carbon Assimilation Through Redox and ROS Linked Metabolite Shuttles

K. Padmasree and Ch. Dinakar

Department of Biotechnology, School of Life Sciences, University of Hyderabad, Hyderabad-500 046

ABSTRACT

The carbon and energy metabolism in mesophyll cells of higher plants is coordinated by the close interaction between chloroplasts and mitochondria. The special features of mitochondrial electron transport chain offer flexibility to the plant to adapt or acclimatize to changes in both carbon and energy metabolism that occur during photosynthesis. Though the occurrence of dark respiration in light was debated constantly, several experimental evidences were in favor of its operation. During illumination, as the rate of photochemical reactions is much greater than the rate at which Benson-Calvin cycle operates, chloroplasts tend to accumulate reducing equivalents. Mitochondria play a significant role in dissipating these excess reducing equivalents through different redox and reactive oxygen species (ROS) linked metabolite shuttles to optimize photosynthetic carbon assimilation.

Keywords: Metabolic shuttles, Reactive oxygen species, Oxidative electron transport.

Abbreviations

SOD: superoxide dismutase, *MAL:* malate, *OAA:* oxaloacetate, *COX:* cytochrome oxidase, *AOX*: alternative oxidase, *SHAM:* salicylhydroxamic acid, *Asc/DHA:* ascorbate/dehydroascorbate, *GSH/GSSG:* reduced glutathione/oxidized glutathione, *UQ*: ubiquinone, *PGA:* 3-phosphoglyceric acid or 3-phosphoglycerate.

INTRODUCTION

Photosynthesis and respiration are the most important metabolic processes associated with carbon and energy metabolism in higher plants (Fig. 12.1). Photosynthesis is a reductive process involving chloroplasts, and respiration is an oxidative process involving mitochondria. During photosynthesis, light energy is transformed through photochemical reactions into biochemically usable chemical potential energy (ATP) and redox potential energy (NADPH) by various electron transport chain (ETC) components of the thylakoid membrane in the chloroplasts. These ATP and NADPH generated during the photochemical reactions are used in the subsequent steps of carbon dioxide fixation and reduction in the Calvin-Benson cycle to generate carbohydrates.

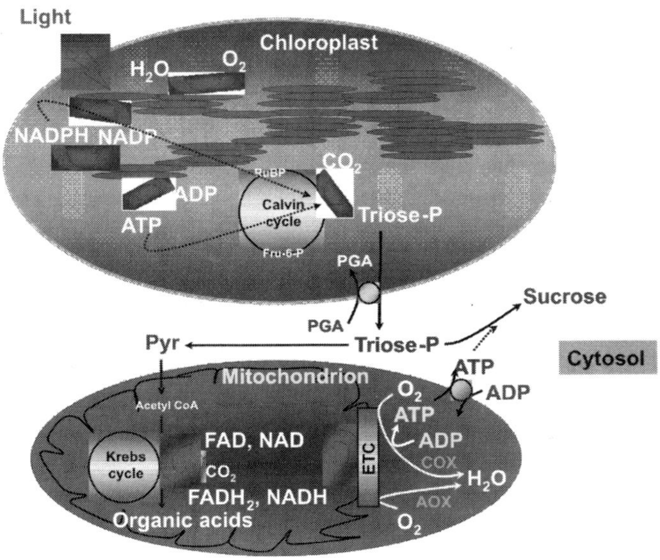

Fig. 12.1 Carbon and energy metabolism of chloroplasts and mitochondria in mesophyll cells of higher plants.

During respiration, the carbohydrates generated through Calvin-Benson cycle are oxidized through glycolysis and tricarboxylic acid (TCA) cycle. The CO_2 generated through decarboxylation reactions of the TCA cycle is released into the atmosphere or recycled to chloroplasts for utilisation in Calvin-Benson cycle. The carbon skeletons of the TCA cycle intermediates are used in various cellular biosynthetic processes, e.g., nitrogen metabolism. On the other hand, NADH generated through TCA cycle is further oxidised by ETC components of the mitochondria and ATP is generated. Though 40-50% of the net carbon gain through photosynthesis is ultimately lost from the plant through respiration, both these processes contribute significantly to the cellular demands of ATP and NAD(P)H.

SPECIAL FEATURES OF PLANT MITOCHONDRIAL ELECTRON TRANSPORT CHAIN

Plant mitochondria possess several non-energy conserving pathways of electron transport, both for reduction and oxidation of ubiquinone along with the standard five respiratory chain complexes, viz., the rotenone sensitive complex I (NADH dehydrogenase), complex II (succinate dehydrogenase), complex III (cytochrome bc1 reductase), the cyanide-sensitive complex IV (cytochrome *c* oxidase) and complex V (ATP synthase) and two mobile electron carriers, ubiquinone and cytochrome c, to mediate electron transport from NAD(P)H to molecular O_2. While the non-energy conserving pathways for the reduction of ubiquinone include external and internal NAD(P)H dehydrogenases, the non-energy conserving pathway for the oxidation of ubiquinone includes the alternative oxidase (AOX) pathway (Fig. 12.2).

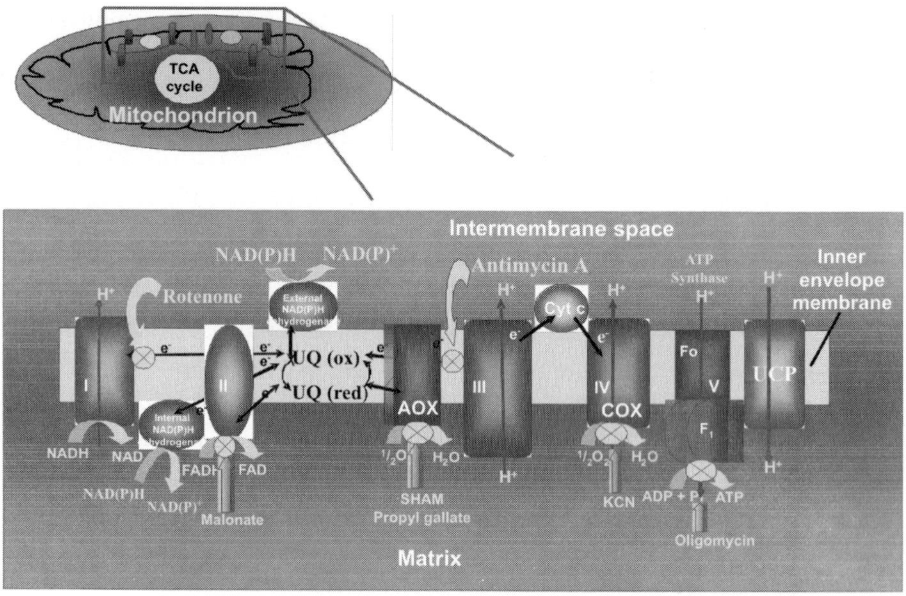

Fig. 12.2 The respiratory chain in higher plant mitochondria. Electron flow through COX pathway to molecular O_2 generates electromotive force for ATP synthesis, and the electron flow through AOX pathway generates heat. Abbreviations: I, complex I or NADH dehydrogenase; II, complex II or succinate dehydrogenase; III, complex III or cytochrome bcI reductase; IV, complex IV or cytochrome oxidase; UQ, ubiquinone; AOX, alternative oxidase pathway and UCP, uncoupling protein. Crosses inside circles indicate sites of inhibition of the miETC by different metabolic inhibitors.

The alternative NAD(P)H oxidizing enzymes are rotenone insensitive and will reduce the ATP-yield of respiration, since they are non-proton pumping and bypass the proton-pumping complex I (Rasmusson *et al.*, 2008). On the other hand, electron transport through AOX pathway is cyanide resistant and the energy generated through this pathway is simply released as heat. Also, the presence of an uncoupling protein (UCP), a homologue to thermogenin, in mammalian brown fat cells on the inner membrane further adds to the complexity of the plant miETC. UCP allow protons to diffuse down their concentration gradient from the intermembrane space into the matrix, circumventing the ATP synthase complex and thus uncoupling electron transport from ATP production. All these non-energy conserving bypass mechanisms provide flexibility to the standard miETC to adapt or acclimatize to the changes in the carbon and energy metabolism during photosynthesis or under stress conditions.

RESPIRATION DURING LIGHT

The nature and occurrence of dark respiration is debated in leaves under light. Different workers reported that respiration is either stimulated or unaffected or inhibited to varying extents in the presence of light (Graham, 1980; Raghavendra *et al.*, 1994; Krömer, 1995; Hoefnagel *et al.*, 1998; Gardeström *et al.*, 2002; Padmasree *et al.*, 2002; Raghavendra and Padmasree, 2003; Møller and Gardeström, 2007; Nunes-Nesi *et al.*, 2007). Radiogasometric method introduced by Parnik and Keerberg (2007) is helpful in measuring the rates of CO_2 fixation, photorespiration and respiration simultaneously during steady state photosynthesis. Using this method, the light inhibition of respiration was found to be larger in species that accumulate starch when compared to the species that do not accumulate starch.

The operation of the Krebs cycle in the light is known to be affected by a combination of atleast two factors: the reversible inactivation of the mitochondrial pyruvate dehydrogenase complex in the light (Tovar-Mendez *et al.,* 2003) and the rapid export of TCA cycle intermediates out of the mitochondria (Hanning and Heldt, 1993) for utilization in glutamate synthesis (Hodges, 2002). It is now clear that while some of the reactions of TCA cycle are inhibited, oxidative electron transport and phosphorylation continue to be active in light and benefit chloroplast metabolism (Padmasree *et al.,* 2002; Raghavendra and Padmasree, 2003). However, variability in mitochondrial O_2 consumption, indicative of mitochondrial oxidative metabolism, in light may depend on several factors, among which metabolic (substrate supply, redox status of the cell and cellular demands for ATP) and environmental (e.g., light intensity, temperature, water and nutrient availability) factors were identified to be important (Hoefnagel *et al.*, 1998; Nunes-Nesi *et al.*, 2007).

Light is also known to regulate the gene expression of several key respiratory enzymes, which include cytochrome oxidase, phosphoenolpyruvate carboxylase, NADP malic enzyme and isocitrate dehydrogenase. The light dependent expression of these mitochondrial genes suggest a close association between photosynthesis and mitochondrial metabolism (Svensson and Rasmusson, 2001; Nunes-Nesi *et al.* 2007; Rasmusson and Escobar, 2007). On the other hand, microarray analysis showed a clear trend of reduced expression of the genes associated with respiratory processes in the light (Urbanczyk-Wochniak *et al.*, 2006). However, this may not necessarily lead to downregulation in the metabolic flux through the pathway. The decreased rate of photosynthesis under photorespiratory conditions in the cytoplasmic male-sterile mutant (CMSII) of *Nicotiana sylvestris*, which exhibits a loss of function of mitochondrial complex I also suggests the close interaction between light and respiration (Dutilleul *et al.*, 2003).

MITOCHONDRIAL RESPIRATION PLAYS AN IMPORTANT ROLE IN OPTIMIZING PHOTOSYNTHETIC CARBON ASSIMILATION BY DISSIPATING EXCESS CHLOROPLASTIC REDUCING EQUIVALENTS

Redox reactions are the fundamental metabolic processes through which cells convert and distribute the energy that is necessary for growth and maintenance. Chloroplasts always have a tendency to get over reduced as the rate of photochemical reaction and utilization of reducing potential in metabolism differ by at least 15 orders of magnitude (Huner *et al.*, 1998). Further, various abiotic stress conditions, for example, sub-optimal CO_2/O_2 or excess light, accumulate redox equivalents in chloroplasts and lead to photoinhibition. Hence, it is essential that these excess reduced equivalents are dissipated quickly to prevent damage to the thylakoid membranes (Niyogi, 1999). Plants have different outlets to dissipate excess reducing equivalents from chloroplasts and minimize the damage caused by photoinhibition. The dissipatory mechanisms include thermal dissipation capacity, operation of xanthophyll cycle, water-water cycle, PS I cyclic electron transport, and rapid turnover of the D1 protein of PS II. Apart from this, mitochondrial oxidative metabolism and photorespiration also play an important role in preventing over-reduction of chloroplastic redox carriers (Raghavendra and Padmasree, 2003). Both redox and ROS linked metabolite shuttles play a significant role in dissipating chloroplastic reducing equivalents and mediating beneficial interactions of mitochondrial metabolism with photosynthetic carbon assimilation (Fig. 12.3).

Importance of redox-linked metabolite shuttles

The malate-OAA shuttle is an important system for reductant transport (Scheibe, 2004). Hatch *et al.* (1984) indicated that a highly active malate-OAA transporter exists in the chloroplasts. Similar to several Calvin cycle enzymes, NADP dependent malate dehydrogenase (NADP-MDH) in the chloroplast stroma is activated by the light-dependent Fd-thioredoxin (Fd-Trx) system (Miginiac-Maslow *et al.*, 2000). The activated NADP-MDH produces malate and $NADP^+$ from OAA and NADPH in chloroplasts. Malate can be exported to the mitochondria in exchange for OAA via dicarboxylate transporters. It has been demonstrated that the acceptor side of PS I was more oxidized in NADP-MDH over-expressing potato leaves than in wild type and NADP-MDH underexpressing ones (Backhausen *et al.*, 1998). In potato leaves under high CO_2 conditions, NADPH is more efficiently consumed by the CO_2 assimilation process, thereby decreasing the NADP-MDH activation state and the reduced fraction of the acceptor side of PSI (Backhausen and Scheibe, 1999). These results support the possibility that NADP-MDH plays a key role in recycling NADPH, especially when its production is in excess of the requirement for the photosynthetic CO_2 assimilation (Scheibe *et al.*, 2005).

The triose phosphate-phosphate transporter (TPT) exchanges triose-P (e.g., dihydroxyacetone phosphate, DHAP) with inorganic phosphate (Pi). In illuminated leaves, triose-P is exported from the chloroplasts to the cytosol and used for sucrose synthesis. TPT imports Pi into the chloroplast, and this is essential for optimal photophosphorylation in the chloroplasts of illuminated leaves. Alternatively, TPT can also exchange triose-P with 3-phosphoglyceric acid (3-PGA). In this case, excess reductants are exported from the chloroplasts in the form of triose-P (Gardeström *et al.*, 2002). The exported triose-P is oxidized by the cytosolic non-phosphorylating NADP-dependent glyceraldehyde 3-phosphate dehydrogenase (NADP-G3PDH), which produces 3-PGA and NADPH or phosphorylating NAD-dependent glyceraldehyde 3-phosphate dehydrogenase (NAD-G3PDH) and phosphoglycerate kinase (PGK), which produces 3-PGA, ATP and NADH (Gardeström *et al.*, 2002; Plaxton and Podesta, 2006). However, NADP-G3PDH has a higher affinity for triose-P than NAD-G3PDH. Although NADP-G3PDH is inhibited by high NAD(P)H concentrations, the cytosolic NAD(P)H pool is more oxidized under light than in dark (Igamberdiev and Gardeström, 2003). Thus any excess reductants in the chloroplasts can be exported to the cytosol via TPT. In leaves of an *Arabidopsis* TPT-lacking mutant, starch accumulated in the chloroplasts, and both photosynthesis as well as the growth of the mutant were decreased (Walters *et al.*, 2004).

Mitochondria play a significant role in maintaining optimal levels of reducing equivalents in the chloroplasts to keep up the Calvin-Benson cycle activity (Woodson and Chory, 2008) by coordinating with peroxisomes and cytosol. The reductants in excess of the requirements of the Calvin-Benson cycle are exported out of chloroplasts through the shuttling of malate-OAA by dicarboxylate translocator and PGA-DHAP by Pi-translocator (Scheibe *et al.*, 2005; Noguchi and Yoshida, 2008). The relative levels of triose-P/PGA and malate/OAA reflect the redox state of cytosol and the cell. Any limitation on the mitochondrial metabolism in the presence of antimycin A and SHAM lead to a marked decrease in bicarbonate dependent photosynthetic O_2 evolution (indicative of Calvin cycle activity) and rise in the redox state of cells, as indicated by the rise in the ratios of malate/OAA or triose-P/PGA. These results support the importance of mitochondrial oxidative electron transport through COX and AOX pathways in dissipating excess chloroplastic reducing equivalents and thereby optimizing chloroplastic photosynthesis (Padmasree and Raghavendra, 1999).

Importance of ROS-linked metabolite shuttles

Generation of ROS from redox reactions of chloroplasts and mitochondria has been identified as an inevitable process of aerobic metabolism (Apel and Hirt, 2004; Møller *et al.*, 2007). Four major types of ROS: singlet oxygen ($^1O_2$), superoxide (O_2^-), hydrogen peroxide (H_2O_2) and hydroxyl radicals (OH^-) are produced in mesophyll protoplasts during active photosynthesis. In chloroplasts, PSII and PSI are the major sites for the production of singlet oxygen and superoxide radicals (Apel and Hirt, 2004). In mitochondria, complex I, ubiquinone and complex III of electron transport chain are the major sites for the generation of superoxide radicals (Møller, 2001; Rhoads *et al.*, 2006; Navrot *et al.*, 2007). In

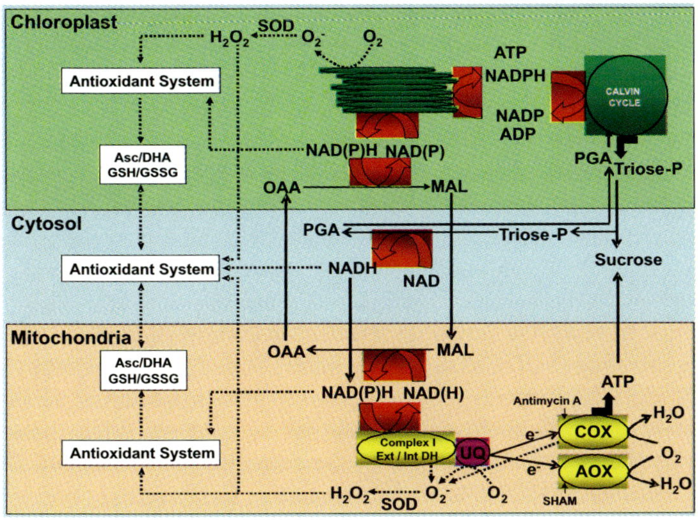

Fig. 12.3 Biochemical crosstalk between chloroplasts, mitochondria and cytosol mediated by metabolites. The excess reducing equivalents generated in chloroplasts are transported to mitochondria through malate-OAA shuttle and Triose-P-PGA shuttle Malate is either oxidized by complex I or Ext/Int DH while triose-P is utilized in sucrose synthesis. Mitochondria supply ATP for sucrose biosynthesis in cytosol. Any over-reduction in COX pathway or AOX pathway generates back pressure on chloroplastic redox carriers leading to generation of ROS in both chloroplasts and mitochondria. These ROS either diffuse directly from one cellular compartment to other or alternatively scavenged by the compartment-specific antioxidant system. The antioxidant system operated in different compartments may crosstalk through exchange of antioxidant metabolites [Asc/DHA] & [GSH/GSSG] related to Asc-GSH cycle. Solid lines indicate the operation of redox-linked metabolite shuttle and dashed line indicate the operation of ROS-linked metabolite shuttle.

both of these organelles, O_2^- radicals are immediately dismutated to H_2O_2, a less toxic form of ROS by superoxide dismutase (SOD). H_2O_2 is also formed as a by-product of photorespiration during the oxidation of glycolate to glyoxylate in peroxisomes (del Rio *et al.*, 2006). Both O_2^- and H_2O_2 participate in a fenton-type reaction with free Cu and Fe ions available in the cell and generate OH^- radicals.

Traditionally, ROS were considered to be toxic by-products of aerobic metabolism, which were disposed of using antioxidants (Asada, 1999; Apel and Hirt, 2004; Noctor *et al.*, 2007; Shao *et al.*, 2008). However, in recent years, it has become apparent that plants actively produce ROS as signaling molecules to control processes such as growth, cell cycle, programmed cell death, abiotic stress responses, pathogen defense and systemic signaling and development (Mittler *et al.*, 2004; Møller *et al.*, 2007). The intensity, duration and localization of different ROS signals are determined by the interplay between the ROS-production and ROS-scavenging pathways of the cell. AOX and uncoupling protein (UCP) of the mitochondrial electron transport were suggested to be proactive enzymes for ROS avoidance in plants. The suggested pathways for ROS scavenging along with catalase (CAT) were: (i) The water-water cycle, (ii) The ascorbate-glutathione (Asc-GSH) cycle and (iii) The glutathione peroxidase (GPX) cycle. In all these pathways, SOD acts as the first line of defense converting O_2^- into H_2O_2. H_2O_2 is subsequently detoxified by CAT, ascorbate peroxidase (APX) and GPX. In contrast to CAT, APX and GPX require an ascorbate (Asc) and/or a glutathione (GSH) regenerating cycle, which use electrons from NAD(P)H. The NAD(P)H-dependent pathways of ROS scavenging mechanisms act either by reducing ROS NAD(P)H directly or by protecting or regenerating the oxidized proteins (Navrot *et al.*, 2007). Using transmission electron microscopy, the localization of ROS scavenging pathways has been detected in all the following compartments of the plant cells: chloroplasts (grana and stroma), peroxisomes, mitochondria and cytosol (Mittler *et al.*, 2004).

The direct role of ROS in signal transduction is ensured only if ROS escape destruction by antioxidants or otherwise consumed in a ROS cascade. Thus, the major low molecular weight antioxidants, Asc and GSH determine the specificity of the signal. Along with ROS, Asc and GSH also act as signal-transducing molecules that can either signal independently or further transmit ROS signals. The specific interplay between ROS and the Asc-GSH cycle constituents generated compartment (chloroplastic, mitochondrial, cytosolic and peroxisomal) specific changes in both the absolute concentrations of ROS and antioxidant compounds, and thereby changes in redox ratio of Asc (Asc/DHA) and GSH (GSH/GSSG) during *Botrytis cinerea* infection in tomato leaves (Kuzniak and Sklodowska, 2005). Further, both AA and SHAM, metabolic inhibitors of cytochrome and alternative oxidase pathways caused marked reduction in photosynthetic carbon assimilation with concomitant rise in total cellular ROS and decrease in redox ratios of both Asc and GSH. Further, in coordination with ROS and redox ratios of Asc and GSH, ROS generating (SOD), ROS scavenging (CAT and APX) and antioxidant (Asc and GSH) regenerating (MDAR and GR) enzymes are also known to participate during the beneficial interactions of mitochondrial metabolism with photosynthetic carbon assimilation. While the activities of SOD and CAT were sensitive to changes in AOX pathway, the activities of MDAR and GR were sensitive to changes in the COX pathway. However, restriction of neither COX pathway nor AOX pathway affected the activity of APX (Dinakar *et al.*, 2010).

FUTURE PERSPECTIVES

Most of the work on the metabolites mediating the beneficial interactions between chloroplasts and mitochondria was done using mesophyll protoplasts. Further, it would be of interest to validate the concept by directly feeding the leaves with redox- and ROS-linked metabolites, followed by monitoring the changes in reducing equivalents (NAD[P]H) and different ROS (hydroxyl, singlet O_2, superoxide, hydrogen peroxide) levels.

REFERENCES

Apel K, Hirt H (2004) Reactive oxygen species: metabolism, oxidative stress, and signal transduction. Annu Rev Plant Biol 55: 373-399

Asada K (1999) The water-water cycle in chloroplasts: scavenging of active oxygens and dissipation of excess photons. Annu Rev Plant Physiol Plant Mol Biol 50: 601-639

Backhausen JE, Emmerlich A, Holtgrefe S, Horton P, Nast G, Rogers JJM, Muller-Rober B, Scheibe R (1998) Transgenic potato plants with altered expression levels of chloroplast NADP malate dehydrogenase: interactions between photosynthetic electron transport and malate metabolism in leaves and in isolated intact chloroplasts. Planta 207: 105-114

Backhausen JE, Scheibe R (1999) Adaptation of tobacco plants to elevated CO_2: influence of leaf age on changes in physiology, redox states and NADP-malate dehydrogenase activity. J Exp Bot 50: 665-675

del Rio LA, Sandalio LM, Corpas FJ, Palma JM, Barroso JB (2006) Reactive oxygen species and reactive nitrogen species in peroxisomes. Production, scavenging and role in cell signaling. Plant Physiol 141: 330-335

Dinakar C, Abhaypratap V, Yearla SR, Raghavendra AS, Padmasree K (2010) Importance of ROS and antioxidant system during the beneficial interactions of mitochondrial metabolism with photosynthetic carbon assimilation. Planta 231: 461-474

Dutilleul C, Driscoll S, Cornic G, De Paepe R, Foyer CH, Noctor G (2003) Functional mitochondrial complex I is required by tobacco leaves for optimal photosynthetic performance in photorespiratory conditions and during transients. Plant Physiol 131: 264-275

Gardeström P, Igamberdiev A, Raghavendra AS (2002) Mitochondrial functions in the light and significance to carbon-nitrogen interactions. In: Foyer C, Noctor G, eds, Photosynthetic Nitrogen Assimilation and Associated Carbon Metabolism, Vol 12. Kluwer Academic Publishers, Dordrecht, pp 151-172

Graham D (1980) Effects of light on dark respiration. In: Davies DD, ed, The Biochemistry of Plants: A Comprehensive Treatise. Academic Press, New York, pp 525-579

Hanning I, Heldt HW (1993) On the function of mitochondrial metabolism during photosynthesis in spinach (*Spinacia oleracea* L) leaves. Partitioning between respiration and export of redox equivalents and precursors for nitrate assimilation products. Plant Physiol 103: 1147-1154

Hatch MD, Droscher L, Flugge UI, Heldt HW (1984) A specific translocator for oxaloacetate transport in chloroplasts. FEBS Lett 178: 15-19

Hodges M (2002) Enzyme redundancy and the importance of 2-oxoglutarate in plant ammonium assimilation. J Exp Bot 53: 905-916

Hoefnagel MHN, Atkin OK, Wiskich JT (1998) Interdependence between chloroplasts and mitochondria in the light and the dark. Biochim Biophys Acta 1366: 235-255

Huner NPA, Oquist GM, Sarhan F (1998) Energy balance and acclimation to light and cold. Trends Plant Sci 3: 224-230

Igamberdiev AU, Gardeström P (2003) Regulation of NAD- and NADP- dependent isocitrate dehydrogenases by reduction levels of pyridine nucleotides in mitochondria and cytosol of pea leaves. Biochim Biophys Acta 1606: 117-125

Krömer S (1995) Respiration during photosynthesis. Annu Rev Plant Physiol Plant Mol Biol 46: 45-70

Kuzniak E, Sklodowska M (2005) Compartment-specific role of the ascorbate–glutathione cycle in the response of tomato leaf cells to *Botrytis cinerea* infection. J Exp Bot 56: 921-933

Miginiac-Maslow M, Johansson K, Ruelland E, Issakidis-Bourguet E, Schepens I, Goyer A, Lemaire-Chamley M, Jacquot J-P, Le Marechal P, Decottignies P (2000) Light-activation of NADP-malate dehydrogenase: A highly controlled process for an optimized function. Physiol Plant 110: 322-329

Mittler R, Vanderauwera S, Gollery M, Breusegem FV (2004) Reactive oxygen gene network of plants. Trends Plant Sci 9: 490-498

Møller IM (2001) Plant mitochondria and oxidative stress: Electron transport, NADPH turnover, and metabolism of reactive oxygen species. Annu Rev Plant Physiol Plant Mol Biol 52: 561-591

Møller IM, Gardeström P (2007) Plant mitochondria-more active than ever. Physiol Plant 129: 1-5

Møller IM, Jensen PE, Hansson A (2007) Oxidative modifications to cellular components in plants. Annu Rev Plant Biol 58: 459-481

Navrot N, Rouhier N, Gelhaye E, Jacquot JE (2007) Reactive oxygen species generation and antioxidant systems in plant mitochondria. Physiol Plant 129: 185-195

Niyogi KK (1999) Photoprotection revisited: genetic and molecular approaches. Annu Rev Plant Physiol Plant Mol Biol 50: 333-359

Noctor G, De Paepe, Foyer CH (2007) Mitochondrial redox biology and homeostasis in plants. Trends Plant Sci 12: 125-134

Noguchi K, Yoshida K (2008) Interaction between photosynthesis and respiration in illuminated leaves. Mitochondrion 8: 87-99

Nunes-Nesi A, Sweetlove LJ, Fernie AR (2007) Operation and function of tricarboxylic acid cycle in illuminated leaf. Physiol Plant 129: 45-56

Padmasree K, Padmavathi L, Raghavendra AS (2002) Essentiality of mitochondrial oxidative metabolism for photosynthesis: Optimization of carbon assimilation and protection against photoinbition. Crit Rev Biochem Mol Biol 37: 71-119

Padmasree K, Raghavendra AS (1999) Response of photosynthetic carbon assimilation in mesophyll protoplasts to restriction on mitochondrial oxidative metabolism: Metabolites related to the redox status and sucrose biosynthesis. Photosynth Res 62: 231-239

Parnik T, Keerberg O (2007) Advanced radiogasometric method for the determination of the rates of photorespiratory and respiratory decarboxylations of primary and stored photosynthates under steady-state photosynthesis. Physiol Plant 129: 34-44

Plaxton WC, Podesta FE (2006) The functional organization and control of plant respiration. Crit Rev Plant Sci 25: 159-198

Raghavendra AS, Padmasree K, Saradadevi K (1994) Interdependence of photosynthesis and respiration in plant cells: interactions between chloroplasts and mitochondria. Plant Sci 97: 1-14

Raghavendra AS, Padmasree K (2003) Beneficial interactions of mitochondrial metabolism with photosynthetic carbon assimilation. Trends Plant Sci 8: 546-553

Rasmusson AG, Escobar MA (2007) Light and diurnal regulation of plant respiratory gene expression. Physiol Plant 129: 57-67

Rasmusson AG, Geisler DA, Moller IM (2008) The multiplicity of dehydrogenases in the electron transport chain of plant mitochondria. Mitochondrion 8: 47-60

Rhoads DM, Umbach AL, Subbaiah CC, Siedow JN (2006) Mitochondrial reactive oxygen species. Contribution of oxidative stress and interorganellar signaling. Plant Physiol 141: 357-366

Scheibe R (2004) Malate valves to balance cellular energy supply. Physiol Plant 120: 21-26

Scheibe R, Backhausen JE, Emmerlich V, Holtgrefe S (2005) Strategies to maintain redox homeostasis during photosynthesis under changing conditions. J Exp Bot 56: 1481-1489

Shao HB, Chu LY, Shao MA, Cheruth AJ, Mi HM (2008) Higher plant antioxidants and redox signaling under environmental stresses. C R Biologies 331: 433-441

Svensson AS, Rasmusson AG (2001) Light-dependent gene expression for proteins in the respiratory chain of potato leaves. Plant J 28: 73-82

Tovar-Mendez A, Miernyk JA, Randall DD (2003) Regulation of pyruvate dehydrogenase complex activity in plant cells. Eur J Biochem 270: 1043-1049

Urbanczyk-Wochniak E, Usadel B, Thimm O, Nunes-Nesi A, Carrari F, Davy M, Blasing O, Kowalczyk M, Weicht D, Polinceusz A, Meyer S, Stitt M, Fernie AR (2006) Conversion of MapMan to allow the analysis of transcript data from *Solanaceous* species: effects of genetic and environmental alterations in energy metabolism in the leaf. Plant Mol Biol 60: 773-792

Walters RG, Ibrahim DG, Horton P, Kruger NJ (2004) A mutant of Arabidopsis lacking the triose-phosphate/phosphate translocator reveals metabolic regulation of starch breakdown in the light. Plant Physiol 135: 891-906

Woodson JD, Chory J (2008) Coordination of gene expression between organellar and nuclear genomes. Nature Rev/Gen 9: 383-395

CHAPTER 13

Leaf Senescence and Abiotic Stresses Share Chloroplast Degradation, ROS Involvement and Transcription Factors

Renu Khanna-Chopra and S. Srivalli

Stress Physiology and Biochemistry Laboratory, Water Technology Centre, Indian Agricultural Research Institute (IARI), New Delhi-110012, India

ABSTRACT

Leaf senescence is a genetically programmed decline in various cellular processes including photosynthesis and involves hydrolysis of macromolecules such as proteins, lipids, etc. It is governed by developmental age and is induced or enhanced by environmental stresses such as drought, heat, salinity and others. Internal factors such as reproductive structures also influence the rate of leaf senescence. ROS generation is one of the earliest responses of plant cells under abiotic stresses and senescence. Chloroplasts are the main targets of ROS linked damage during various environmental stresses and natural senescence as ROS detoxification systems decline with age. Plants adapt to environmental stresses through the process of acclimation, which involves less ROS production coupled with an efficient antioxidant defence. Chloroplasts are a major site of protein degradation and Rubisco is rapidly and selectively degraded during senescence and stress. The process of protein degradation is initiated by ROS and involves the action of proteolytic enzymes such as cysteine and serine proteases. The mechanism of Rubisco degradation still remains to be elucidated. The molecular understanding of the leaf senescence was achieved through the characterization of senescence associated genes (SAGS) and various senescence mutants of *Arabidopsis*, which is a suitable model plant showing monocarpic senescence. The regulation of senescence involves many regulatory elements composed of positive and negative elements to fine tune the initiation and progression of senescence. The regulatory factors include transcription factors, receptors and signaling components for hormones and stress responses and regulators of metabolism. Considerable crosstalk is observed in the regulatory network involving hormones, ROS and transcription factors, both in natural and stress related senescence and abiotic stress responses. This review gives an overview on chloroplast protein degradation during leaf senescence and abiotic stresses and also highlights the role of ROS management and transcription factors in both the processes.

Keywords : Senescence, Abiotic stresses, Transcription factors, Detoxification, Proteolysis.

Abbreviations

PS II: Photosystem II, *MeJA:* Methyl Jasmonate, *PQ:* Plastoquinone *ABA:* Abscisic Acid, *JA:* Jasmonic Acid

LHC II: Light Harvesting Complex, *NO*: Nitric Oxide, *APX:* Ascorbate Peroxidase, *GR:* Glutathione Reductase.

INTRODUCTION

Environmental stresses such as drought, extremes of temperature, salinity etc. cause considerable damage to crop productivity. Such stresses, in turn can also lead to senescence, a genetically programmed decline in various cellular processes involving loss in photosynthetic activities and hydrolysis of macromolecules such as proteins, lipids etc. The oxidative burst, during which large quantities of reactive oxygen species (ROS), like superoxide, hydrogen peroxide, hydroxyl radicals, and singlet oxygen are generated, is one of the earliest responses of plant cells under various abiotic and biotic stresses and natural course of senescence (Bhattacharjee, 2005). Thus, ROS are a central component in stress and senescence process.

Chloroplasts with their strong photooxidative potential are the most likely organelles to be affected by a significant production of ROS. Thylakoids univalently photoreduce oxygen to $O_2^{-\cdot}$ due to the presence of a high concentration of oxygen and an electron transport chain (Navari-Izzo *et al.,* 1999). Consequently, they are the main targets of ROS linked damage during various environmental stresses. Chloroplasts are also considered to be the target organelles of age-associated oxidative stress in plants (Munné-Bosch and Alegre, 2002). As such specific ROS detoxification systems located in both the thylakoid and stroma have been evolved (Noctor *et al.,* 2000). ROS are scavenged in the chloroplasts mainly by the enzymes of the ascorbate-glutathione pathway in conjunction with the reduced non-enzymic ascorbate and glutathione molecules. The enzymes of this pathway include ascorbate peroxidase (APX, EC 1.11.1.11), which utilizes reduced ascorbate (AsA) to reduce H_2O_2 to water; monodehydroascorbate reductase (MDHAR, EC 1.1.5.4), which catalyses the conversion of monodehydroascorbate (MDHA), an intermediary radical, to AsA, dehydroascorbate reductase (DHAR, EC 1.8.5.1), which catalyses the conversion of the oxidized form of ascorbate, dehydroascorbate (DHA) to AsA, using reduced glutathione (GSH) as the substrate. Glutathione, in turn, is maintained in the reduced state by glutathione reductase (GR, EC 1.6.4.2).

Plant species vary in their response and sensitivity to stresses depending on the location, growth stage and the intensity of the stress. Plants can, for instance, adjust to the changes in the environment through the process of acclimation. Acclimation is a physiological change, which occurs within a single life cycle in many crop species, i.e., exposure to a mild form of stress makes the plant less sensitive to a second cycle of stress of a more severe nature. Acclimation responses have been studied for various types of stresses such as cold, anoxia, light etc. Of these, responses during cold acclimation are well documented (Fryer *et al.,* 1998; Prasad *et al.,* 1994; Wise, 1995). Information on oxidative stress management in relation to drought acclimation has been generated in wheat (Selote and Khanna-Chopra, 2006; Khanna-Chopra and Selote, 2007) and rice (Srivalli *et al.,* 2003).

Leaf senescence is basically governed by the developmental age, but is influenced by both internal and environmental signals. The internal factors include reproductive structures and hormones, while environmental factors include biotic and abiotic stresses, which generally enhance senescence. The molecular understanding of the leaf senescence was achieved through the characterization of senescence associated genes (SAGS) and various senescence mutants of *Arabidopsis*, which is a suitable model plant showing monocarpic senescence. The regulation of senescence involves many regulatory elements composed of positive and negative elements to fine tune the initiation and progression of senescence. The regulatory factors include transcription factors, receptors and signaling components for hormones and stress responses and regulators of metabolism. Recent technological advances such as microarray analysis using ESTs and genes have allowed investigation of the SAGs during leaf senescence at the genome wide scale in *Arabidopsis* and other plants (Lim *et al.,* 2007). These studies have been carried out using transcription factors also at different developmental stages and under biotic and abiotic stresses, providing a clue to the transcriptional regulatory network during leaf senescence and stresses (Chen *et al.,* 2002). Considerable crosstalk is observed in the regulatory network involving hormones, ROS and transcription factors, both in natural and stress related senescence and abiotic stress responses. This review gives an overview on chloroplast protein degradation during leaf senescence and abiotic stresses and also highlights the role of ROS management and transcription factors in both the processes.

DEGRADATION OF CHLOROPLAST PROTEINS DURING SENESCENCE AND ABIOTIC STRESSES

In mesophyll cells, chloroplasts are dismantled early during senescence and stromal enzymes are degraded leading to the decline in photosynthetic activity, whereas mitochondria remain functional (Hörtensteiner and Feller, 2002). Enzymes involved in carbon and nitrogen assimilation are lost and the amino acids derived from their catabolism may be exported via the phloem with or without prior modification (e.g., production of amides from other amino acids). Glutamine synthetase, the key enzyme in the assimilation of ammonia, is present in the chloroplasts and cytosol. The plastidial form is lost in cereal leaves in an early phase of senescence, while the cytosolic form remains active longer. Glutamine synthetase is rather susceptible to proteolysis in isolated chloroplasts and it is degraded far more rapidly than Rubisco or other enzymes involved in carbon assimilation (Thoenen and Feller, 1998). Hence, a shift in ammonia assimilation from the chloroplast to the cytosol of mesophyll cells of tobacco during senescence was suggested (Brugie`re *et al.*, 2000).

Chloroplasts are a major site of protein degradation during senescence and stress. Ribulose-1, 5-bisphosphate carboxylase/oxygenase (Rubisco) is the most abundant protein on earth and contributes a high percentage to the total leaf nitrogen in plants (Feller *et al.*, 2008a, b). Rubisco is rapidly and selectively degraded during senescence and stress (Moreno *et al.*, 1995). The nitrogen demand of the growing sink is fulfilled mainly by the degradation of Rubisco, which is also the prime enzyme involved in CO_2 fixation, both during senescence and stress (Demirevska-Keopva *et al.*, 2005; Thoenen *et al.*, 2007). This degradation is mediated by the action of proteases, and the acidic endoproteases have been found to be responsible for the loss in flag leaf nitrogen in hexaploid wheat (Dalling *et al.*, 1976). Rubisco degradation can occur under conditions of both stress and senescence (Khanna-Chopra *et al.*, 1999). However, under abiotic stresses, inducing a rapid net degradation of Rubisco fragments of the large subunit may become detectable. Rubisco degradation product, 52 kDa was visible at later stages of leaf senescence under heat stress in wheat leaves (Chauhan *et al.*, 2009).

It has been shown earlier that degradation of Rubisco along with several stromal proteins may be initiated by ROS (Roulin and Feller, 1998). Protein oxidation is widespread and is often used as a diagnostic marker for oxidative stress. Oxidative damage to proteins may be a prerequisite to protein loss, a characteristic feature of stress and senescence as ROS levels are known to increase and proteins are 'marked' before their degradation. Within the chloroplast, oxidation of critical cysteine residues enhances the binding of the Rubisco protein to the chloroplast envelope membranes, marking the protein for degradation (Marín-Navarro and Moreno, 2006).

Cysteine proteases are involved in a variety of proteolytic functions in higher plants. Increase in cysteine proteases activities has been shown in senescing leaves of cowpea (Srivalli *et al.*, 2001), wheat (Martínez *et al.*, 2007) and *Arabidopsis* and soybean (Otegui *et al.*, 2005). They are also induced in response to stresses such as cold and drought (Schaffer and Fischer, 1988; Srivalli and Khanna-Chopra, 2006). The relatively acidic pH optima of many of the endogenous plant cysteine proteases indicate that they are localized in the vacuole (Srivalli *et al.*, 2001). In wheat, five bands of cysteine proteases were detected in the senescing flag leaf and four of these were found to be localized in the vacuoles (Martínez *et al.*, 2007). Papain-like cysteine proteinases are often found in senescing organs, particularly leaves (Lohman *et al.*, 1994; Khanna-Chopra *et al.*, 1999). In general, vacuolar cysteine proteases are the predominant proteases in senescing leaves, and as the vacuole plays an important role in the defense against pathogens, this seems to be its prime role than remobilization at least in the initial stages (Müntz, 2007). In fact, plastoglobuli containing photosynthetic components have been observed emerging from the chloroplast envelopes of senescing soybean leaves and moving towards the vacuole where they are degraded (Guiamét *et al.,* 1999). In addition, transfer of catabolites from senescent chloroplasts to the vacuole is mediated by ATP-dependent transport systems (Hörtensteiner, 2006). However, key questions have remained regarding the extent to which Rubisco is degraded outside the chloroplast and how Rubisco degradation is controlled (Feller *et al.,* 2008 b).

Prins *et al.* (2008) have presented evidence that chloroplast proteins, particularly Rubisco and Rubisco activase, are the major targets of leaf cysteine proteases. The degradation of Rubisco involves interactions with the cytosol, as it can be blocked by endogenous cytosolic cysteine protease inhibitors such as rice cystatin, OC-1. The demonstration of Rubisco vesicular bodies (RVB), which are vacuole type compartments outside the chloroplast, suggested a link between the protein turnover machinery in the chloroplasts, cytosol and vacuoles. They have formulated a new model for Rubisco degradation (Fig. 13.1). According to the model, the binding of oxidized Rubisco protein to the chloroplast envelope might facilitate the accumulation of the proteins at the sites where vesicles (RVB) form. RVBs might be formed as a part of the autophagositic process and engulf cytosolic as well as chloroplastic proteins. Vesicles containing cysteine proteases in the cytosol continuously interact with those from functional chloroplasts to remove proteins marked for degradation. Thereafter Rubisco degradation products appear in the vacuole. The occurrence of RVBs has been observed even in young leaves, thereby indicating the involvement of the above process in Rubisco protein turnover. Transgenic tobacco overexpressing cysteine proteinase inhibitor cystatin exhibited slower senescence and delay in decline of photosynthesis and Rubisco activity (Prins *et al.,* 2008). The plants had extended lifespan, accumulated more leaf area and biomass before flowering and could avoid stress induced premature senescence. This study reemphasizes the importance of cysteine proteases both in natural and stress induced senescence.

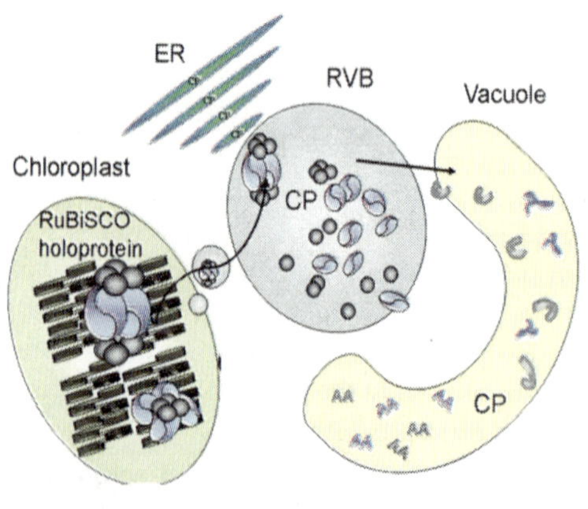

Fig. 13.1 A hypothetical model for Rubisco degradation via autophagy and the plant vesicle trafficking system. AA, amino acids; CP, cysteine proteinase; ER, endoplasmic reticulum; OC-1, oryzacystatin-1; RVB, Rubisco vesicular body (Prins *et al.* 2008).

The degradation of chlorophyll to colourless tetrapyrroles during senescence is an effort by the plants to remobilize 20% of cellular nitrogen fixed in chlorophyll apoproteins, as chlorophyll breakdown is tightly linked to the breakdown of chlorophyll-binding proteins (Hörtensteiner, 2006). Chlorophyll breakdown during senescence is achieved with the help of a novel plastid localized enzyme, pheophytinase, rather than chlorophyllase as was assumed earlier (Schelbert *et al.,* 2009). Therefore, chlorophyll synthesis and breakdown are metabolically separated during leaf senescence. The catabolism of chlorophyll is accompanied by the loss of green colour to nonfluorescent catabolites. It occurs within the vacuoles and is catalyzed non-enzymatically by the acidic vacuolar pH (Oberhuber *et al.,* 2003). The expression of the genes involved in chlorophyll degradation is under the control of hormones such as ethylene and methyl jasmonic acid, which accelerate senescence. Proteases involved in the degradation of chlorophyll-binding proteins are largely unknown, except the chloroplast-localized DegP and FtsH involved in the degradation of D1 protein of the PS II reaction center (Adam and Clarke 2002). In stay green genotypes, the degradation of soluble proteins is close to normal, however, light harvesting and reaction center thylakoid membrane proteins are more stable (Thomas *et al.,* 2002).

REGULATION OF SENESCENCE BY ROS AND TRANSCRIPTION FACTORS

Senescence is characterized by a wide array of anabolic and catabolic processes that are driven primarily by endogenous factors but are strongly influenced by environmental conditions. External stimulus such as infection, drought, cold, UV radiation etc. and developmental changes can lead to accelerated production of ROS and the subsequent onset of cell death (Ahmed *et al.,* 2009). ROS are important signaling intermediates that regulate cell death in plants. Recent genetic evidence suggests that ROS do not trigger PCD or senescence by causing physicochemical damage to the cell, but rather these metabolites act as signals that activate genetically programmed pathways of gene expression, which leads to regulated cell suicide events (Foyer and Noctor, 2005). ROS can act synergistically in concert with nitric oxide (NO) to trigger rapid cell death in soybean cells (Delledonne *et al.,* 1998). During senescence, however, NO appears to work as an antioxidant and slows degeneration. Nitric oxide synthase 1 (NOS 1), which produces NO in mitochondria, reduces ROS accumulation and the resulting oxidative damage in entire leaves during dark induced senescence in *Arabidopsis* (Guo and Crowford, 2005).

Dark-induced senescence in the leaves of *Pelargonium* was manifested in chlorophyll breakdown, and an increase in reactive oxygen species (ROS) levels followed by a subsequent induction of two senescence-associated gene (SAG) transcripts: senescence-related transcription factor PeWRKY6-1 and cysteine protease homolog PeSAG12-1(Rosenvasser *et al.,* 2006). The expression levels of other SAGs in *Arabidopsis* have been found to increase because of ROS induction (Navabpour *et al.,* 2003). ROS generated lipid peroxidation is an inherent feature of senescing cells and a source of ROS, especially alkoxy, peroxy radicals and singlet oxygen, which are highly toxic. The increase in leaf H_2O_2 levels occurs in parallel with increase in lipid peroxidation and protein oxidation in senescent leaves (Vanacker *et al.,* 2006). Peroxidation of lipid during plant cell senescence can be triggered either by ROS or lipoxygenase, as it has been shown by some tissues where lipoxygenase activity increases with advancing senescence (Dhindsa *et al.,* 1981).

Our current understanding of the relationship between environmental responses and leaf senescence mostly comes from the study of senescence response to the phytohormones such as abscisic acid (ABA), jasmonic acid (JA), ethylene, and salicylic acid (SA) that are extensively involved in response to various abiotic and biotic stresses. A recent report argued that ABA induces accumulation of H_2O_2 in senescing rice leaf, which in turn accelerates leaf senescence (Hung and Kao, 2004). ABA also induces the expression of antioxidant genes and enhances the activities of antioxidative enzymes such as SOD, APX and CAT, thus protecting the cellular functions required for progression and completion of senescence (Hung and Kao, 2003). It appears that ABA controls both cellular protection activities and senescence activities. The balance between these two activities seems to be important in controlling the progression of leaf senescence and may be adjusted by other senescence-affecting factors such as age.

Methyl jasmonate (MeJA) and its precursor, JA, promote senescence in detached oat and *Arabidopsis* leaves. Exogenously applied MeJA to detached *Arabidopsis* leaves leads to a rapid loss of chlorophyll content and photochemical efficiency of photosystem II (PS II) and increased expression of *SAG*s such as *SEN4, SEN5,* and γ *VPE.* JA-dependent senescence is defective in the JA-insensitive mutant *coronatine insensitive 1(coi1)*, implying that the JA signaling pathway is required for JA to promote leaf senescence (He *et al.,* 2002). Functional studies on a nuclear localized CCCH-type zinc finger protein, OsDOS (*Oryza sativa:* Delay of the Onset of Senescence), also support the involvement of MeJA in leaf senescence. The expression of the *OsDOS* gene was down-regulated during leaf senescence. It appears that OsDOS acts as a negative regulator of leaf senescence in integrating the JA signaling pathway into age-dependent senescence (Kong *et al.,* 2006).

Ethylene, another stress hormone, is also found to interact with ROS in their signaling network. O_3-induced superoxide radical accumulation and cell death have been demonstrated to be substantially reduced in ethylene-insensitive *Arabidopsis* mutant ein-2, whereas ethylene-sensitive *Arabidopsis* mutant eto-1 is hypersensitive to ozone (Overmyer *et al.,* 2000). Plants that were exposed to stresses such as ozone or UV-B irradiation, each resulting in increased levels of ROS, responded by showing symptoms of premature senescence followed by necrotic cell death (Pell *et al.,* 1997). SA is the hormone involved in pathogen response and pathogen-mediated cell death. Recent studies have shown the role of SA in age-dependent leaf senescence. The concentration of endogenous SA is four times higher in senescing leaves of *Arabidopsis*. The higher SA level in senescing leaves appears to be involved in upregulation of several *SAG*s during leaf senescence. Transcriptome analysis showed that the change of transcriptome mediated by the SA pathway was highly similar to that mediated by age-dependent senescence (Lim *et al.,* 2007).

NAC domain transcription factors represent a large fraction of the senescence-regulated genes in many plant species including crops, suggesting that they play an important role in the control of senescence. In *Arabidopsis,* a regulatory role in senescence has been demonstrated for ANACO29 and ANACO92 which is regulated by micro-RNA and involves ethylene in age related senescence (Kim *et al.,* 2009). In wheat the senescence-controlling NAC transcription factor NAM-B1 was identified by positional cloning (Uauy *et al.,* 2006), which was shown to improve protein, zinc and iron content in the grain. WRKY transcription factors perform central functions in the regulation of developmental processes and senescence. Expression profiling in *Arabidopsis* revealed that WRKY transcription factors are the second largest family of transcription factors in the senescence transcriptome. WRKY genes that regulate senescence include AtWRKY6, AtWRKY53, AtWRKY70 and OsWRKY23. WRKY53 is known to function in early stages of senescence (Miao and Zentgraf, 2007). WRKY53 can be induced by H_2O_2, and epigenetic programming has also been implicated in the mechanism whereby AtWRKY53 regulates senescence. AtWRKY6 is strongly up-regulated during senescence, and analysis of AtWRKY6 target genes identified the SENESCENCEINDUCEDRECEPTOR KINASE/FLG22-INDUCED RECEPTOR-LIKE KINASE (SIRK/FRK1). SIRK/FRK1encodes a receptor-like protein kinase whose expression is strongly and specifically induced during leaf senescence. AtWRKY70 acts as a negative regulator of senescence, while OsWRKY23 acts as a positive regulator of senescence. Thus, members of the WRKY transcription factor family, both positively and negatively, regulate this process (Rouston *et al.,* 2010).

Chloroplastic antioxidant defense is highly important during oxidative stress induced senescence. Triple null mutants of *Arabidopsis* in stromal APX (sAPX), thylakoidal APX (tAPX) and low cellular ascorbate levels showed bleaching and necrosis accompanied by accumulation of H_2O_2 during light stress (Giacomelli *et al.,* 2007). The ubiquitin-dependent protein degradation pathway, which selectively targets a diverse range of substrates for degradation by the 26S proteosome, is important in the degradation of leaf proteins during senescence (Stone *et al.,* 2005).The activity of this pathway is influenced by cellular reductants such as glutathione (Theriault *et al.,* 2000). High GSH/GSSG ratios can prevent the binding of target proteins, and regulation of the reduction state of protein thiols might participate in the regulation of this reaction cascade (El Yahyaoui *et al.*, 2004). The mitogen-activated protein kinase (MAPK) cascades play diverse roles in intracellular and extracellular signaling in plants. Steroid-induced expression of constitutively active MAPK KinaseE3 (MKK3) or MKK4 in transgenic *Arabidopsis* plants results in the accumulation of H_2O_2 and also the stress- and senescence-associated plant hormone ethylene (Takahashi *et al.*, 2007).

Loss of photosynthetic competence during senescence is attributable, at least in part, to impairment of photosynthetic electron transport. Plants can sense and translate the ROS signal into cellular responses by redox-sensitive proteins (RSPs), which can undergo reversible oxidation / reduction and may switch 'on' or 'off' depending on the cellular redox state. ROS can oxidize the RSPs directly or indirectly via glutathione or thioredoxin (redox sensitive molecules). The best-studied example of involvement of RSP in redox signaling is the cytochrome-bf complex of photosynthetic electron transport chain located in chloroplast of higher plants where when PS II receives excess light, plastoquinone (PQ) becomes predominantly reduced, resulting in the activation of LHC II kinase via structural changes around Riske centre (2Fe–2S) protein of cytochrome bf (Cyt bf) complex (Vener *et al.*, 1998). The kinase in turn phosphorylates LHC II and PS II, resulting in migration of LHC II away from PS II and further reducing ligand absorption by PS II. Oxidation of PQ reverses structural changes of Cytbf and leads to kinase inactivation. Phosphatase-mediated dephosphorylation of mobile LHC II leads to reassociation of PS II and further increase in absorbation of light energy by PS II. Alternately, redox state of PQ controls adjustment of stoichiometry of PS I and PS II by transcriptional regulation of chloroplastic genes that encode apoprotein of PS I (psA and psB proteins) and PS II (D1 proteins) reaction centers (Karuse, 1994; Pfannschmidt *et al.*, 1999). ROS produced under excess light may cause physical separation of PS II from LHC II by degrading D1 proteins, thus reducing light energy absorption. ROS are central regulators of senescence in plants. The literature reveals a strong correlation between ROS and phytohormone signaling during abiotic stresses and development of senescence (Love *et al.*, 2008). However, there could be additional levels of cell death control that extend beyond those of hormones and ROS.

OXIDATIVE STRESS AND ITS MANAGEMENT DURING SENESCENCE AND ABIOTIC STRESSES

Senescence is the final stage in a plant's life ultimately leading to death. Senescence, as in the case of other PCD processes, is also associated with an increased production of ROS such as hydrogen peroxide, superoxide and its more toxic derivative hydroxyl radical (Breusegem and Dat, 2006). These toxic ROS oxidize proteins, lipids and DNA resulting in lipid peroxidation, cellular damage and cell death as the antioxidant status of the leaf is also reduced (Kukavica and Veljovic-Jovanovic, 2004). The increase in oxidative stress occurs to increase in plant age especially in chloroplasts (Munne-Bosch and Alegre, 2002). The biosynthesis and content of ascorbate decreases (Queval *et al.*, 2007; Srivalli and Khanna-Chopra, 2009) and senescence is often accompanied by a decrease in the activities of some antioxidant enzymes and pyridine nucleotides (Jimenez *et al.*,1998, Orendi *et al.*, 2001, Srivalli *et al.*, 2001; Srivalli and Khanna-Chopra, 2009; Queval *et al.*, 2007). Development dependent and stress-dependent changes in the expression of APX and CAT genes have been described as well as altered sensitivities of APX1 to H_2O_2 (Zimmermann and Zentgraf, 2005; Zimmermann *et al.*, 2006). The reduction in antioxidant capacity might be important in the activation of proteases such as Cys proteases, which tend to be inhibited by reductants such as GSH (Groten *et al.*, 2006).

Studies on mutants provide direct evidence that redox factors and antioxidants play important roles in plant senescence. Novel roles of NAD in signaling and regulation of cell longevity are being elucidated in plants (Noctor *et al.*, 2006; Ying, 2008). *Arabidopsis* mutant (old5) showing disruption in NAD synthesis show early leaf senescence (Jing *et al.*, 2005). The timing of senescence is also altered in the ascorbate-deficient *Arabidopsis* mutant vtc1 (Barth *et al.*, 2004), but this may be due to the enhanced susceptibility of individual ascorbate-deficient cells to PCD (Pavet *et al.*, 2005).Transgenic inhibition of the protein degradation machinery by constitutive cystatin expression in tobacco favors delayed senescence and alters lifespan (Prins *et al.*, 2008). Similarly, *Arabidopsis* mutants such as sag101, which is defective in an acyl hydrolase and ore9 that lacks a functional F-box protein that is part of the ubiquitin-dependent proteasome system shows delayed senescence and extended longevity (Woo *et al.*, 2002). The delayed leaf senescence mutants of *Arabidopsis thaliana*, *ore*1, *ore*3 and *ore*9 were found to be more tolerant of oxidative stress and hence provided genetic evidence that oxidative stress tolerance is linked to control of leaf longevity in plants (Woo *et al.*, 2004).

Abiotic stresses, such as drought, can increase the production of reactive oxygen species (ROS) in plants. An increase in ROS levels can provoke a partial or severe oxidation of cellular components inducing redox status changes, so continuous control of ROS and therefore of their metabolism is decisive under stress conditions.While under non-stress conditions, most cellular compartments maintain a reducing environment, stress conditions generally lead to an increase in ROS and induce oxidative stress, which is also defined as a disruption of redox signaling and redox control. Since, a majority of abiotic stress studies are performed under controlled conditions in the laboratory, they do not reflect the actual conditions that occur in the field. Plant acclimation to a particular abiotic stress condition requires a specific response that is tailored to the precise environmental conditions the plant encounters (Mittler, 2006). For instance, ROS are associated with different abiotic stress conditions, however, different genes of the ROS network of *Arabidopsis* were found to respond differently to different stress treatments (Mittler *et al.*, 2004). An experimental set up where the plant faces multiple cycles of stress intensities or combinations of different stresses would probably better simulate the existing conditions in the field. Such an experiment was conducted in wheat and rice where the plants were subjected to multiple cycles of increasing water deficit through withholding of water (Srivalli *et al.*, 2003, Selote and Khanna-Chopra, 2006). Each cycle was interrupted by rewatering the plants in order to simulate the naturally occurring water deficiency and sufficiency cycles occurring under field conditions. The differential response of the antioxidant enzymes with increased stress intensity followed by recovery highlighted the different role of each in the drought acclimation process of upland rice (Srivalli *et al.* 2003). Co-ordinated defense helped the plants to recover in terms of growth on rewatering after stress cycles. In this regard, the increased activities of some of the existing isoforms of SOD, APX, CAT and POX and the induction of new isoform of CAT, which was retained during recovery in third stress cycle, underlines the importance of these enzymes in the drought acclimation process (Figs. 13.2 and 13.3). However, chloroplasts were susceptible to ROS linked damage during drought stress, as the enzymes of the Halliwell-Asada pathway were lower in severe stress. In addition, it was observed that stromal APX activity was reduced significantly in the second stress cycle, while all thylakoidal APX isoforms showed considerable activity during stress. Hence, stromal APX was sensitive to stress, thus contributing to the susceptibility of chloroplast to drought stress induced oxidative stress.

Studies on drought stress with and without acclimation in a drought resistant wheat variety clearly showed that acclimation reduced oxidative stress and oxidative damage by enhancing antioxidant defense. There was a 7-fold increase in superoxide radicals in leaves of wheat seedlings experiencing severe stress for the first time as compared to the 3-fold increase in leaves of acclimated plants (Selote *et al.*, 2004). This oxidative stress tolerance of acclimated plants was due to a co-ordinated antioxidant defense at both the cellular and subcellular levels (Selote and Khanna-Chopra, 2006). Up-regulation of SOD and APX protected the leaf chloroplasts from toxic ROS. Acclimated plants also had higher GSH levels and GR

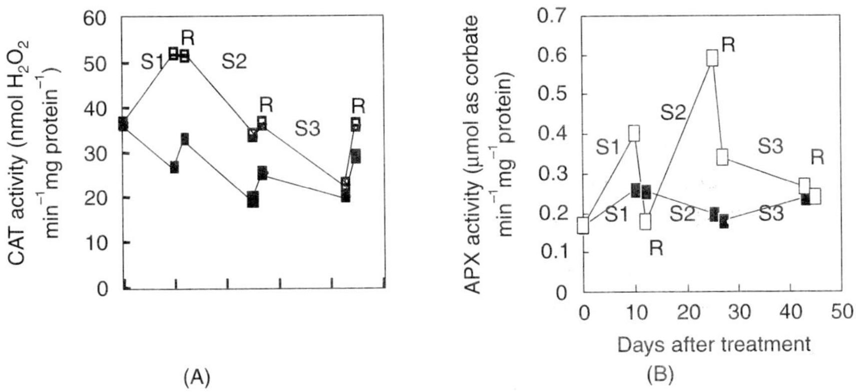

Fig. 13.2 Effect of three successive water stress cycles of increasing intensity on the antioxidant defense in leaves of upland rice cultivar Tulsi (drought tolerant). Each stress cycle was terminated by a 48-h period of rehydration. (A) CAT activity and (B) APX activity (mol ascorbate min^{-1} mg^{-1} protein) in Control ■; Stress □; R, Recovery; S, Stress. (Srivalli *et al.*, 2003).

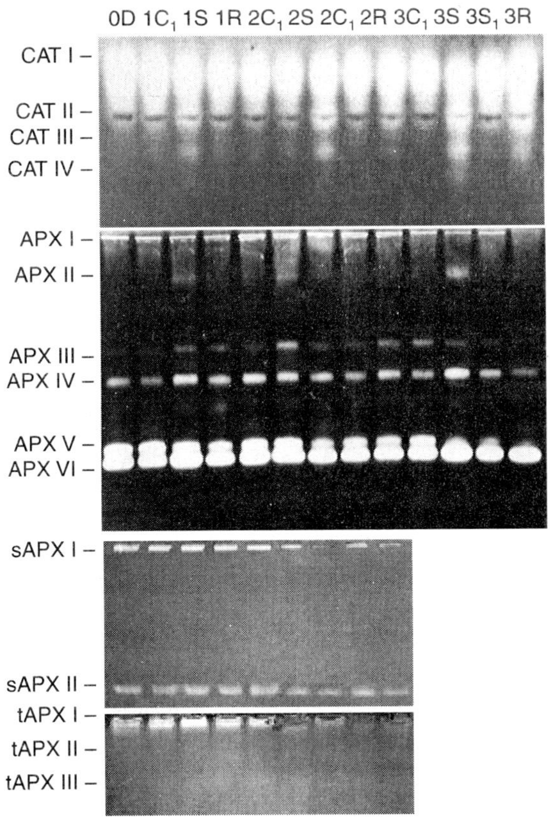

Fig. 13.3 Effect of three successive water stress cycles of increasing intensity on the antioxidant defense in leaves of upland rice cultivar Tulsi (drought tolerant). Each stress cycle was terminated by a 48-h period of rehydration. (A) CAT isoforms. (B) APX isoforms. (C) APX activity in stroma and (D) APX activity in thylakoid fraction. 1, 2 and 3 stand for the first, second and third cycle, respectively. C1 and C2 are the controls for the stress (S) and recovery (R), respectively (Srivalli *et al.*, 2003).

activity in crude extracts and in the chloroplasts. In non-acclimated plants, the failure in the induction of APX and ascorbate-glutathione cycle enzymes made chloroplasts susceptible to oxidative damage (Fig. 13.4). Genotypic differences too play a crucial role in the development of acclimatory stress response. The drought-resistant wheat genotype (C306) showed better acclimation induced oxidative stress management than the susceptible genotype (Moti) under field conditions. It was observed that resistant genotype had higher tolerance to oxidative stress with higher GR activity and GSH/GSSG ratio under stress than the susceptible genotype (Fig. 13.5, Khanna-Chopra and Selote, 2007).

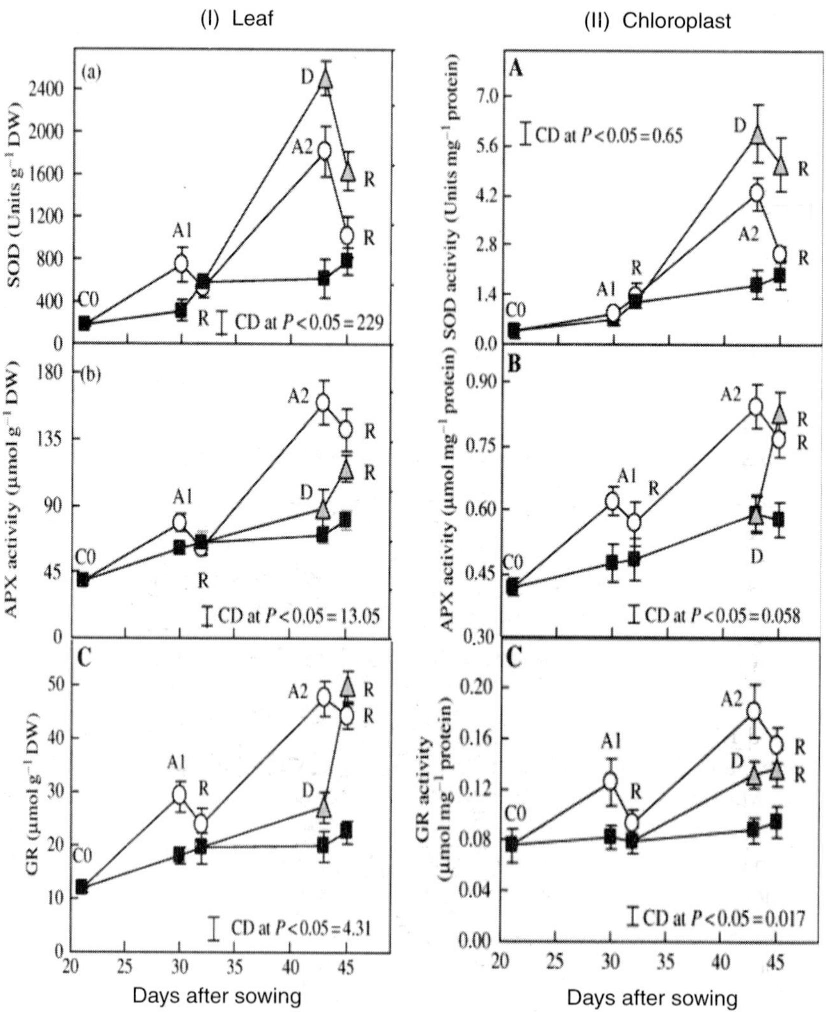

Fig. 13.4 Activity of antioxidant enzymes – (A) SOD, (B) APX and (C) GR in the leaves (I) and chloroplasts (II) of wheat seedlings (cv. C306, drought tolerant) subjected to acclimation and drought stress (non-acclimation) treatments. A group of plants were subjected to acclimation treatment (A) – i.e. two cycles of water stress, first of mild intensity (A1) and subsequently a second one of severe intensity (A2). Each cycle is terminated by rewatering (R) the plants. Another group of plants (non-acclimated) were exposed directly to single water stress of severe intensity (D) followed by 48 h of rewatering (R) period. DAS indicates days after sowing. Control ■, acclimated water stress O, and non-acclimated water stress Δ. (Selote and Khanna-Chopra, 2006).

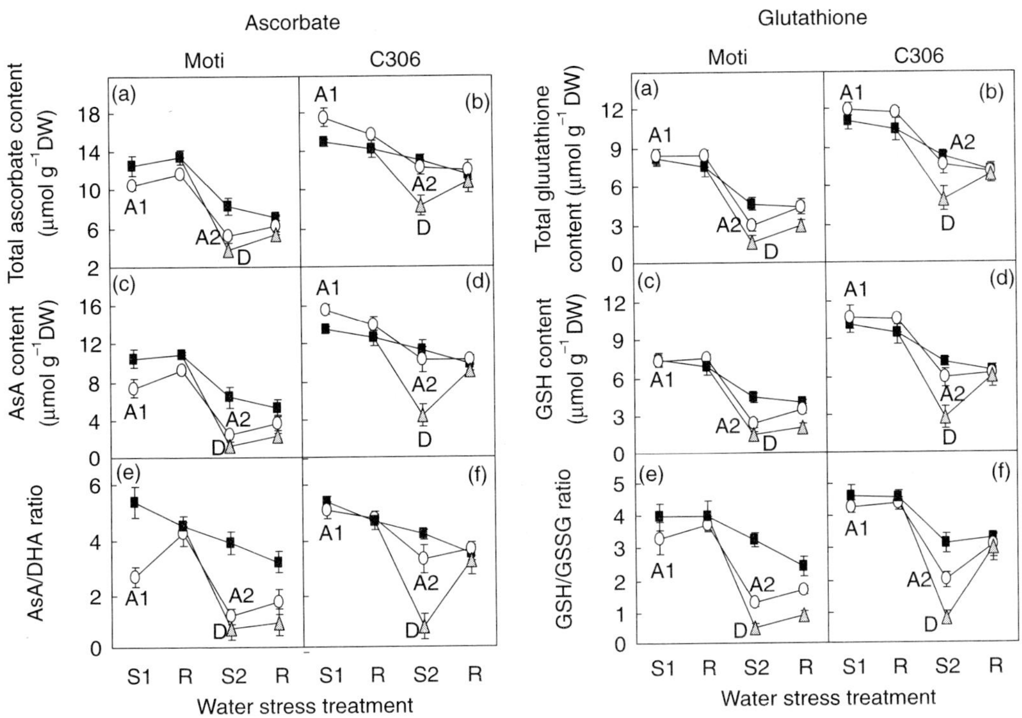

Fig. 13.5 Changes in the (a,b) total ascorbate content, total glutathione content, (c,d) reduced ascorbate (AsA) content, reduced glutathione (GSH) content and (e, f) ratio AsA to DHA, GSH to GSSG in the flag leaf of two wheat cultivars C306 (drought tolerant) and Moti (drought susceptible) during monocarpic senescence subjected to drought stress during post-anthesis period.A group of plants of each cultivar were subjected to acclimation treatment (A) i.e., two cycles of water stress, first of mild intensity during vegetative growth period (A1) and subsequently a second one of severe intensity during post-anthesis period (A2). Rewatering (R) period of 48 h terminates each stress cycle. Another group of plants (non-acclimated) of each genotype were made to experience directly single water stress of severe intensity (D) followed by 48 h of rewatering (R) period. DAS indicates days after sowing. Control (acclimated water stress 0, and non-acclimated water stress Δ. (Khanna-Chopra and Selote, 2007).

Drought induced ROS production has been observed in a number of plant species growing under different conditions, and the ability of plants to control oxidant levels is highly correlated with stress tolerance (Jubany-Mari *et al.*, 2010). Information on the relationship between drought, oxidative stress and antioxidant defence in field grown plants is lacking. Under natural climatic conditions, different abiotic stresses can occur simultaneously and plants need to cope with all of them. On the other hand, plants may have more resources to resist because their roots have more room to spread and look for water in deeper layers of the soil.

Heat stress is also among the primary causes of reduced crop yields. Improved understanding of plant responses to temperature can lead to more reliable predictions of crop yield and may help in designing crops that are better matched to the environment (Atkinson and Porter, 1996). Heat stress damages cellular structure and metabolic pathways and contributes to secondary water stress. There have been several reports on the oxidative stress and the response of antioxidant defense mechanism in heat stressed plants (Anderson and Padhye, 2004; Dat *et al.*, 1998; Larkindale and Knight, 2002). Heat stress was shown to cause impairments in mitochondrial functions and result in the induction of oxidative damage (Vacca *et al.*, 2004). In addition, acquired thermotolerance was shown to be mediated in part by enhancing cellular mechanisms that prevented oxidative damage under heat stress (Bergmüller *et al.*, 2003; Larkindale and Huang, 2004). Accumulation of GSH has also been observed in heat stressed tomato seedlings (Revero *et al.*, 2004). In fact, heat stress also increased GSH levels in the flag leaf of two wheat genotypes with contrasting behaviour in heat tolerance at all the stages during grain development (Chauhan, 2005). Overall, GSH levels and GSH/GSSG ratio declined during grain development in flag leaves of wheat in both control and heat stressed plants. The rate of decline was, however, higher under heat stress environment. Reducing the rate of ROS production in cells is likely to be as important as active scavenging of ROS during stress (Suzuki and Mittler, 2006). Studies on heat-acclimated versus non-acclimated cool season turfgrass species suggested that the former had lower production of ROS as a result of enhanced synthesis of ascorbate and glutathione (Xu *et al.*, 2006).

REGULATION OF PHOTOSYNTHEIC METABOLISM UNDER ENVIRONMENTAL STRESSES BY TRANSCRIPTION FACTORS

Abiotic stresses such as drought, temperature and salinity limit the growth and productivity. These effects are caused by altered morphology and physiology resulting from changes to processes such as inhibition of cell division and effects on metabolism including photosynthesis. Responses to abiotic stress require the production of important metabolic proteins (such as those involved in synthesis of osmoprotectants) and regulatory proteins operating in the signal transduction pathways such as kinases and transcription factors (TFs). TFs play a critical role in the abiotic stress response by controlling gene expression (Chaves and Oliveira, 2004).

TFs are proteins with a DNA domain that bind to the cis-acting elements present in the promoter of a target gene. They induce or repress the expression. TFs can be grouped into families according to their DNA-binding domain. A group of genes controlled by a certain type of TF is known as a regulon. In the plant response to abiotic stresses, both ABA dependent and ABA independent regulon are identified to play a role. These are CBF/DREB regulon, the NAC and ZF-HD (Zinc-finger homedomain) regulon, the AREB/ABF (ABA-responsive clement-binding protein/ABA-binding factor) and the MYC/MYB regulon.

Photosynthesis plays a central role as an energy source for plant metabolism, and its efficiency may be drastically reduced due to abiotic stresses. Plants have developed many strategies to acclimate to adverse environmental conditions with the main aim to maintain photosynthetic efficiency at as high a level as possible thereby avoiding the energy imbalance and the consequent photo-oxidative damage. Abiotic stresses reduce the rate of photosynthesis by reducing CO_2 diffusion, PSII efficiency, electron transport, ROS formation, RUBP content, Rubisco activity and photorespiration. Droughts can lead to a limitation in the supply of ATP in the chloroplast as a consequence of decreased ATP synthase activity (Lawlor and Tezara, 2009). Stomatal closure is the earliest response to drought, and the dominant limitation in photosynthesis at mild to moderate stress but under severe stress progressive reduction in the biochemical processes is the main cause of reduced photosynthetic CO_2 assimilation (Tezara *et al.*, 1999).

Stomatal closure/opening through both ABA dependent and independent pathways does involve transcriptional regulation. The first indication that TFs could be involved in stomatal regulation came from mutants defective in ABA signaling (Parcy and Giraudat 1997). Subsequently several TFs belonging to MYB family, i.e., MYB61, MYB60 were shown to express in guard cells and control stomatal movements through ABA independent pathways (Liang *et al.*, 2005). NAC TFs may also be involved in controlling stomatal aperture. Transgenic rice plants overexpressing SNACl are more sensitive to ABA and lose water more slowly by closing more stomata, leading to improvements in performance under drought and salinity (Saibo *et al.*, 2009).

An emerging area of research is the effect of abiotic stresses on stomatal development, since stomatal number affects both photosynthesis and WUE and thereby contributes to the long term adaptation to adverse environmental conditions. Stomatal development is influenced by environmental conditions such as CO_2 concentration, light and UV-B irradiation (Gitz *et al.*, 2005) and ABA (Franks and Farquhar, 2001). Stomatal development requires a strict control over differentiation and cell division so that stomatal morphology, distribution and behaviour respond properly to the environment ensuring plant performance. Several TFs including some from MYB family have been shown to control stomatal development at various steps in *Arabidopsis* (Barton, 2007). In addition, mitogen-activated kinases have been shown to influence stomatal development and also play an important role in environmental stress response (Wang *et al.,* 2007). Hence, studies are required to clarify the regulatory mechanisms underlying environmental control of stomata formation.

In addition to stomatal regulation, abiotic stress conditions reduce the photosynthetic rate by controlling the expression of genes involved in non-stomatal processes associated with photosynthesis and carbohydrate metabolism. Environmental cues control the expression of genes encoding chlorophyll *a/b* binding proteins, PS I, PS II subunits, oxygen evolving complex and Rubisco subunits (Saibo *et al.,* 2009). Most genes associated with photosynthesis are under the control of a transcriptional regulatory network evolved to control plant respose to external stimuli. Transcriptional profiling studies have shown that although some genes are up-regulated, many photosynthesis related genes are down-regulated under abiotic stress conditions, thus accounting for growth retardation. Those genes may be target for TFs such as the Cys2/His2- type zinc finger proteins STZ and AZF2, which were shown to function as transcription repressors under drought, cold and high salinity stress conditions resulting in growth retardation (Sakamoto *et al.,* 2006). TFs acting as activators are also involved in the modulation of photosynthesis under abiotic stress. A good example is the regulation of the genes encoding CAB proteins of PSII. Two MYB-like TFs from barley, HvMCB1 and HvMCB2 were shown to bind specifically to defined regions of CAB promoters derived from barley and wheat. These TFs have characteristic features of transcriptional activators and are required for maximal CAB gene expression. Interestingly transcription of both genes HvMCB1 and HvMCB2 is regulated by environmental factors. When barley leaves were subjected to salt, osmotic and oxidative stress, the transcripts of both MYB-encoding gene decreased significantly (Churin *et al.,* 2003). In *Brassica napus,* it was shown that the gene expression of both GLK1 and GLK2, belonging to MYB TF family, is regulated by cold stress (Savitch*et al.,* 2005). The involvement of several MYB-type TFs in the regulation of photosynthesis-related genes in response to abiotic stress indicates that this TF family is highly implicated in this process. Other TFs, such as LONG HYPOCOTYL 5 (HY5), are known to be involved in the regulation of CAB gene expression and other photosynthesis-related genes, such as the ribulose bisphosphate carboxylase small subunit (RbcS1A). HY5 appears to regulate the expression of several *Arabidopsis* genes known to respond to abiotic stress conditions e.g., CBF1, DREB2A, RD20 and MYB59 (Lee *et al.,* 2007). Hence, it is inferred that HY5 may also be involved in the regulation of photosynthesis by adverse environmental conditions, as extreme cold conditions also induce HY5 expression.

The regulation of photosynthesis-related genes by an abiotic stress response can also involve an indirect action of specific TFs. For instance, it is known that over-expression of OsMYB4 enhances compatible solute e.g., glycine betaine(GB)accumulation and increases stress tolerance of *Arabidopsis thaliana* (Mattana *et al.,* 2005). Vannini *et al.* (2004) have reported that *Arabidopsis* plants over-expressing OsMYB4 show improved PSII stability and higher tolerance to photoinhibition. In addition, GB stabilizes Rubisco conformation under high salinity stress (Sakamoto and Murata, 2002). OsMYB4 gene transcription is known to be regulated by abiotic stresses, namely cold, in an ABA-independent way (Vannini *et al.*, 2004). Thus, it is possible that OsMYB4 may play a role in the regulation of photosynthesis by environmental cues. Another interesting point is the transcriptional regulation of plastid-encoded photosynthesis genes by environmental stimuli. The expression of these genes is controlled through nuclear-encoded transcription regulators, as none of the genes in the higher plant chloroplast encodes TFs. It would be particularly important to know how the transcription of the chloroplast-encoded genes is regulated by abiotic stresses.

Hence, knowledge about the TFs involved in regulation of photosynthesis related genes, both stomatal and non-stomatal, by the different abiotic stresses is still limited. Many transgenic plants overexpressing TFs show improved abiotic stress tolerance related to enhanced photosynthetic parameters (Table 13.1). However, these results must be carefully analysed, as few photosynthetic parameters are usually evaluated and the growth/stress conditions are normally different from natural field conditions.

Table 13.1 Some transgenic plants overexpressing abiotic stress-related transcription factors show stress tolerance and improved photosynthesis.

Transgenic plant	TF family	Abiotic Stress Tolerance	Effect of Photosynthesis	Reference
Brassica napus	AP2/EREPB	Freezing	Increased CO_2 assimilation Increased photochemical efficiency	Savitch *et al.* (2005)
Rice	AP2/EREPB	Drought/High salt/ low temperature	Improved maximum quantum efficiency of PSII	Oh *et al.* (2007)
Rice	AP2/EREPB	Drought/High Salt	Lower stomatal conductance/ enhanced photosynthesis assimilation and efficiency	Karaba *et al.* (2007)
Tobacco	AP2/EREPB	Low Temperature	Higher chlorophyll fluorescence/ higher net photosynthetic rate	Shan *et al.* (2007)
Arabidopsis	bZIP	Drought and Heat Shock	Improved photosynthetic capacity	Zhang *et al.* (2008)
Rice	NAC	Drought/High salt	Loses water more slowly by closing more stomatal pores/ no effect on photosynthesis rate	Hu *et al.* (2006)
Arabidopsis, Maize	NF-Y (HAP)	Drought	Higher chlorophyll index and Higher photosynthesis rates	Nelson *et al.* (2007)
Arabidopsis	MYB	Cold/Freezing	Improved PSII stability. Tolerance to photoinhibition	Vannini *et al.*(2004)

CONCLUSION AND FUTURE RESEARCH

Senescence is an age dependent slow form of cell death that primarily allows plants to remobilize materials and nutrients to other parts of plants where growth might be occurring (Lim *et al.*, 2007). In annual crops such as wheat, rice and legumes, monocarpic senescence plays an important role in mobilizing nutrients from leaves to developing seeds/grains in the post flowering phase (Khanna-Chopra and Srivalli, 2008). Reproductive sink size, biotic and abiotic stresses influence the rate of senescence. Our studies have clearly shown that reproductive sink acts as a stress leading to higher oxidative damage to proteins and drives the mobilization of nitrogen to the developing seeds (Srivalli and Khanna-Chopra, 2009). Rubisco is the major protein degraded in which ROS and cysteine proteases play a major role. The mechanism of Rubisco degradation is being worked out and most probably involves the SAVs in addition to the vacuoles (Prins *et al.*, 2008).

Combining the tolerance of abiotic stresses with high yield is a cherished goal in agriculture since abiotic stresses, specially drought and heat, reduce crop yield considerably (Vinocur and Altman, 2005). Acclimation is an important component of abiotic stress tolerance and needs to be understood at the physiological, biochemical and molecular level. Enhanced ROS generation is one of the earliest responses of plant cells under abiotic stresses and senescence. ROS play a key signaling role in plants and control growth, development and abiotic stress resposes (Bailey-Serres and Mittler, 2006, Foyer and Noctor, 2009). The steady state level of ROS in cells, as well as the duration, subcellular localization and intensity of ROS signal are thought to be controlled in cells by the ROS gene network, which comprises 150 genes in *Arabidopsis* and includes genes encoding ROS generating proteins such as NADPH oxidases and ROS-scavenging enzymes including SODs, CATs, thio- and glutaredoxins and APXs (Mittler *et al.,* 2004). The functions of genes in ROS network and their role in sensing and transducing ROS signals in different processes are virtually unknown.

Studies on drought stress with and without acclimation in wheat and rice on oxidative stress and antioxidative defense have clearly shown that highly coordinated antioxidant defense and specifically some enzymes such as APX, CAT and enzymes of Halliwell-Asada pathway play an important role in the process of acclimation (Selote and Khanna-Chopra, 2006). In non-acclimated plants, the failure in the induction of APX and ascorbate-glutathione cycle enzymes made chloroplasts susceptible to oxidative damage. Other studies in *Arabidopsis* have also highlighted the importance of APX, specially the cytosolic APX1, in protecting cells from light stress induced oxidative damage (Develetova *et al.,* 2005). In fact, cytosolic APX1 protected chloroplast and was essential for the proper functioning of thylakoidal and stromal APX during light stress. Such studies highlight the need to understand the ROS homeostasis and signaling between neighboring compartments in cells, as ROS generation and defense is spread over the entire cell.

Transcription factors regulate leaf senescence and abiotic stress response (Nakashima *et al.,* 2009). In fact, considerable crosstalk exists between the gene expression pattern in the transcription factor genes that are induced in both the processes. For example, among the 43 transcription factor genes that are induced during senescence, 28 genes are also induced during various stresses (Lim *et al.,* 2007). The transcription factors WRKY 53 and NAC act as positive regulators in leaf senescence. Nac regulon also plays an important role in abiotic stress responsive gene expression. Overexpression of the SNAC1 and SNAC2 enhanced drought and salt tolerance in rice without growth retardation and yield penalty (Hu *et al.,* 2006). WRKY transcription factors play pivotal roles in regulating many stress reactions in plants, however, unravelling their roles in abiotic stress responses needs more study. Overexpression studies have shown that WRKY transcription factors enhance salt, heat, cold and drought tolerance and hence form part of the signaling processes associated with transcriptional reprogramming when plants encounter abiotic stresses (Rouston *et al.,* 2010). The potential functions of most leaf senescence associated transcription factors remain to be elucidated. Functional characterization of these genes including the signaling pathways they are involved in and the target genes they regulate will be invaluable in understanding the complex molecular pathways regulating leaf senescence.

Leaf senescence may reduce yield in crop plants by limiting the growth phase and may also cause post-harvest spoilage in vegetable crops. Manipulation of leaf senescence using biotechnological approaches requires in depth knowledge of the novel regulatory elements specific to leaf senescence using senescence mutants and stress responsive mutants in *Arabidopsis* and crop plants.

ACKNOWLEDGEMENTS

This research was supported by the grants of the National Fellow and ICAR-NPTC scheme of Indian Council of Agricultural Research, New Delhi, India.

REFERENCES

Adam Z, Clarke AK (2002) Cutting edge of chloroplast proteolysis. Trends Plant Sci 7: 51-456

Ahmed P, Jaleel CA, Azooz MM, Nabi G (2009) Generation of ROS and nonenzymatic antioxidants during abiotic stresses in plants. Bot Res International 2: 11-20

Anderson JA, Padhye SR (2004) Protein aggregation, radical scavenging capacity, and stability of hydrogen peroxide defense systems in heat-stressed *Vinca* and sweet pea leaves. J Am Soc Hort Sci 129: 54-59

Atkinson D, Porter JR (1996) Temperature, plant development and crop yields. Trends Plant Sci 1: 199-124

Bailey-Serres J, Mittler R (2006) The roles of reactive oxygen species in plant cells. Plant Physiol 141: 311

Barth C, Moeder W, Klessig DF, Conklin PL (2004) The timing of senescence and response to pathogens is alteredin the ascorbate-deficient mutant vitamin C-1. Plant Physiol 134: 178-192

Barton MK (2007) Making holes in leaves: promoting cell state transitions in stomatal development. Plant Cell 19: 1140-1143

Bergmüller E, Porfirova E, Dörmann P (2003) Characterization of an *Arabidopsis* mutant deficient in γ-tocopherol methyltransferase. Plant Mol Biol 52: 1181-1190

Bhattacharjee (2005) Reactive oxygen species and oxidative burst: Roles in stress, senescence and signal transduction in plants. Curr Sci 89: 1113-1121

Breusegem FV, Dat JF (2006) Reactive oxygen species in plant cell death. Plant Physiol 141: 384-390

Brugie`re N, Dubois F, Masclaux C, Sangwan RS, Hirel B (2000) Immunolocalization of glutamine synthetase in senescing tobacco (*Nicotiana tabacum* L.) leaves suggests that ammonia assimilation is progressively shifted to the mesophyll cytosol. Planta 211: 519 527

Chauhan S (2005) Physiological and molecular basis of heat tolerance with emphasis on oxidative stress metabolism in wheat. PhD Thesis. HNB Garhwal University, Srinagar, Uttarakhand, India

Chauhan S, Srivalli S, Nautiyal AR, Khanna-Chopra R (2009) Wheat cultivars differing in heat tolerance show a differential response to monocarpic senescence under high temperature stress and the involvement of serine proteases. Photosynthetica 47: 536-547

Chaves MM, Oliveira MM (2004) Mechanisms underlying plant resilience to water deficits: prospects for water-saving agriculture. J Exp Bot 55: 2365-2384

Chen W, Provart NJ, Glazebrook J, Katagiri F, Chang HS (2002) Expression profilematrix of *Arabidopsis* transcription factor genes suggests their putative functions in response to environmental stresses. Plant Cell 14: 559-74

Churin Y, Adam E, Kozma-Bognar L, Nagy F, Borner T (2003) Characterization of two Myb-like transcription factors binding to CAB promoters in wheat and barley. Plant Molecular Biology 52: 447-462

Dalling MJ, Boland G, Wilson JH (1976) Relation between acid proteinase activity and redistribution of nitrogen during grain development in wheat. Aust J Plant Physiol 3: 721-730

Dat JF, Lopez-Delgado H, Foyer CH, Scott IM (1998) Parallel changes in H_2O_2 and catalase during thermotolerance induced by salicylic acid or heat acclimation in mustard seedlings. Plant Physiol 116: 1351-1357

Delledonne M, Xia Y, Dixon RA, Lamb C (1998) Nitric oxide functions as a signal in plant disease resistance. Nature 394: 585-588

Demirevska-Kepova K, Hölzer R, Simova-Stoilova L, Feller U (2005) Heat stress effecects on Rubisco, Rubisco binding protein and Rubisco activase in wheat leaves. Biol Plant 49: 521-525

Devletova S, Rizhsky L, Liang H, Shengqiang Z, Oliver DJ, Coutu J, Shulaev V, Schlauch K, Mittler R (2005) Cytosolic ascorbate peroxidase 1 is a central component of the reactive oxygen gene network of *Arabidopsis*. Plant Cell 17: 268-281

Dhindsa RJ, Dhindsa PP, Thorpe TA (1981) Leaf senescence: correlated with increased level of membrane permeability and lipid peroxidation and decreased level of superoxidedismutase and catalase. J Exp Bot 32: 93-101

El Yahyaoui F, Kuster H, Ben Amor B (2004) Expression profiling in *Medicago truncatula* identifies more than 750 genes differentially expressed during nodulation, including many potential regulators of the symbiotic program. Plant Physiol 136: 3159-3176

Feller U, Anders I, Demirevska K (2008a) Degradation of Rubisco and other chloroplast proteins under abiotic stresses. Gen Appl Plant Physiol 34: 5-18

Feller U, Anders I, Mae T (2008b) Rubiscolytics: fate of Rubisco after its enzymatic function in a cell is terminated. J Exp Bot 59: 1615-1624

Foyer CH, Noctor G (2005b) Redox homeostasis and antioxidant signalling: a metabolic interface between stress perception and physiological responses. Plant Cell 17: 1866-1875

Foyer CH, Noctor G (2009) Redox regulation in photosynthetic organisms: signaling, acclimation, and practical implications. Antioxidants and Redox Signalling 11: 861-905

Franks PJ, Farquhar GD (2001) The effect of exogenous abscisic acid on stomatal development, stomatal mechanisms, and leaf gas exchange in *Tradescantia verginiana*. Plant Physiol 125: 935-942

Fryer MJ, Andrews JR, Oxborough K, Blowers DA, Baker NR (1998) Relationship between CO_2 assimilation, photosynthetic electron transport and active O_2 metabolism in leaves of maize in the field during periods of low temperature. Plant Physiol 116: 571-580

Giacomelli L, Masi A, Ripoll DR, Lee MJ, van Wijk KJ (2007) *Arabidopsis thaliana* deficient in two chloroplast ascorbate peroxidases shows accelerated light-induced necrosis when levels of cellular ascorbate are low. Plant Mol Biol 65: 627-644

Gitz DC, Liu-Gitz L, Britz SJ, Sullivan JH (2005) Ultravoilet-B effects on stomatal density, water use efficiency, and stable carbon isotope discrimination in four glass house grown soyabean (*Glycine max*) cultivars. Environ Exp Bot 53: 343-355

Groten K, Dutilleul C, van Heerden PDR, Vanacker H, Bernard S, Finkemeier I, Dietz KJ, Foyer CH (2006) Redoxregulation of peroxiredoxin and proteinases by ascorbate and thiols during pea root nodule senescence. FEBS Letters 580: 1269-1276

Guiamét JJ, Pichersky E, Nooden LD (1999) Mass exodus from senescing soybean chloroplasts. Plant Cell Physiol 40: 986-992

Guo FQ, Crawford NM (2005) *Arabidopsis* nitric-oxide synthase I is targeted to mitochondria and protects against oxidative damage and dark-induced senescence. Plant Cell 17: 3436-3450

He Y, Fukushige H, Hildebrand DF, Gan S (2002) Evidence supporting a role of jasmonic acid in *Arabidopsis* leaf senescence. Plant Physiol 128: 876-84

Hörtensteiner S (2006) Chlorophyll degradation during senescence. Annu Rev Plant Biol 57: 55-77

Hörtensteiner S, Feller U (2002) Nitrogen metabolism and remobilization during senescence. J Exp Bot 53: 927-937

Hu H, DaiM, Yao J, Xiao B, Li X, Zhang Q, Xiong L (2006) Overexpressing a NAM, ATAF, and CUC (NAC) transcription factor enhances drought resistance and salt tolerance in rice. Proc Natl Acad Sci USA 35: 12987-12992

Hung KT, Kao CH (2003) Nitric oxide counteracts the senescence of rice leaves induced-by abscisic acid. J Plant Physiol 160: 871-79

Hung KT, Kao CH (2004) Hydrogen peroxide is necessary for abscisic acid induced senescence of rice leaves. J Plant Physiol 161: 1347-57

Jime´nez A, Hernandez JA, Pastori G, del Rio LA, Sevilla F (1998) Role of the ascorbate-glutathione cycle of mitochondria and peroxisomes in the senescence of pea leaves. Plant Physiol 118: 1327-1335

Jing HC, Schippers JHM, Hiller J, and Djikwel DP (2005) Ethylene-induced leaf senescence depends on age-related changes and OLD genes in *Arabidopsis*. J Exp Bot 56: 2915-2923

Jubany-Marí T, Munné-Bosch S, Alegre L (2010) Redox regulation of water stress responses in field-grown plants: Role of hydrogen peroxide and ascorbate. Plant Physiology and Biochemistry 48: 351-358

Karaba A, Dixit S, Greco R, Aharoni A, Trijatmiko KR, Martinez NM, Krishnan A, Nataraja KN, Udayakumar M, Pereira A (2007) Improvement of water use efficiency in rice by expression of HARDY, an *Arabidopsis* drought and salt tolerance gene. Proc Nat Acad Sci USA 104: 15270-15275

Karuse GH (1994) The role of oxygen in photoinhibition and photosynthesis, In: Causes of Photooxidative Stress and Amelioration of Defense Systems in Plants (eds Foyer, CH and Mullineauxe, PM), CRC Press, Boca Raton, FL, USA pp 43-76

Khanna-Chopra R, Srivalli B, Ahlawat YS (1999) Drought induces many forms of cysteine proteases not observed during natural senescence. Biochem Biophys Res Comm 225: 324-327

Khanna-Chopra R, Selote DS (2007) Acclimation to drought stress generates oxidative stress tolerance in drought-resistant than susceptible wheat cultivar under field conditions. Environ Exp Bot 60: 276-283

Khanna-Chopra R, Srivalli S (2008) Interaction of reproductive sink and abiotic stress with monocarpic senescence: molecular analysis. Indian Agriculturist 135-149

Kim JH, Woo HR, Kim J, Lim PO, Lee IC, Choi SH, Hwang D and Nam HG (2009) Trifucate feed-forward regulation of age-dependent cell death involving *mi*R164 in *Arabidopsis*. Science 323: 1053-1057.

Kong Z, Li M, Yang W, Xu W, Xue Y (2006) A novel nuclear-localized CCCH-type zinc finger protein, OsDOS, is involved in delaying leaf senescence in rice (*Oryza sativa* L.). *Plant Physol* 141: 1376-88

Kukavica B, Veljovic-Jovanovic S (2004) Senescence-related changes in the antioxidant status of ginkgo and birch leaves during autumn yellowing. Physiol Plant 122: 321-327

Larkindale J, Knight MR (2002) Protection against heat stress-induced oxidative damage in *Arabidopsis* involves calcium, abscisic acid, ethylene and salicylic acid. Plant Physiol 128: 682-695

Larkindale J, Huang B (2004) Thermotolerance and antioxidant systems in *Agarostis stolonifera*: involvement of salicyclic acid, abscisic acid, calcium, hydrogen peroxide and ethylene. J Plant Physiol 161: 405-413

Lawlor DW, Tezara W (2009) Causes of decreased photosynthetic rate and metabolic capacity in water-deficient leaf cells: a critical evaluation of mechanisms and integration of process. Annals Bot 103: 561-579

Lee J, He K, Stolc V (2007) Analysis of transcription factor HY5genomic binding sites revealed its hierarchical role in light regulationof development. The Plant Cell 19: 731-749

Liang YK, Dubos C, Dodd IC, Holroyd GH, Hetherington AM, Campbell MM (2005) AtMYB61, an R2R3-MYB transcription factor controlling stomatal aperture in *Arabidopsis thaliana*. Current Biol 15: 1201-1206

Lim PO, Kim HJ, Nam HG (2007) Leaf senescence. Ann Rev of Plant Biol 58: 115-136

Lohman KN, Gan S, Manorama CJ, Amasino R (1994) Molecular analysis of natural leaf senescence in *Arabidopsis thaliana*. Physiol Plant 92: 322-328

Love AJ, Milner JJ, Sadanandom A (2008) Timing is everything: regulatory overlap in plant cell death. Trends Plnat Sci 13: 589-595

Marín-Navarro J, Moreno J (2006) Cysteines 449 and 459 modulate the reduction–oxidation conformational changes of ribulose 1,5-bisphosphate carboxylase/oxygenase and the translocation of the enzyme to membranes during stress. Plant Cell Environ 29: 898-908

Martínez DE, Bartoli CG, Grbic V, Guiamet JJ (2007) Vacuolar cysteine proteases of wheat (*Triticum astivum* L.) are commen to leaf senescence induced by different factors. J Exp Bot 58:1099-1107

Mattana M, Biazzi E, Consonni R, Locatelli F, Vannini C, Provera S, Coraggio I (2005) Overexpression of Osmyb4 enhances compatible solute accumulation and increases stress tolerance of *Arabidopsis thaliana*. Physiol Plant 125 :212 -223

Miao Y, Zentgraf U (2007) The antagonist function of *Arabidopsis* WRKY53 and ESR/ESP in leaf senescence is modulated by the jasmonic and salicylic acid equilibrium. Plant Cell 19: 819-830

Mittler R (2006) Abiotic stresses, the field environment and stress combination. Trends Plant Sci 11: 15-19

Mittler R, Vanderauwera S, Gollery, Van Breusegem F (2004) The reactive oxygen gene network of plants. Trends Plant Sci 9: 490-498

Moreno J, Penarubbia L, Gracia-Ferris C (1995) The mechanism of redox regulation of ribulose 1,5-bisphosphate carboxylase/ oxygenase turnover a hypothesis. Plant Physiol Biochem 33: 121-127

Munné-Bosch S, Alegre L (2002) Plant ageing increases oxidative stress in chloroplasts. Planta 214: 608-615

Müntz K (2007) Protein dynamics and proteolysis in plant vacuoles. J Exp Bot 58: 2391-2407

Nakashima K, Ito Y, Yamaguchi-Shinozaki K (2009) Transcriptional regulatory response to abiotic stresses in *Arabidopsis* and grasses. Plant Physiol 149: 88-95

Navabpour S, Morris K, Allen R, Harrison E, AH-Mackerness S, Buchanan-Wollaston V (2003) Expression of senescence enhanced genes in response to oxidative stress. J Exp Bot 54: 2285-2292

Navari-Izzo F, Pinzino C, Quartacci MF, Sgherri CLM (1999) Superoxide and hydroxyl radical generation, and superoxide dismutase in PS II membrane fragments from wheat. Free Radic Res 31: S3-S9

Nelson DE, Repetti PP, Adams TR, Creelman RA, Wu J, Warner DC, Anstrom DC, Bensen RJ, Castiglioni PP, Donnarummo MG, Hinchey BS, Kumimoto RW, Maszle DR, Canales RD, Krolikowski KA, Dotson SB, Gutterson N, Ratcliffe OJ, Heard JE (2007) Plant nuclear factor Y (NF-Y) B subunits confer drought tolerance and lead to improved corn yields on water-limited acres. Proc Nat Acad Sci USA 104: 16450-16455

Noctor G, Veljovic-Jovanovic S, Foyer CH (2000) Peroxide processing in photosynthesis: antioxidant coupling and redox signalling. Phil Trans Roy Soc B 355: 1465-1475

Noctor G, Queval G, and Gakie`re B. (2006) NAD(P) synthesis and pyridine nucleotide cycling in plants and their potential importance in stress conditions. J Exp Bot 57: 1603-1620

Oberhuber M, Breghold J, Breuker K, Hörtensteiner S, Kräutler B (2003) Breakdown of chlorophyll: A nonenzymatic reaction accounts of the formation of colourless "nonfluorescent" chlorophyll catabolites. Proc Natl Acad Sci USA 100: 6910-6915

Oh SJ, Kwon CW, Choi DW, Song SI, Kim JK (2007) Expression of barley HvCBF4 enhances tolerance to abiotic stress in transgenic rice. Plant Biotech J 5: 646-656

Orendi G, Zimmermann P, Baar C, and Zentgraf U (2001) Loss of stress-induced expression of catalase3 during leaf senescence in *Arabidopsis thaliana* is restricted to oxidative stress. Plant Sci 161: 301-314

Otegui MS, Noh YS , Martinez DE, Petroff MGV, Staehelin LA, Amasino RM, Guiamet JJ (2005) Senescence-associated vacuoles with intense proteolytic activity develop in leaves of Arabidopsis and soyabean. Plant J 41: 831- 844

Overmyer K, Tuominen H, Kettunen R, Betz C, Langebartels C, Sandermann H, Jr. Kangasjärvi J (2000) Ozone-sensitive *Arabidopsis* rcd1 mutant reveals opposite roles for ethylene and jasmonate signaling pathways in regulating superoxide-dependent cell death. Plant Cell 12: 1849-1862

Parcy F, Giraudat J (1997) Interactions between the ABI1 and the ectopically expressed ABI3 genes in controlling abscisic acid resposes in *Arabidopsis* vegetative tissue, Plant J 11: 693-702

Pavet V, Olmos E, *Kiddle* G, Mowla S, Kumar S, Antoniw J,Alvarez ME, Foyer CH (2005) Ascorbic acid deficiency activatescell death and disease resistance responses in *Arabidopsis*. Plant Physiol 139: 1291-1303

Pell EJ, Schlagnhaufer CD, Arteca RN (1997) Ozone-induced oxidative stress: mechanisms of action and reaction. Physiol Plant 100: 264-273

Pfannschmidt T, Nilsson A, Allen JF (1999) Photosynthetic control of chloroplast gene expression. Nature 397: 625-629

Prasad TK, Anderson MD, Martin BA, Stewart CR (1994) Evidence for chilling-induced oxidative stress in maize seedlings and a regulatory role for hydrogen peroxide. Plant Cell 6: 65-74

Prins A, van Heerden PDR, Olmos E, Kunert KJ, Foyer CH (2008) Cysteine proteinases regulate chloroplast protein content and composition in tobacco leaves: a model for dynamic interactions with ribulose-1,5-bisphosphate carboxylase/oxygenase (Rubisco) vesicular bodies. J Exp Bot 59: 1935-1950

Queval G and Noctor G (2007) A plate-reader method for the measurement of NAD, NADP, glutathione and ascorbate in tissue extracts. Application to redox profiling during *Arabidopsis* rosette development. Anal Biochem 363: 58-69

Revero RM, Ruiz JM, Remero LM (2004) Importance of N source on heat stress tolerance due to the accumulation of proline and quaternary ammonium compounds in tomato plants. Plant Biol 6: 702-707

Rosenvasser S, Mayak S, Friedman H (2006) Increase in reactive oxygen species (ROS) and in senescence-associated gene transcript (SAG) levels during dark-induced senescence of *Pelargonium* cuttings and the effect of gibberellic acid. Plant Sci 170: 873-879

Roulin S, Feller U (1998) Dithiothreitol triggers photooxidative stress and fragmentation of the large subunit of ribulose-1,5-bisphosphate carboxylase/oxygenase in intact pea chloroplasts. Plant Physiol Biochem 36: 849-856

Rouston P J, Somssich I E, Ringler P, Shen Q J (2010) WRKY transcription factors. Treands Plant Sci 15: 247-258

Saibo NJM, Lourenco T, Oliveira MM (2009) Transcription factors and regulation of photosynthetic and related metabolism under environmental stresses. Annals Bot 103: 609-623

Sakamoto A, Murata N (2002) The role of glycine betaine in the protection of plants from stress: clues from transgenic plants. Plant Cell and Environ 25: 163-171

Sakamoto H , Maruyama K, Sakuma Y, Meshi T, Iwabuchi M, Shinozaki K Yamaguchi-Shinozaki K (2006) Arabidopsis Cys2/His2-type zinc-finger proteins function as transcription repressors under drought, cold, and high-salinity stress conditions. Plant Physiol 136 : 2734-2746

Savitch LV, Allard G, Motoaki Seki, Robert LS, Tinker NA, Huner N P A, Shinozaki K, Singh J (2005) The effect of overexpression of two *Brassica* CBF/DREB1-like transcription factors on photosynthetic capacity and freezing tolerance in *Brassica napus*. Plant Cell Physiol 46 :1525-1539

Schaffer MA, Fischer RL (1988) Analysis of mRNAs that accumulate in response to low temperature identifies a thiol protease gene in tomato. Plant Physiol 87: 431-436

Schelbert S, Aubry S, Burla B, Agne B, Kessler F, Krupinska K, Hörtensteiner S (2009) Pheophytin Pheophorbide Hydrolase (Pheophytinase) is involved in chlorophyll breakdown during leaf senescence in *Arabidopsis* Plant Cell 21: 767-785

Selote D, Khanna-Chopra R (2004) Drought induced spikelet sterility is associated with an inefficient antioxidant defense in rice panicle. Physiol Plant 121: 462-471

Selote D, Khanna-Chopra R (2006) Drought-acclimation confers oxidative stress tolerance by inducing coordinated antioxidant defense at cellular and sub-cellular level in leaves of wheat seedlings. Physiol Plant 127: 494-506

Shan DP, Huang JG, Yang YT, Guo,YH, Wu CA, Yang, GD Gao, Z Zheng, CC (2007) Cotton GhDREB1 increases plant tolerance to low temperature and is negatively regulated by gibberellic acid. New Phytologist 176 : 70-81

Srivalli B, Bharti S, Khanna-Chopra R (2001) Vacuolar cysteine proteases and ribulose-1,5-bisphosphate carboxylase/oxygenase degradation during monocarpic senescence in cowpea leaves. Photosynthetica 39: 87-93

Srivalli B, Sharma G, Khanna-Chopra R (2003) Antioxidative defense system in an upland rice cultivar subjected to increasing intensity of water stress followed by recovery. Physiol Plant 119: 503-512

Srivalli S, Khanna-Chopra R (2006) Drought stress during pod development in cowpea induces loss in ribulose-1,5 bisphosphate carboxylase/oxygenase and an increase in cysteine proteases. J Plant Biol 33: 22-31

Srivalli S, Khanna-Chopra R (2009) Delayed wheat flag leaf senescence due to removal of spikelets is associated with increased activities of leaf antioxidant enzymes, reduced glutathione/oxidized glutathione ratio and oxidative damage to mitochondrial proteins. Plant Physiol Biochem 47: 663-670

Stone SL, Hauksdottir H, Troy A, Herschleb J, Kraft E, Cullis J (2005) Functional analysis of the RING-type ubiquitin ligase family of *Arabidopsis*. Plant Physiol 137: 317-327

Suzuki N, Mittler R (2006) Reactive oxygen species and temperature stresses: A delicate balance between signaling and destruction. Physiol Plant 126: 45-51

Takahashi F, Yoshida R, Ichimura K, Mizoguchi T, Seo S, Yonezawa M, Maruyama K, Yamaguchi-Shinozaki K, Shinozaki K (2007) The mitogen-activated protein kinase cascade MKK3-MPK6 is an important part of the jasmonate signal transduction pathway in *Arabidopsis*. Plant Cell 19: 805-818

TezaraW, Mitchell VJ, Driscoll SD, Lawlor DW (1999) Water stress inhibits plant photosynthesis by decreasing coupling factor and ATP. Nature 401: 914-917

Theriault A, Wang Q, Van Iderstine SC, Chen B, Franke AA, Adeli K (2000) Modulation of hepatic lipoprotein synthesis and secretion by taxifolin, a plant flavonoid. Journal of Lipid Research 41: 1969-1979

Thoenen M, Feller U (1998) Degradation of glutamine synthetase in intact chloroplasts isolated from pea (*Pisum sativum*) leaves. Australian Journal of Plant Physiology 25: 279-286

Thoenen M, Herrmann B, Feller U (2007) Senescence in wheat leaves: is a cysteine endopeptidase involved in the degradation of the large subunit of Rubisco? Acta Physiol Plant 29: 339-350

Thomas H, Ougham H, Canter P, Donnison I (2002) What stay-green mutants tell us about nitrogen remobilization in leaf senescence. J Exp Bot 53: 801-808

Uauy C, Distelfeld A, Fahima T, Blechl A, Dubcovsky J (2006) A NAC gene regulating senescence improves grain protein, zinc and iron content in wheat. Science 314: 1298-1301

Vacca RA, de Pinto MC, Valenti D, Passarella S, Marra E, De Gara L (2004) Production of reactive oxygen species, alteration of cytosolic ascorbate peroxidase and impairment of mitochondrial metabolism are early events in heat shock induced programmed cell death in tobacoo bright-yellow 2 cells. Plant Physiol 134: 1100-1112.

Vanacker H, Sandalio LM, Jimenez A, Palma JM, Corpas FJ, Meseguer V, Gomez M, Sevilla F, Leterrir M, Foyer CH, del Rio LA (2006) Role of redox regulation in leaf senescence of pea plants grown in different sources of nitrogen nutrition. J Exp Bot 57: 1735-1745

Vannini C, Locatelli F, Bracale M, Magnani E, Marsoni M, Osnato M, Mattana M, Baldoni, Coraggio I (2004) Overexpression of the rice Osmyb4 gene increases chilling and freezing tolerance of *Arabidopsis thaliana* plants. The Plant Journal 37(1): 115-127

Vener AV, Ohad I, Anderson B (1998) Protein phosphorylation and redox sensing in chloroplast thylakoids. Curr Opin Plant Biol 1: 217-229

Vinocur B, Altman A (2005). Recent advances in engineering plant tolerance to abiotic stresses: achievements and limitations. Curr Opin Biotehnol 16: 123-132

Wang H, Ngwenmyama N, Liu Y, Walker JC, Zhang S (2007) Stomatal development and patterning are regulated by environmentally responsive mitogen-activated protein kinase in *Arabidopsis*. Plant Cell 19: 63-73

Wise RR (1995) Chilling-enhanced photooxidation: The production, action and study of reactive oxygen species produced during chilling in the light. Photosyn Res 45: 79-97

Woo HR, Goh CH, Park JH, de la Serve BT, Kim JH, ParkYI, Nam HG (2002) Extended leaf longevity in the ore 4-1mutant of *Arabidopsis* with a reduced expression of plastid ribosomal protein gene. Plant J 31: 331-340

Woo HR, Kim JH, Nam HG, Lim PO (2004) The delayed leaf senescence mutants of *Arabidopsis*, *ore1*, *ore3*, and *ore9* are tolerant to oxidative stress. Plant Cell Physiol 45: 923-932

Xu S, Li J, Zhang X, Wei H, Cui L (2006) Effects of heat acclimation pretreatment on changes of membrane lipid peroxidation, antioxidant metabolites, and ultrastructure of chloroplasts in two cool-season turfgrass species under heat stress. Environ Exp Bot 53: 274-285

Yahyaoui F, Kuster HK, Amor BB, Hohnjec N, Puhler A, Becker A, Gouzy J, Vernie T , Gough C, Niebel A, Godiard L, Gamas P (2004) Expression Profiling in Medicago truncatula Identifies More than 750 Genes differentially expressed during nodulation, including many potential regulators of the symbiotic program. Plant Physiol 136: 3159-3176

Ying W (2008) NAD/NADH and NADP/NADPH in cellular functions and cell death: regulation and biological consequences. Antiox Red Signal 10: 179-206

Zhang X, Wollenweber B, Jiang D, Liu F, Zhao J (2008) Water deficits and heat shock effects on photosynthesis of a transgenic *Arabidopsis thaliana* constitutively expressing ABP9, a bZIP transcription factor. J Exp Bot 59: 839-848

Zimmermann P, Zentgraf U (2005) The correlation between oxidative stress and leaf senescence during plant development. Cell Mol Biol Lett 10: 515-534

Zimmermann P, Heinlein C, Orendi G, Zentgraf U (2006) Senescence-specific regulation of catalase in *Arabidopsis thaliana* (L.) Heynh. Plant Cell Environ 29: 1049-1056

CHAPTER 14

Inactivation and Repair of Photosynthetic Machinery under Heat Stress

Suleyman I. Allakhverdiev[1, 2, 4, *], Vladimir D. Kreslavski[2], Irina R. Fomina[2, 3], Dmitry A. Los[1], Vyacheslav V. Klimov[2], Mamoru Mimuro[4], Prasanna Mohanty[5] and Robert Carpentier[6]

[1]*Institute of Plant Physiology, Russian Academy of Sciences, Botanicheskaya 35, Moscow 127276, Russia*

[2]*Institute of Basic Biological Problems, Russian Academy of Sciences, Pushchino, Moscow Region 142290, Russia*

[3]*Biosphere Systems International Foundation, Tucson, Arizona 85755, USA*

[4]*Graduate School of Human and Environmental Studies, Kyoto University, Kyoto 606-8501, Japan*

[5]*Regional Plant Resource Center, Nayapalli, Bhubaneswar 751015, India and Jawaharlal Nehru University, New Delhi-11006, India*

[6]*Groupe de Recherche en Biologie Végétale, Université du Québec à Trois-Rivières, C.P. 500, Québec, Canada, G9A 5H7*

ABSTRACT

There are several target sites for elevated temperature-induced damage in plants such as the oxygen evolving complex along with the associated cofactors in photosystem II (PS II), the ATP generating system and carbon fixation by Rubisco due to Rubisco activase. The impairment of the photosynthetic machinery (PM) leads to the generation of reactive oxygen species (ROS), such as singlet oxygen in PS II, superoxide radical ($^{\cdot}O_2^-$) in PS I, and others. Moderately high temperatures do not cause direct damage to PS II but through ROS. The balance between (pro) oxidant and antioxidant levels, light intensity during heating and ATP supply are the crucial factors for the extent of recovery of PM from heat induced damage. Most recent investigations on heat stress responses are focused on the role of heat shock transcription factors (HSFs), chaperones, and heat-induced genes important for plant cell heat acclimation, with little information about chloroplastic (photosynthetic) heat tolerance. HSFs can be sensitive sensors for the effect of heat stress on thylakoid membranes, although, the mechanism of their involvement is still obscure. In this chapter, we summarize the recent progress in the study of molecular mechanisms of damage and acclimation of PM, especially PS II, during moderate heat stress.

Keywords: Acclimation, Heat stress, Photosynthesis, Photosystem II.

Abbreviations

APX: ascorbate peroxidase, *CAT*: catalase, *Chl:* chlorophyll, *Cytb:* cytochrome b, *EPR:* electron paramagnetic resonance, *ETC:* electron transport chain, *Fo:* basal fluorescence, *Fm:* maximum fluorescence, *Fv:* variable fluorescence, *GB:* glycinebetaine, *HSFs:* heat shock transcription factors, *HSPs:* heat shock proteins, *NMR:* nuclear magnetic resonance, *OEC:* oxygen-evolving complex, *3-PGA:* 3-phosphoglyceric acid, *PM*: photosynthetic machinery, *PQ:* plastoquinone, *PS I:* photosystem I, *PS II:* photosystem II, *QTL:* quantitative trait loci, *QA:* the first stable quinone electron acceptor of PS II, *QB:* the secondary quinone electron acceptor, *RC:* reaction center, *ROS:* reactive oxygen species; Rubisco, ribulose-1,5-bisphosphate carboxylase/oxygenase, *SOD:* superoxide dismutase, *WT:* wild type.

INTRODUCTION

Threats of global warming and climate change make heat stress a general concern especially for tropical agriculture, as they limit plant productivity (Boyer, 1982; Hall, 2001). Under *in vivo* conditions, heat-stress linked alterations depend on the stage of growth of photosynthetic tissue, where the young developing and the old senescing tissues exhibit different sensitivity (Kalitulo *et al.*, 2003).

Photosynthesis, especially PS II activity, is one of the most heat sensitive cellular-functions (Berry and Björkman, 1980; Yordanov *et al.*, 1986; Carpentier, 1999). Thermal stress leads to an increase in the relative fraction of Q_B-nonreducing RC (Havaux *et al.*, 1991), loss of thylakoid membrane integrity, especially destacking of the thylakoid membranes (Gounaris *et al.*, 1984; Semenova, 2004), and dysfunction in the system of CO_2 assimilation (Sharkey, 2005). Heat-stress decreases the activities of heat-sensitive enzymes and the level of expression of most of the genes involved in energy and lipid metabolism, pigment biosynthesis, and photosynthesis (Allakhverdiev *et al.*, 2001, 2002; Suzuki *et al.*, 2005).

The extent of damage *in vivo* under any type of stress depends on the balance between the damage and repair processes during the stress. This is particularly true for PS II and it provides the basis for acclimation and photosynthetic recovery processes (Adir *et al.*, 2003; Allakhverdiev and Murata, 2004, 2008; Allakhverdiev *et al.*, 2002, 2003, 2004, 2005a,b, 2007; Mohanty *et al.*, 2007; Murata *et al.*, 2007). Moderate heat stress enhances photoinhibition, induced by strong light, but photosynthetic inhibition is usually reversible and reduced activity of the PM is recovered. This fact is explained by the hypothesis that most types of environmental stressors act primarily by inhibiting the repair of PS II due to formation of ROS, while light can damage PS II directly (Nishiyama *et al.*, 2001, 2005; Allakhverdiev and Murata, 2004). It might be suggested that CO_2 limitation and moderate heat stress generate ROS such as H_2O_2 due to suppression of the fixation of CO_2 and decreased levels of 3-PGA. These ROS, in turn, inhibit protein synthesis in photosynthesis, and, thus, the repair of the PS II reaction center complex.

The *de novo* synthesis of photosynthetic proteins demanding energy in the form of ATP and light is likely to play a crucial role in the repair of heat-induced photoinhibition of PM (Adir *et al.*, 2003; Allakhverdiev *et al.*, 2005a, 2008). Additionally, exposure to low or moderate light during short-time heat treatment is also shown to be beneficial to plants due to its diminishing effect on the extent of thermal damage to PM, especially PS II (Havaux *et al.*, 1991; Havaux, 1993). Strong light exposure, however, increases the magnitude of inhibition of PM induced by heat stress (Al-Khatib and Paulsen, 1989; Kreslavski and Khristin, 2003). Photosynthetic organisms are able to overcome the damaging effects of light and other oxidative stresses via the rapid and efficient repair of PS II (Aro *et al.*, 1993; Andersson and Aro 2001; Allakhverdiev and Murata, 2004). The capacity of PS II for recovery in the light can be one of the main mechanisms for elimination of the inhibitory action of heat stress on PS II (Nishiyama *et al.*, 2001, 2005, 2006; Kreslavski *et al.*, 2008).

It is widely accepted that the co-ordination of plastid protein complex formation involves bidirectional signaling pathways, including signals from the nucleus to the plastids (anterograde signals) and vice versa (retrograde signals) (Fey *et al.,* 2005). Changes in the redox state of the chloroplast electron transport chain or pools of photosynthesis coupled redox-active compounds (thioredoxin, glutathione) exert regulation over both plastid- and nuclear encoded chloroplast-expressed proteins (Dat *et al.,* 2000). The mechanisms of redox sensing and signal transmission, the identification of regulatory targets, and the understanding of adaptation mechanisms have been the focus of many recent researches (Strand, 2004; Fey *et al.,* 2005; Pshybytko *et al.,* 2008).

Thylakoid membrane fluidity might act as a sensor for temperature-induced functional changes as well (Horvath *et al.,* 1998; Los and Murata, 2004). In general, a transient elevation in temperature, usually 10–15°C above ambient, is considered as heat shock or heat stress (Wahid *et al.,* 2007). The severity of heat-induced damage depends not only on the tested systems and the potential for temperature tolerance, but also on the temperature gradient and its mode of heat application (Bukhov and Mohanty, 1999).

In this review, we focus on moderate high temperature stress effects *in vivo*, heat stress targets and factors regulating the mechanisms of stress impairments, acclimation, and the recovery of photosynthetic function.

HEAT DAMAGE AND SIGNALING

Targets of Heat Stress

There are several target sites for elevated temperature-induced damage such as the CO_2 fixation system, photophosphorylation, the electron transport chain, and the OEC (Nash *et al.,* 1985; Feller *et al.,* 1998; Bukhov and Mohanty, 1999; Carpentier, 1999; Sharkey, 2005).

Both low and high temperature stress cause multi-step injuries to the PM, although the early steps of impairment, caused by cold and heat, seem to be different. The enzymes of the Calvin-Benson cycle are heat labile. Thus the carbon assimilation system is sensitive to elevated temperature and gets strongly inhibited at moderate thermal stress (Berry and Björkman, 1980; Weis, 1981; Feller *et al.,* 1998; Sharkey, 2005). The decline in Rubisco activity by moderate thermal stress correlated with the loss of photosynthesis (Law and Crafts-Brandner, 1999). The enzyme Rubisco of higher plants is heat stable but the loss in activity at elevated temperature is due to changes in the activity of Rubisco activase, which is extremely sensitive to elevated temperature (Salvucci and Crafts-Brandner, 2004; Sharkey, 2005).

Among the various components of PM, the OEC of PS II is most sensitive to heat as revealed by *in vivo* and *in vitro* studies (Katoh and San Pietro, 1967; Berry and Björkman, 1980; Mamedov *et al.,* 1993; Allakhverdiev *et al.,* 1994, 1996; Havaux and Tardy, 1996; Carpentier, 1999). This appears to be due to a heat-induced loss of two out of four Mn from the Mn-containing cluster of the OEC (Nash *et al.,* 1985; Enami *et al.,* 1994, 1998). However, it has been demonstrated in recent years that PS II activity is not inhibited at temperatures where leaf CO_2 assimilation is reduced due to the decrease in the activity of Rubisco activase (Salvucci and Crafts-Brandner, 2004; Tang *et al.,* 2007).

The maximum quantum yield of PS II photochemistry (F_v/F_m), where F_v are photoinduced changes in fluorescence of Chl *a* and F_m – maximal fluorescence, is used as a conventional index to estimate heat damage to PS II. However, several studies have shown that F_v/F_m is not sensitive to certain stresses under particular conditions (Law and Crafts-Brandner, 1999; Crafts-Brandner and Salvucci, 2000; Wen *et al.,* 2005). Some authors (Strasser, 1997; Strasser *et al.,* 2000) pointed out that this parameter does not accurately reveal alterations in PS II behavior in detail. Li *et al.* (2009) explored the heterogeneity of PS II function in soybean leaves when F_v/F_m was identical under different heat treatments and found heterogeneous behavior of PS II. In the last decades, the Chl *a* fluorescence O-J-I-P transient has become one of the most popular tools in photosynthesis research being very sensitive to environmental stresses, in particular, to high temperature (Srivastava *et al.*, 1997; Strasser, 1997; Tóth *et al.*, 2007; Chen *et al.*, 2008; Yordanov *et al.*, 2008). Besides, the most convenient test for monitoring heat-induced changes remains the rise in minimum Chl *a* fluorescence, F_0 level, as it indicates the critical temperature for PS II inactivation (Havaux, 1993; Yamane *et al.*, 2000). Yamane *et al.* (2000) found that the main causes of F_0 increase differ from species to species. In some species (potato, tobacco, etc.) the high-temperature induced F_0 increase is not due to the inactivation of PS II photochemical reactions, but due to the dark reduction of Q_A via PQ (Sazanov *et al.*, 1998; Yamane *et al.*, 2000; Bukhov and Carpentier,s 2000).

Early Events of Heat Stress Signaling

Low concentrations of oxidants are required for cell function under physiological conditions and generation of ROS, particularly H_2O_2, during abiotic stresses has been proposed as part of the signaling cascade leading to protection from these stresses and acclimation (Foyer *et al.*, 1997; Dat *et al.*, 1998; Mittler *et al.*, 2004). Miller *et al.* (2007) found that a ROS-related signal, generated in the chloroplast, can enhance basal thermotolerance and that this signal might require an intact retrograde signaling pathway. Both peroxide and superoxide can induce the synthesis of HSPs as well as glutathione peroxidases and ascorbate peroxidase (Vranova *et al.*, 2002), important for protection of PM against heat stress.

Development of Oxidative Stress

There are two principal modes of stress-induced impairment of photosynthesis: first, a direct damage induced by the stress factor and second, inhibition of *de novo* protein synthesis by ROS. The extent of damage of PM depends on the pool of H_2O_2 and other ROS, activities of antioxidant enzymes and content of low-molecular antioxidants (Asada, 2006). Usually, enhanced level of ROS stimulates the development of impairment to the PM (Gong *et al.*, 2001; Pastori and Foyer, 2002). Both decline in CO_2 assimilation and O_2 evolution due to heat stress are assumed to enhance ROS production and the consequent loss in the light-dependent repair processes (Sairam *et al.*, 2000).

Heat Induced Changes in the Structure of Thylakoid Membranes, Lipid Composition and Fluidity

The heat-induced changes in the structure of Chl-protein complexes and thylakoid membranes, as well as the inactivation of enzyme activity and change in membrane fluidity, constitute the early effects of thermal stress (Vani *et al.*, 2001; Semenova, 2004; Kreslavski *et al.*, 2008). The integrity of thylakoid membranes is compromised. A decreased membrane stacking and a general reorganization of the thylakoid membranes have been observed with temperatures in the interval of 35-45°C (Gounaris *et al.*, 1983, 1984; Semenova, 2004). The heat-induced structural transformations lead to ion leakage from heat-stressed leaves (Inaba and Grandall, 1988; Wahid and Shabbir, 2005) and changes in energy distribution between the two photosystems *in vivo* (Mohanty *et al.*, 2002).

It is suggested that heat-induced destacking of thylakoid membranes occurs as a two-step process. The first step of membrane disintegration is located in the appressed region of thylakoids and involves destacking of thylakoid membranes. Thylakoid membrane structure undergoes strong disintegration during the second step observed at high temperatures (> 45°C) and relates to general vesicle formation of thylakoid membranes (Gounaris *et al.*, 1984). Besides temperature, the effect depends on the duration of heat treatment. Indeed, no destacking of thylakoid membranes was observed in chloroplasts isolated from intact primary wheat leaves heat-treated at 42°C for 20 min in the dark, whereas heating at 42°C for 40 min demonstrated the destacking that was mainly reversible after 6 h of light exposure of leaves (Kreslavski *et al.*, 2008). The destacking can be associated with the reversible disconnection of light-harvesting complex II (Georgieva, 1999). It might be suggested that heating influences the surface membrane charge due to changes in charge screening characteristics of the membranes.

The extent of molecular disorder and molecular motion within a lipid bilayer is referred to as the fluidity of the membrane (Los and Murata, 2004). A number of studies focused on the effects of low temperature and demonstrated clearly that membrane fluidity decreases with a decrease in temperature (Inaba *et al.,* 2003; for reviews see Murata and Los, 1997; Los and Murata, 2004). Conversely, high temperatures cause an increase in the fluidity of membranes, which can lead to disintegration of the lipid bilayer (Carratu *et al.,* 1996; Vigh *et al.,* 1998; Semenova, 2004; Kreslavski *et al.,* 2008).

It was suggested that the expression of heat-shock genes depends on the fluidity of the membrane in yeast cells (Carratu *et al.,* 1996; Vigh *et al.,* 1998). It is supposed that the increase of membrane fluidity inherent to the increased temperatures can activate the expression of HSP genes, some of which code proteins preventing membrane disintegration (Vigh *et al.,* 1998). Evidence to support this possibility for photosynthetic cells was obtained in studies of the responses of *Synechocystis* cells to heat shock (Horvath *et al.,* 1998; Török *et al.,* 2001). At the optimal growth temperature, the membrane fluidizer, benzyl alcohol, activated the transcription of the *hspA* gene as efficiently as heat stress. It was suggested that thylakoid membranes act as a cellular thermometer that senses thermal stress (Horvath *et al.* 1998). However, other studies did not support the putative involvement of membrane fluidity in heat-induced gene expression and heat sensing (Inaba *et al.,* 2003). Thus, the detection of heat-stress-induced membrane fluidization by cellular protein sensors still remains a controversial issue (Los and Murata, 2004).

Rigidification of thylakoid membranes appears to invoke altered expression profiles of heat shock genes, suggesting that the temperature sensing mechanism may be located on the thylakoid membrane (Horvath *et al.,* 1998). The prospect of the thylakoid membrane acting as a heat sensor is physiologically crucial, because it is susceptible to temperature upshift, owing to its highly unsaturated character, and the presence of photosystems, which are fragile to temperature changes (Sung *et al.,* 2003).

Alterations in lipid composition can affect heat stress signaling either through global effects on the physical state of membrane, or via specific interactions of lipids with proteins (Vigh *et al.,* 2007). It was suggested that membrane fluidity is linked to the degree of unsaturation of membrane lipids (Gombos *et al.,* 1994; Vigh *et al.,* 1998; Los and Murata, 2004). The mechanism of lipid desaturation in the acclimation to high temperature is still unclear and data are contradictory. In *Chlamydomonas*, saturation of membrane lipids was reported to enhance the thermal stability of PS II (Sato *et al.,* 1996), while in cyanobacteria desaturation of membrane lipids was shown to have no effect on the thermal stability of PS II (Gombos *et al.,* 1991, 1994; Wada *et al.,* 1994). Gombos *et al.* (1994) monitored the changes in heat tolerance after the elimination of dienoic and trienoic fatty acids from membrane lipids of *Synechocystis*. Heat resistance of photosynthetic oxygen evolution was not associated with fatty acid unsaturation. On the other hand, on the basis of recent observations of the differences in the temperature dependence and thermosensitivity of PS II in *Synechocystis* sp. PCC 6803 grown at 25 and 35°C (Aminaka *et al.,* 2006), it was suggested that the changes in the lipid desaturation levels of plasma and thylakoid membranes are amongst major factors responsible for the change in the thermosensitivity of PS II.

Redox State of Chloroplast Components

There is an abundant evidence in the literature that redox state of the ETC is involved in the expression of photosynthetic genes in prokaryotes (Oh and Kaplan, 2001; Wilson *et al.,* 2006). In eukaryotes, a consensus is emerging that expression of chloroplastic genes, as well as retrograde regulation of nuclear genes encoding photosynthetic and nonphotosynthetic proteins can occur via the modulation of the redox state of the membrane-bound quinone pools associated with photosynthetic and respiratory electron transport (Allen *et al.,* 1995; Pfannschmidt, 2003).

Components of the photosynthetic ETC exhibit significant changes in their reduction/oxidation (redox) state depending on the photosynthetic electron flow and therefore serve as signaling parameters which report environmental influences on photosynthesis. Such redox signals control chloroplast and nuclear gene expression events and play an important role in the co-ordination of both genetic compartments (Pfannschmidt and Liere, 2005; Dietzel *et al.,* 2008). Signaling pathways are triggered by the redox state of the PQ-pool, the thioredoxin system, and the acceptor availability at PS I, in addition to control by oxolipins, tetrapyrroles, carbohydrates, and abscisic acid. They provide a functional feedback control, which couples the expression of chloroplast genes to the actual function of photosynthesis, and, by this means, help to acclimate the photosynthetic process to environmental cues, especially to heat stress.

The photosynthetic adjustment at the level of the light-harvesting complex associated with PS II (LCHII) in *Dunaliella salina* to different irradiances and high temperatures was investigated by Maxwell *et al.* (1995). The authors concluded that photosynthetic adjustment to temperature mimics adjustment to high irradiance through a common redox sensing/signaling mechanism. Both temperature and light modulate the redox state of the first, stable quinone electron acceptor of PS II, Q_A, which reflects the redox poise of the intersystem electron transport chain. Changes in redox poise signal the nucleus to regulate cab mRNA abundance, which, in turn, determines the accumulation of light-harvesting apoprotein. This redox mechanism may represent a general acclimation mechanism for photosynthetic adjustment to environmental stimuli (Maxwell *et al.,* 1995).

Using HPLC analysis, the effect of heat stress conditions on the redox state of the PQ pool was examined in the study of Pshybytko *et al.* (2008). Heat stress (40°C, 3 h, and illumination at 100 μmol m^{-2} s^{-1}) had influence on the redox state of the PQ pool in thylakoids isolated from heat-treated barley seedlings. The authors suggested that a thermoinduced change of the redox state of the PQ pool and a redistribution of PQ molecules between photoactive and non-photoactive pools are the mechanisms which reflect and regulate the response of the photosynthetic apparatus under heat stress conditions.

Recent structural studies suggest that redox-dependent changes in the secondary structure of the cytochrome b_6f components may be involved in the signaling process (Allen, 2004). However, strong genetic and biochemical proof about a link between heat signaling and cytochrome b_6f is still lacking.

It is also important to note that low-molecular cytosolic and nuclear-encoded heat shock stress proteins can be imported into chloroplasts (Lubben and Keegstra, 1986; Vierling *et al.,* 1986).

Role of Cofactors and Extrinsic Proteins of PS II in Thermal Damage

Among the various components of the PS II, the OEC located at the donor side of PS II is often the most sensitive to heat stress (Katoh and San Pietro, 1967; Berry and Björkman, 1980; Mamedov *et al.,* 1993; Havaux and Tardy, 1996). Heating of PS II to moderate temperatures inactivates oxygen evolution (Shutilova *et al.,* 1995). Even at not so high temperatures (> 35°C) and short-time stress exposure, a significant inactivation of OEC can be observed (Nash *et al.,* 1985; Enami *et al.,* 1994). Usually the inactivation is reversible as the temperature is lowered to room temperature.

It was demonstrated that the oxidation of two bound water molecules to molecular oxygen is catalyzed by a metal center at the lumenal side of PS II, which contains four Mn atoms and, presumably, one calcium as well as one chloride ion as essential cofactors (Dau *et al.,* 2001; Ferreira *et al.,* 2004; for a review see McEvoy and Brudvig, 2006). Besides, bicarbonate has been suggested to play a role in the assembly of the OEC (Allakhverdiev *et al.,* 1997), and in the protection of the donor side of PS II against thermo- and photoinactivation (Klimov *et al.,* 1997, 2003; Villarejo *et al.,* 2002; Zharmukhamedov *et al.,* 2007; Shutova *et al.,* 2008). As many as five extrinsic proteins associate in different combinations depending on the type of organism (Roose *et al.,* 2007). The extrinsic proteins are labeled as O (PsbO), P (PsbP),

Q (PsbQ), R (PsbR), U (PsbU), and V (PsbV). Only the PsbO protein (the 33 kDa or manganese stabilizing protein) is ubiquitous to all oxygenic photosynthetic organisms. Three most spread extrinsic proteins of molecular mass 33, 23, and 17 kDa (PsbO, PsbP, PsbQ, respectively) are bound at the luminal side of PS II in many types of photosynthetic organisms (Seidler, 1996; Roose *et al.*, 2007). The 33 kDa protein is known to stabilize the Mn cluster (Nash *et al.*, 1985; De Las Rivas and Heredia, 1999; Roose *et al.*, 2007), and the 23 kDa protein has been shown to enhance the binding of calcium and chloride in addition to Mn (Bondarava *et al.*, 2005; Roose *et al.*, 2007). The extrinsic proteins and cofactors seem to counteract the negative action of ROS as well. Thus, protection of the oxygen-evolving machinery by the extrinsic proteins of PS II was shown essential for the development of cellular thermotolerance in *Synechocystis* sp. PCC 6803 (Kimura *et al,*. 2002).

Several studies have demonstrated that, in higher plants, the heat-induced inactivation of oxygen evolution in PS II complex is coupled to the unbinding of the three extrinsic polypeptides with molecular masses of 17, 23, and 33 kDa (Nash *et al.*, 1985; Enami *et al.*, 1994, 1998). Loss of cofactors, especially PsbU, induced the inactivation of PS II, thermal damage of D1 protein and production of ROS, both in the light and the dark when thermal stress persists. It was suggested that PsbU provides protection from reactive oxygen species, as PsbU mutants have enhanced mechanisms to detoxify exogenously applied H_2O_2 (Balint *et al.*, 2006).

Assays of the Mn content of heated PS II, by atomic absorption spectroscopy, EPR, and NMR, have shown the release of Mn from its binding site (Nash *et al.*, 1985; Enami *et al.*, 1994, 1998). In the absence of the extrinsic proteins, the release of Mn has been proposed to be the consequence of its reduction to Mn^{2+} by external reductants, namely by OH^- ions, in a reaction which may involve the production of peroxide (Thompson *et al.*, 1989, Yocum, 1992). The intermediate formed after a 10-min heat treatment of PSII membranes contains a binuclear Mn unit with the two Mn(III) ions separated by 2.7 Å and connected by a di-μ-oxo bridge (Pospiŝil *et al.*, 2003). For comparison, the intact Mn_4Ca complex in its S_1 state seems to contain two di-μ–oxo bridges between high-valence Mn.

Among intrinsic components of PS II the D1 protein is the most vulnerable. Moderate heat exposure (40°C for 30 min) of spinach thylakoids could cleave the D1 protein producing a 9 kDa C-terminal and 23 kDa N-terminal fragments. Only the D1 protein was cleaved in the thylakoids but not D2, CP43 or CP47 (Yoshioka *et al.*, 2006). Degradation of phosphorylated D1 was effected by a FtsH protease located in the stroma which was assumed to diffuse to the granal area as a result of heat triggered swelling of thylakoids (Yamamoto *et al.*, 1999; Komayama *et al.*, 2007). It is assumed that degradation of thermally damaged (heat inhibited) and high light damaged (photoinhibited) D1 follows the same route (Yoshioka *et al.*, 2006). It is still unknown if elevated temperature treatment of intact leaves and mesophyll cells, protoplasts, or algal cells would exhibit similar or identical results; no data is available at the moment to support this proposition.

ACCLIMATION AND RECOVERY OF PHOTOSYNTHETIC MACHINERY

It is well-known that increasing the temperature at which plants are grown causes an upward shift of the optimal temperature of photosynthesis in different plant species and renders the photosynthetic apparatus more tolerant to heat stress (Berry and Björkman, 1980). This phenomenon is termed acclimation. There are rapid and long-time mechanisms of acclimation of PM to elevated temperatures. The energy distribution between the two photosystems is one of the fast protective mechanisms against heat stress (Mohanty *et al.*, 2002). The distribution of absorbed light energy between the two photosystems is dynamically balanced and regulated by a process termed "state transitions". Redistribution of PQ molecules from the photoactive to the non-photoactive pool could be considered a protective mechanism for the prevention of PS II inactivation on the acceptor side (Pshybytko *et al.,* 2008). Efforts were undertaken to define cell components that stabilize the PM and assist in its tolerance to heat stress. The content of carotenoids, which participate in the xanthophyll cycle (Tefler and Barber, 1995; Tefler, 2005), HSPs (Preczewski *et al.,* 2000), extrinsic proteins (Nishiyama *et al.,* 1997; Tanaka *et al.,* 2000; Kimura *et al.,* 2002), isoprene (Sharkey, 2005), protective compatible solutes (Yang *et al.,* 2005), and changes in fatty acid composition (Yordanov, 1993; Gombos *et al.,* 1994; Murakami *et al.,* 2000; Zaharieva *et al.,* 2001) have been considered to contribute to the photosynthetic acclimation. Havaux (1996) has already reported that by heating, PQ is reduced in darkness by stromal reductants, and could be oxidized by fast illumination of far red light, preferentially exciting PS I. Lowered stromal redox status can lead to significant stimulation of PS I-mediated cyclic electron flow (Joët *et al.,* 2002). Acclimation to heat stress seems to operate through the sensing of redox components, especially PQ/plastoquinol (Pshybytko *et al.,* 2008). It is also suggested that HSPs, which are formed during heating in animal and plant organisms, are also involved in the PS II resistance (Heckathorn *et al.,* 1998, 2002) and their synthesis is correlated with the acquisition of thermotolerance (Vierling, 1991). Further, it has been proposed that deep oxidation of violaxanthine to zeaxanthine favours ΔpH-induced conformational changes in the light harvesting system of PS II, converting PS II to a state of high thermal energy dissipation (Horton *et al.,* 1994). A xanthophyll-modulated increase in heat losses is believed to protect the PS II reaction centers from overexcitation and subsequent photoinhibition. The role of these components is discussed in detail in some reviews (Yordanov, 1993; Bukhov and Mohanty, 1999; Sharkey, 2005) and therefore we will not develop this aspect further.

Role of Oxidant/Antioxidant Balance in Recovery from Heat Stress

ROS play not only damaging but also key signaling roles in plants, and their cellular concentration is controlled by a complex network of ROS metabolizing enzymes found in several different cellular compartments (Foyer and Noctor, 2005). ROS, especially H_2O_2 can be generated even during moderate heat stress. The generation of ROS is observed in the electron transport systems of chloroplast, both in PS I and PS II, and mitochondria (Asada, 1999, 2006; Mubarakshina *et al.,* 2008), as well as in the Calvin-Benson cycle (Kim and Portis, 2004). The pathways of ROS generation have been investigated by many researchers, but the data are often contradictory (Allakverdiev *et al.,* 2008). *In vitro* experiments from several laboratories suggested that thermal stress at high enough temperatures produces ROS such as superoxide radicals, hydroxyl radicals and hydrogen peroxide at the PS II RC. Photosynthetic systems have evolved defense mechanisms against ROS, involving both low-molecular antioxidants and antioxidant enzymes. The presence of antioxidant enzymes saves the organisms by limiting the formation of singlet oxygen, an highly toxic ROS, also produced under strong light (Krieger-Liszkay, 2005). Like singlet oxygen, hydroxyl radicals are also toxic for plant cells, and their conversion to H_2O_2 reduces their damaging effect (Asada, 2006). There are observations that formation of ROS and their scavenging by antioxidants also occurs *in vivo* (El-Shitinawy *et al.,* 2004; Kreslavski *et al.,* 2007, 2010). Major ROS-scavenging mechanisms include SOD, CAT and APX (Allen, 1995; Mittler, 2002; Asada, 2006).

Components of cell detoxification mechanisms have been investigated in specific experiments to improve thermotolerance using transgenic plants. Overexpression of Cu/Zn SOD is noted to protect plants from high temperature stress (Tang *et al.,* 2007). Peroxidases, especially APX, are one of the basic elements participating in photosynthetic response to heat stress and acclimation of PM (Foyer *et al.,* 1997; Dat *et al.,* 1998; Chen *et al.,* 2004). Chen *et al.* (2004) overexpressed a tomato gene encoding for glutathione peroxidase in tobacco. Transient expression of transgene protected transgenic leaves from heat stress. To understand how different ROS signals initiated in the chloroplast and the cytosol are integrated in the whole cell, Miller *et al.* (2007) generated a double mutant lacking the genes encoding the thylakoid APX (*tylapx*) and cytosolic APX1 (*apx*1). The study suggested that two different signals are generated in plants lacking cytosolic APX1 or tylAPX. The lack of a chloroplastic hydrogen peroxide removal enzyme, tylAPX, triggered a specific signal in cells that results in enhanced tolerance to heat stress, whereas the lack of a cytosolic hydrogen peroxide removal enzyme, APX1, triggered a different signal, which resulted in stunted growth and enhanced sensitivity to oxidative stress. Further investigation of the enhanced heat tolerance in plants lacking tylAPX, using mutants deficient in chloroplast-to-nuclei retrograde signaling, suggests the existence of a chloroplast-generated stress signal that enhances basal thermotolerance in plants. Activity of cytosolic antioxidant enzymes seems to be important for the protection of PM from heat-induced photoinhibition. For instance, the study of Davletova *et al.* (2005) demonstrated that in the absence of the cytosolic H_2O_2-scavenging enzyme APX1, the entire chloroplastic H_2O_2-scavenging system of *Arabidopsis thaliana* collapses, H_2O_2 levels increase, and protein oxidation occurs.

There is a link between H_2O_2 production and antioxidant enzyme activities, however, many details are unknown. The extent of damage caused to PM, its capacity for post-stress recovery and the possible roles of H_2O_2, SOD, CAT, and APX in the recovery process were investigated in wheat seedlings exposed to mild short-time heat stress (40°C and 42°C) (Kreslavski *et al.*, 2008, 2009). The damage to PM and its recovery depended on temperature and light intensity during seedling recovery (Kreslavski *et al.*, 2008). High temperature led to imbalance, in which the redox steady state of the cell is altered in the direction of pro-oxidants (H_2O_2) and increase in the activities of antioxidant enzymes (Fig. 14.1).

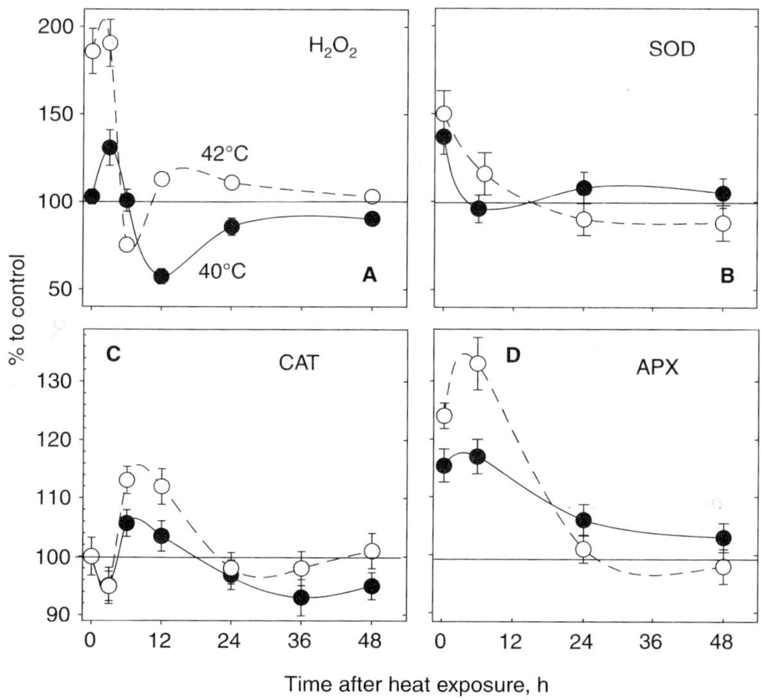

Fig. 14.1 Effect of heat pretreatment on the level of H_2O_2 (A), activities of superoxide dismutase (SOD) (B), catalase (CAT) (C), and ascorbate peroxidase (APX) (D) in leaf extracts during post-stress seedling recovery at 20°C and I = 40 μE m^{-2} s^{-1}. Initial activities of SOD, CAT, and APX were 430±40 unit g^{-1} (FW), 2500 μmol H_2O_2 g^{-1} (FW) min^{-1}, and 6.3 ± 1 μmol^{-1} (FW) min, respectively. ○-42°C; ●-40°C ▼. Adapted from Los *et al.*, 2010.

However, during next light-stimulated recovery of the PM, the balance was restored. It is suggested that the balance between (pro)oxidant and antioxidant levels during heat stress in the dark and heat-induced photoinhibition is the crucial factor for the extent of recovery of the PM from heat induced damage.

ROS Inhibit the Repair of Photodamaged PS II

The ROS are generated in all cell compartments and their formation is required for functioning of photosynthetic organisms (Suzuki and Mittler, 2006). The question about the role of different ROS in heat-induced inhibition of the activity of the PA is still open (Thompson *et al.,* 1989; Sharkey, 2005), whereas generation of ROS during photoinhibition is more evident (Aro *et al.,* 1993; Andersson and Aro, 2001). Recent studies in the cyanobacterium *Synechocystis* have revealed that ROS inhibit *de novo* synthesis of D1 protein via the inhibition of translation of the *psbA* gene that encodes the precursor of D1 (Nishiyama *et al.,* 2001). Since the production of ROS is enhanced through the inhibition of the Calvin cycle (Asada, 1999; Sharkey, 2005) sensitive to elevated temperatures (Salvucci, and Crafts-Brandner, 2004; Sharkey, 2005), extra ROS may be involved in the inhibition of the repair of PS II at high temperature (Takahashi *et al.,* 2004; Nishiyama *et al.,* 2006). Therefore, one of the targets of the action of moderate heat stress under light may be the system of the heat-induced repair of photodamaged PS II. On another hand, ROS may be involved in signal transduction chain that engages the protective and acclimation systems of the PM (Foyer *et al.,* 1997; Demmig-Adams and Adams, 2002, El-Shitinawy *et al.,* 2004; Hung *et al.,* 2005), the detail of which are yet to be known.

It may also be concluded that there is a close relationship between the mode of recovery of PM and oxidant/antioxidant balance in seedlings exposed to heating and then to light for recovery. High oxidant level seems to be one of the basic reasons of heat-induced photoinhibition, which retards recovery from heat stress, whereas light intensity, energy support and prooxidant-antioxidant balance are crucial for the recovery of PM.

Recovery of Photosynthetic Machinery after Short-term Heat Stress

The damage of photosynthetic apparatus by application of elevated temperatures in dark has been studied in higher plants, green algae as well as in cyanobacteria. In each case, the thermal inactivation sites were found to be numerous, but it is difficult to describe each step of the damage as the deteriorations can be sequential or simultaneous. However, it is possible to weigh the relative sensitivity of the target sites. Such studies involved monitoring time and extent of repair after the removal of thermal stress (Mohanty *et al.,* 2007). Studies recently done on wheat plants on the post dark heat stress recovery exhibited the complexity of the recovery process. In dark, the recovery of carbon assimilation and PS II performance depended on mitochondrial phosphorylation, while in light the recovery was influenced by temperature and light intensity (Kreslvaski *et al.,* 2008). At least two temporally separated steps were identified in the path of recovery from heat stress at 40°C and 42°C in the plants growing in high and moderate/high light, respectively. Both photochemical activity of PS II and the activity of CO_2 assimilation system were lowered during the first step in comparison with the corresponding activities immediately after heat treatment. During the second step, the photosynthetic activities were completely or partly recovered. Recovery from heat stress at 40°C was accompanied by the appreciably higher rate of cyclic photophosphorylation in comparison with control non-heated seedlings. Combination of moderate heat stress and low or moderate light intensity during post-stress cultivation of seedlings led to almost complete recovery of photosynthesis.

The complete recovery of photosynthetic activity reduced by moderate heat treatment supports the argument that in nature moderate heat stress provides a mechanism for dynamic regulation of the activity of the PM rather than for destructive processes (Sharkey, 2005). At high irradiance, the recovery of photosynthesis was limited as compared with the photosynthesis of plants exposed to low irradiance (Kreslavski *et al.,* 2008). We indicated dark recovery after both heat treatments at 40°C and 42°C. Our data suggest that mitochondrial ATP support plays a role in repair and recovery.

Low intensity light seems to be the most optimal for the recovery from heat-induced photoinhibition (Kreslavski *et al.,* 2009).

Quaternary Ammonium Compounds (Glycinebetaine and Choline Chloride) Enhance Protection and Recovery from Heat Stress

Osmolyte accumulation, particularly GB, is one of the key factors to enhance plant tolerance to elevated temperatures. GB, proline and choline chloride help to conserve enzyme activity as well as to stabilize membrane integrity against extreme temperatures, reduce membrane lipid peroxidation, and protect PS II electron transport (Gorham,1995; Caldas *et al.,* 1999; Chen and Murata, 2002; Ohnishi and Murata, 2006; Sheng *et al.,* 2006; Allakhverdiev *et al.,* 2007; Khan *et al.,* 2009). Exogenous application of GB in order to stabilize PS II catalyzed functions has been reviewed earlier (Papageorgiou and Murata, 1995). It has been shown that GB enhances the stability of photochemical performance of the RCs in thylakoid membranes (Mamedov *et al.,* 1993), PS II particles (Allakhverdiev *et al.,* 1996), and D1/D2/Cytb559 complexes (Allakhverdiev *et al.,* 2003). Exogenic GB decreased the temperature of 50% inhibition ($T_{1/2}$) of various processes in PS II and demonstrated a significant effect of GB on almost all photosynthetic characteristics measured *in vitro* (Allakhverdiev *et al.,* 1996).

Plant genetic engineering may be one of the answers to the global warming phenomenon in terms of producing high temperature-tolerant crops. A number of plants have been engineered to contain genes of the GB biosynthetic pathway, which confers enhanced tolerance to a range of abiotic stresses during various plant developmental stages. Unlike natural accumulators, the transgenic plants accumulate very low GB concentrations, insignificant in terms of coping with osmotic stress. The GB accumulation in these transgenic plants varies depending upon their capacity for endogenous choline uptake, the type of gene that catalyzes the GB biosynthetic pathway, and the localization of the transgene product in a particular cellular compartment (Khan *et al.,* 2009). Murata's laboratory in Japan introduced the bacterial *codA* gene, for choline oxidase that converts choline to GB. Transformed plants such as *Arabidopsis,* rice, and tomato were endowed with property for tolerance to various types of stresses: low temperature stress, freezing, highlight or salt stress (Allakhverdiev *et al.,* 1999, 2001, 2002, 2004; for reviews on the subject see Sakamoto and Murata, 2002; Chen and Murata, 2002; Murata *et al.,* 2007; Allakhverdiev and Murata, 2008).

The genetically engineered line of tobacco (*Nicotiana tobacum*) for synthesis of GB did accumulate GB in chloroplasts; the transformed seedlings grew more rapidly at higher temperatures than the wild type. Carbon fixation in transgenic plants was significantly more tolerant to elevated temperatures (Yang *et al.,* 2005). It was argued that synthesis of GB in chloroplasts maintained the activation of Rubisco by sequestering Rubisco activase near thylakoids and preventing its thermal inactivation (Yang *et al.,* 2005, 2007). The accumulation of GB in BADH-transgenic tobacco increased the tolerance of PSII to very high temperatures (up to 50°C) and improved the thermostability of the OEC and the reaction center of PSII (Yang *et al.,* 2007). It is thought that the increased tolerance of PS II to heat stress is associated with the prevention of the heat-induced photoinhibition of PS II.

Precursors of GB-chlorocholine chloride and choline chloride can also protect PM from various environmental stress factors (Kreslavski *et al.,* 2001; Sheng *et al.,* 2006). Root application or spraying of choline chloride or chlorocholine chloride improved the heat tolerance and enhanced the activities of antioxidative enzymes such as SOD and APX in bean leaves and provided heat stability for PS II

(Kreslavski *et al.,* 2001). Enhanced activity of peroxidases was also demonstrated immediately after heat stress in first wheat leaves (Kreslavski *et al.,* 2010). Formation of betaine during decomposition of compounds such as choline chloride and enhancement of antioxidant enzyme activities in treated plants (Kreslvaski *et al.,* 2001, 2010; Sheng *et al.,* 2006) contribute to heat tolerance.

These observations suggest that cells with elevated ability for choline synthesis could be more heat stress resistant than ordinary cells. However, the cyanobacterium *Synechococcus* lacks the ability to synthesize choline. *Synechococcus* cells transformed with gene *codA* for choline oxidase, named PAMCOD acquired heat tolerance ability when these cells were supplemented exogenously with choline chloride, but not its non-transformant PAM cells. In transformed PAMCOD cells, the heat induce half inactivation temperature ($T_{½}$) of Hill activity for O_2 evolution was shifted from 46°C to 54°C and $T_{½}$ for PS II photochemistry was shifted from 51°C to 58°C, exhibiting a significant increase in heat tolerance ability in transformant PAMCOD cells over non-transformant PAM cells (Allakhverdiev *et al.,* 2007). Heat stress combined with light stress caused the inactivation of PS II activity faster than when the two stresses were applied separately as heat stress enhanced the inhibition of PS II repair. The accumulation of GB in PAMCOD cells alleviated such inhibitory effect of heat stress on the repair of PS II by accelerating the synthesis of D1 protein of PS II RC. Therefore, GB accumulation gave to the transformant cells the ability to sustain combined heat and light stresses (Allakhverdiev *et al.,* 2007).

One of the explanations for the positive action of GB on thermotolerance was received on the basis of *in vitro* experiments with D1/D2/Cytb559 complexes isolated from PS II membranes, which indicated that the presence of GB enhanced the thermotolerance of such complexes with a shift in temperature for 50% inactivation from 29°C to 35°C (Allakhverdiev *et al.,* 2003). It has been suggested that elevated temperatures increase the distance between D1 and D2 proteins resulting in changes in impairment of its functions (Braun *et al.,* 1990). The stabilization and protective effect of GB against heat inactivation in isolated D1/D2/Cytb559 and PS II complexes can be explained with minimization of the water-protein interfacial area that leads to preferential hydration of the proteins and, hence, to their tighter packaging and stabilization (Allakhverdiev *et al.,* 1996, 2003).

Role of Heat Shock Proteins in Photosynthetic Machinery Acclimation

HSPs are important for protecting plants against high temperature and other stresses (Barua *et al.* 2003). Addition of purified chloroplast-localized HSPs conferred heat tolerance to the photosynthetic electron transport chain in isolated chloroplasts. *In vivo*, experiments demonstrated that small HSPs could associate with thylakoids and protect O_2 evolution and OEC proteins of PS II against heat stress. Evidence for the significance of chloroplast-localized HSPs for thermotolerance was obtained in tomato species (Heckathorn *et al.,* 1998, 2002). Neta-Sharir *et al.* (2005) demonstrated that the chloroplasts small heat shock protein, HSP21, induced by heat treatment in tomato leaves protected PS II from temperature-dependent oxidative stress. It is considered that chloroplast HSPs do not participate in the repair of stress-related damage, but rather they function to prevent damage (Downs *et al.,* 1999; Heckathorn *et al.,* 2002). HSPs can function in protection as molecular chaperons to prevent but no reverse protein denaturation and aggregation, as membrane stabilizers and possibly, as site-specific antioxidants (Barua *et al.,* 2003). HSP101 has been shown to be essential for thermotolerance by genetic analysis (Hong and Vierling, 2001; Myouga *et al.,* 2006). By genetic manipulation, the role and sub-cellular localization of the small (16 kDa) HSP - HspA was investigated comparing the cyanobacterium *Synechococcus* strain ECT16-1, with constitutively expressed HspA, with the reference strain (Nitta *et al.,* 2005). The protein possesses the unique property to associate with thylakoid membranes during heat stress and support stability of thylakoid membranes.

Interesting results were obtained from proteomic analysis of the heat shock response in *Synechocystis* (Slabas *et al.*, 2006). The examination of heat-induced proteins in *Synechocystis*, which had been grown at 34°C and transferred to 44°C for 60 min indicated changes in 65 different proteins. The most remarkable changes that occurred upon heat shock corresponded to components involved in pathways for the folding, breakdown of aggregates, and degradation of proteins. The small heat-shock protein HspA showed the greatest change, with an increase in level of almost 20-fold. Smaller changes in levels were observed for GroES, GroEL1, GroEL2, HtpG, DnaK, and ClpB, which is an ATP-dependent Clp protease. The PS II complex and, in particular, the oxygen-evolving machinery is the most heat-sensitive component of the photosynthetic apparatus. After heat shock, there is a twofold increase in the level of the manganese-stabilizing protein, PsbO. Levels of phycobillisome proteins fall upon heat shock and the level of RbcL, the large subunit of Rubisco falls to less than half of the control level (Los *et al.*, 2008).

Recently Salvucci (2008) studied the association of Rubisco activase with chaperonin-60b, the chloroplast GroEL homologue. Using an affinity capture procedure, Cpn60b was identified as a protein that was bound to Rubisco activase from leaf extracts prepared from heat-stressed, but not control plants. Thus, this is a protein that potentially interacts with Rubisco activase during heat stress. The observations suggest that the association of Rubisco activase with Cpn60b during heat stress could function to protect Rubisco activase from thermal denaturation, thus providing a mechanism to protect and acclimate photosynthesis to heat stress.

GENETIC BASIS OF HEAT TOLERANCE

Thermotolerance is a complex cellular feature, which depends on the function of many biochemical pathways, as well as on the current physiological state of cells and the whole plant. These aspects of thermotolerance and set of genes implicated to impart heat tolerance have been reviewed (Inaba *et al.*, 2003; Maestri *et al.*, 2002; Los and Murata, 2004; Singh and Grover, 2008). Besides the genetic stock diversity and analysis of QTL, the following factors can be important in providing heat tolerance in plant cells: 1) HSPs; 2) antioxidants; 3) membrane lipid unsaturation; 4) gene expression and translation; 5) protein heat stability; 6) accumulation of solutes. Conventional breeding for high temperature stress tolerance has not been much successful due to several reasons like lack of suitable source of genes in sexually compatible gene pools, complex nature of the HS trait, lack of understanding of the genetic mechanisms of the high temperature tolerance response, etc. (Singh and Grover, 2008).

Genetic engineering for heat tolerance in phototrophic organisms progressed significantly during the last decades. Apart from heat shock proteins, thermotolerance has also been altered by elevating levels of osmolytes, increasing levels of cell detoxification enzymes, and through altering membrane fluidity. It is suggested that HSPs may be directly implicated in thermotolerance as agents that minimize damage to cell proteins. In last years, detailed comprehension has been gained on various components of the system of heat stress response in phototrophic organisms including heat shock genes/proteins, heat shock promoters and heat shock elements, heat shock transcription factors (HSFs), possible receptors of the heat shock response (Fig. 14.2), signaling components and chromatin remodeling aspects (Grover, 2002, Wahid *et al.*, 2007; Los *et al.*, 2008; Singh and Grover, 2008).

Heat Stress-Inducible Genes in *Synechocystis* and Plants

Response to stress begins with perception of stress and transduction of the stress signal. A combination of the systematic mutation of potential sensors and transducers and DNA microarray analysis has led to significant progress in understanding the mechanisms of perception of and reaction to environmental stress, especially heating, in cyanobacteria. The recent progress in the identification of stress-inducible

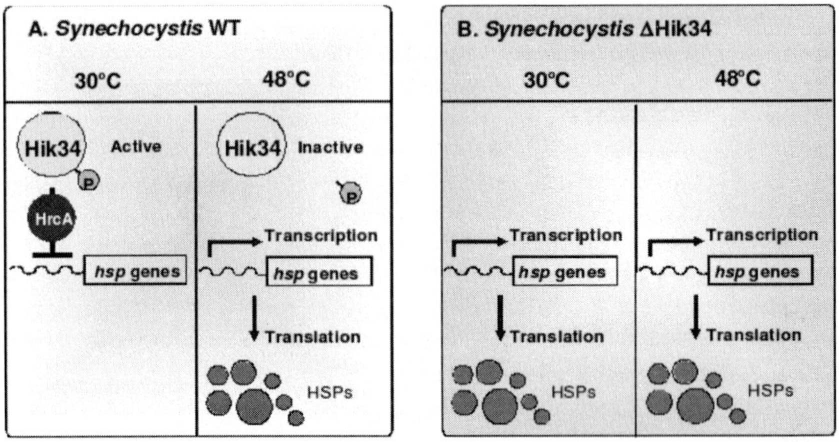

Fig. 14.2 A scheme for heat-induced expression of genes that are controlled by the histidine kinase Hik34. In wild-type cells (A), Hik34 is autophosphorylated at normal temperature (30°C), and it represses transcription of the genes for HSPs. Hik34 may operate through transcription factor HrcA, which also negatively regulated transcription for *hsp* genes. At high temperatures (48°C) Hik34 is no longer phosphorylated, and expression of genes for HSPs is allowed. In cells that are defective in Hik34 (B), transcription of genes for HSPs and translation of HSPs run constitutively and independent of temperature.

genes and systems that regulate responses to stress has been made using DNA microarrays both in plants (Wahid *et al.*, 2007; Singh and Grover, 2008) and cyanobacteria (Suzuki *et al.*, 2005; Los *et al.*, 2008).

Several reports demonstrated that plants can be genetically engineered for high temperature tolerance by altering HSPs either directly or through regulatory chains regulating HSPs levels, levels of osmolytes, components of the cell detoxification mechanisms, and components regulating membrane fluidity (Burke and Chen, 2006; Singh and Grover, 2008). Details of genetic engineering for heat tolerance in plants are reviewed by Singh and Grover (2008) and in *Synechocystis* sp. PCC 6803 (Los *et al.*, 2008; Los *et al.*, 2010).

Heat-induced gene expression was studied in *Synechocystis* at the genome-wide level by Inaba *et al.*, (2003) and Suzuki *et al.* (2005). Upon heat stress, achieved by exposing *Synechocystis* cells, which had been grown at 34°C, to 44°C for 20 min, the levels of expression of 59 genes rose more than threefold and those of 58 other genes fell more than threefold (Suzuki *et al.*, 2005). Most of the 59 genes were strongly induced within 10–20 min after an upward shift in temperature from 34°C to 44°C. However, 1 hour after the onset of heat stress, the levels of their transcripts declined significantly.

Genes, whose expression is rapidly and strongly induced by heat stress are listed in Table 14.1 and a general scheme for heat-induced expression of genes that are controlled by the histidine kinase Hik34 is illustrated in Fig. 14.2. The transiently induced genes include *hspA, groES, groEL1, groEL2, dnaJ, htpG,* and *dnaK2,* which encode proteins that are mostly chaperones. Heat stress also induces the expression of the *clpB1* and *htrA* genes for proteases, the *sigB* gene for a group 2 sigma factor of RNA polymerase, and the *hik34* gene, as well as the *sodB* gene for SOD and the *hliC* gene for a high light-inducible protein. HSPs are shown to prevent the aggregation or denaturation of proteins during heat stress (Low *et al.*, 2000). Stressed cells tend to return to their unstressed state by minimizing damage to protein, primarily through chaperoning activity of the HSPs (Fig. 14.2).

Transcription Factors

As mentioned above, eukaryotic cells respond to elevated temperature or heat shock (HS) by inducing the transcription of genes encoding proteins such as molecular chaperones. Many of these proteins

Table 14.1

Heat-inducible genes in *Synechocystis* sp. PCC 6803.

ORF gene encoded protein ratio of transcript levels

slr1641 clpB1 ClpB protease 92.5 ± 9.1

sll1514 hspA Small heat-shock protein 68.1 ± 2.9

slr2075 groES 10-kDa chaperonin 44.7 ± 2.3

sll0416 groEL2 60-kDa chaperonin 2 42.1 ± 7.7

slr0093 dnaJ Heat-shock protein 40 36.9 ± 3.6

sll0430 htpG Heat-shock protein 90 34.6 ± 5.6

slr0095 o-Methyltransferase 24.4 ± 4.6

sll0306 sigB RNA polymerase group 2 s-factor 22.9 ± 0.7

slr1963 Water-soluble carotenoid protein 22.8 ± 0.0

sll0170 dnaK2 Heat-shock protein 70 22.3 ± 1.5

slr1285 hik34 Histidine kinase 34 18.9 ± 2.6

sll1621 AhpC/TSA family protein 11.0 ± 0.6

ssl3044 Ferredoxin-like protein 10.3 ± 0.0

slr2076 groEL1 60-kDa chaperonin 1 8.7 ± 1.5

ssl1633 hliC CAB/ELIP/HLIP superfamily 6.9 ± 0.5

slr1204 htrA Serine protease 6.0 ± 0.7

slr1516 sodB Superoxide dismutase 5.6 ± 0.3

sll1483 Salt-inducible periplasmic protein 5.1 ± 1.1

sll1009 frpC Iron-regulated protein 4.9 ± 0.5

slr1675 hypA1 Hydrogenase component protein 4.7 ± 0.1

sll1507 Salt-induced periplasmic protein 4.0 ± 0.6

Cells grown in standard growth medium were incubated in the presence of 0.5 M NaCl for 20 min. Genes with ratios of transcript levels greater than 4.5 are listed. Data adapted from Shoumskaya *et al.,* 2005.

are involved in preventing or repairing the damage caused by heat stress and thus confer increased thermotolerance (Vierling, 1991). This phenomenon, known as heat shock responses, is initiated by the activation of the heat shock transcription factors (HSFs), which are the central regulators of the heat shock stress response in all eukaryotic organisms (Baniwal *et al.,* 2004). The heat shock response was shown to be under the control of HSFs (Volkov *et al.,* 2006). The review of Miller and Mittler (2006) describes the recent studies, which support the hypothesis that HSFs can function as molecular sensors that are able to directly sense ROS, for instance H_2O_2, and control the expression of oxidative stress response genes during oxidative stress. Hence, HSFs can be sensitive sensors for heat stress, however, the mechanism involved is still to be unveiled.

Other Factors of Thermal Stability of Photosynthetic Machinery

HSPs are key factors in heat tolerance. However, other compounds different from HSPs can also contribute to heat tolerance (Nishiyama *et al.,* 1999; Tanaka *et al.,* 2000). Although the extrinsically associated proteins are not necessary for oxygen evolution activity *in vitro*, they are required to enhance oxygen evolution and play important roles *in vivo* (reviewed in Seidler, 1996). In the cyanobacterium *Synechococcus* sp. PCC 7002, the oxygen-evolving machinery is stabilized against heat-induced inactivation by PsbU, which is an extrinsic protein of the PS II complex (Nishiyama *et al.,* 1999). Inactivation of *psbU* gene lowered the thermal stability of OEC in the mutant in comparison with the WT cells. However, the levels of HSPs, namely the homologs of HSP70, HSP60, and HSP17 remained unaffected by mutation in *psbU* gene, suggesting that HSPs are not involved in cell acclimation to high temperature (Nishiyama *et al.,*

1999). Similarly, Tanaka *et al.* (2000) showed that the crucial role in heat resistance of OEC of the PS II is played by proteins encoded by nuclear and chloroplast genomes rather than by 70, 60 and 22 kDa HSP. Conceivably, their stabilization requires the synthesis of extrinsic proteins of PS II (e.g., PsbU) (Nishiyama *et al.,* 1999) manifesting cooperative activity during stabilization of OEC of the PS II. It is possible that OEC thermal stability and acclimation is mostly regulated by *psbO* gene that produces the 33 kDa extrinsic protein (Enami *et al.,* 1998; Pueyo *et al.,* 2002, Ferreira *et al.,* 2004).

INTERACTION BETWEEN HEAT STRESS AND OTHER STRESS FACTORS

Heat Stress and Photoinhibition

Heat stress and strong light usually occur together in nature, and their joint manifestation can lead to strengthening of their damaging effects to the PM. On the contrary, exposure to low or moderate irradiance during heat treatment is often beneficial to plants diminishing the extent of damage to the PM (Havaux *et al.,* 1991). Light is often required to tolerate PM to heat stress, especially for the repair of photosystems, using weak light for protein photophosphorylation and stimulation of the activity of many enzymes. In contrast, exposure to strong light increases the magnitude of PM inhibition induced by heat stress. This might be linked to the formation of extra ROS during heat-induced photoinhibition, which affect the repair of PS II (Al-Khatib and Paulsen, 1989; Kreslavski *et al.,* 2008, 2009). This finding is consistent with the hypothesis according to which moderate heat stress is accompanied by the PS II repair without any effect on its photodamage (Nishiyama *et al.,* 2005, 2006; Murata *et al.,* 2007).

Some studies report that exposure to elevated temperatures higher than growth temperature ensures an advantage for plant acclimation to strong light (Havaux, 1993; Kreslavaski *et al.,* 2008). This type of resistance to stress factors is called "cross-tolerance" (Allen, 1995; Pastori and Foyer, 2002), but the mechanism of such cross adaptation is not fully clear. Short-time heat pre-treatment increased the tolerance of PS II to light and elevated temperatures, and short-time exposure to strong light increased the thermostability of PS II in *L. esculentum* leaves (Havaux and Tardy, 1996). Short-time pre-treatment of wheat seedlings at 40°C enhanced their tolerance to following high light exposure (Kreslavski *et al.,* 2008) (Fig. 14.3) that can be explained by the accumulation of HSPs, protecting PM from photodamage. The enhancement might be also due to the higher rate of CO_2 assimilation indicated in hardening plants and, likely, higher rate of cyclic photophosphorylation indicated in chloroplasts isolated from heat-pre-treated seedlings. Accumulation of HSPs (Schroda *et al.,* 1999) seems to contribute to protecting the PA from photoinhibition as well.

It is known that high temperatures inactivate Rubisco activase (Feller *et al.,* 1998; Crafts-Brandner and Salvucci, 2000), which is essential for the activity of Rubisco. Under such conditions, inhibition of the repair of photodamaged PS II as a response to limitations in fixation of CO_2 might accelerate photoinhibition. According to this hypothesis, heat accelerates photoinhibition by inhibiting the repair of photodamaged PS II (Yang *et al.,* 2007). Although these various stresses might inhibit the repair of photodamaged PS II directly, their inhibitory effects might be attributable, in part, to limitation of the fixation of CO_2 in the Calvin cycle (Murata *et al.,* 2007; Mohanty *et al.,* 2007; Takahashi and Murata, 2008).

Moderate heat stress can stimulate dark reduction of plastoquinone and cyclic electron flow in the light (Bukhov *et al.,* 1998, Sharkey, 2005). The assessment of cyclic and non-cyclic photophosphorylation in preparations of thylakoid membranes isolated from leaves of wheat seedlings damaged by heat stress and recovered in light demonstrated that cyclic electron flow was enhanced in the preparations of thylakoid membranes from recovered plants (Kreslavski *et al.,* 2008). These observations are in line with earlier

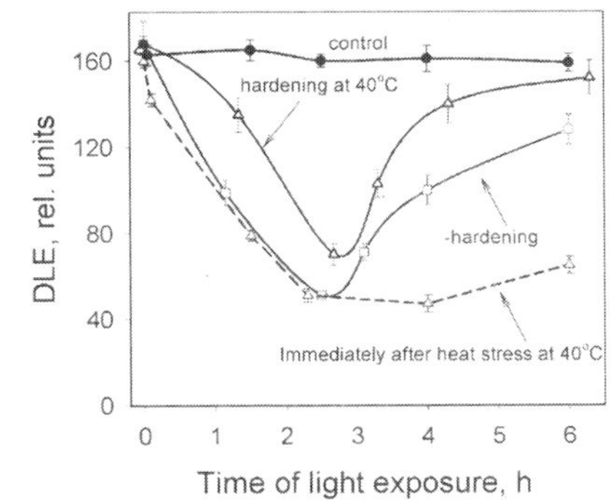

Fig. 14.3 Effect of heat hardening at 40°C on photoinhibition. Seedlings grown for 48 h at 120 μmol m^{-2} s^{-1} were either pre-heated (△) or not (◆, □) and then exposed to strong light (I = 900 μmol m^{-2} s^{-1}) either 24 hr after (◆, △, —) or immediately after (△, ----) heat treatment at 40°C for 20 min ◇. Other, non-pre-treated seedlings (● control) were grown for 48 h at 120 μmol m^{-2} s^{-1} and then exposed to light at I = 120 μmol m^{-2} s^{-1} for 6 h. Means ± SE (*n* = 4). Adapted from Kreslavski *et al.* 2008.

suggestions that cyclic electron flow could provide additional ATP required for protein synthesis and protect PS II from photoinhibition (Havaux *et al.,* 1991; Miyake and Okamura, 2003; Allakhverdiev *et al.,* 2005a).

In spite of complexities, it is clear that post temperature stress recovery needs the protein biosynthetic machinery to repair the heat or photodamage of PS II (Takahashi *et al.,* 2004, Yang *et al.,* 2007). A number of studies also showed that weak light, energy support and suitable balance of oxidants and antioxidants are necessary for the effective repair of PM (Allakhverdiev *et al.,* 2005a,b; Kreslavski *et al.,* 2007, 2009). It may also demonstrated that heat hardening improves the recovery of PM from photoinhibition (Fig. 14.3), and that one of the reasons for the improved recovery from photoinhibition in hardening plants might be the higher activity of antioxidant enzymes and increased content of low-molecular antioxidants in the chloroplasts of hardened plants.

Effects of High Temperature and Salt Stress

There is a cross-tolerance between high salinity and heat stress. Salinity treatment showed no effects on photosystem II photochemistry, but increased the resistance of PS II to heat stress in the halophytes *Suaeda salsa* (Lu *et al.,* 2003) and *Artemisia anethifolia* (Wen *et al.,* 2005). High salinity (100-400 mM NaCl) had no effects on PS II photochemistry either in a dark-adapted state or in a light-adapted state, however, heat tolerance of PS II was increased after salt treatment. These results suggest that the increased thermostability of PS II seems to be associated with the increased resistance of the OEC and the reaction centres of PS II to high temperature (Lu *et al.,* 2003; Wen *et al.,* 2005) as well as the light-harvesting complex (Wen *et al.,* 2005). In addition to this, in leaves of salt-adapted plants, the CO_2 assimilation system was more thermoresistant than in the control (Lu *et al.,* 2003).

Effects of High Temperature and Drought

The most destroying stress combination for crop productivity, which frequently occurs in the field, is drought and heat stress. The proteomic and metabolic analysis of plants subjected to joint action of drought and elevated temperatures indicated 45 different proteins that accumulated in response to stress combination (Koussevitzky *et al.,* 2008). Cytosolic APX1 protein and mRNA accumulated under stress combination. The data obtained with various *Arabidopsis* mutants deficient in APX1, as well as thylakoid or stromal/mitochondrial peroxidases demonstrated that APX-deficient mutant (*apx1*) as opposed to other mutants was significantly more sensitive to stress combination than the wild type. The authors suggested that APX1 plays a key role in response to joint combination of heat stress and drought.

CONCLUSION

The recent advancements in the studies of the effect of heat stress on photosynthesis lead to the following outcome: the primary target sites of thermal impairment or inactivation are the OEC with associated cofactors and suppression of carbon fixation by Rubisco, most likely due to inactivation of Rubisco activase. These two sites of impairment occur together under moderately elevated temperatures for short duration. The reduction in carbon fixation and oxygen evolution activities is followed by disruption of the linear electron flow, which induces the generation of ROS and ROS inhibit the repair of damaged PS II, which largely depends on *de novo* synthesis of proteins of PM (Allakhverdiev *et al.* 2008). The repair system for stress-induced damage is complex and involves multistep mechanisms, whose nature is not completely understood.

Last investigations on heat stress responses are focused on the role of the HSFs, chaperones and heat-induced genes important for plant cell heat acclimation. However, there is little information about chloroplastic (photosynthetic) heat tolerance. HSFs can be sensitive sensors for heat stress in chloroplast, although, the mechanism needs to be examined (Singh and Grover, 2008).

The following areas may remain in focus for future research

(i) The relationship between generation of ROS during heat stress and regulation of activity of genes responsible for the synthesis of stress proteins important for protection and acclimation of the PM.

(ii) The role of thylakoid membrane changes and redox-potential of components of ETC, first PQ pool, in regulation of the photosynthetic tolerance during heat stress.

(iii) The molecular mechanism of stress-induced inhibition of PS II recovery in relation to suppression of protein synthesis by ROS.

In summary, there are multiple effects of elevated temperature on the PA involving reduction in activities or even damage of all parts of the photosynthetic machinery. However, their heat sensitivity and tolerance are different. It may be suggested that under natural conditions photosynthetic regulation during heat stress occurs mainly by inhibition of majority of photosynthetic activities, especially inhibition of the repair of the PS II. At the same time, other activities such as the structural rearrangements in chloroplast thylakoid membranes, enhanced heat dissipation, etc. protect the PM from damage. Prooxidant/antioxidant levels, light intensity during heating and ATP supply are the crucial factors for the extent of the recovery of PM from heat induced damage. Finally, the rate of repair of photosynthetic proteins, first of all, D1 protein, is a key element for recovery of heat damaged PS II. The thylakoid membrane integrity as well as thermostability of photosynthetic structures, redox-potential of components of ETC of the chloroplast and a relationship between antioxidants and oxidants play a key role in resistance and acclimation of the PM to heat stress.

ACKNOWLEDGEMENTS

This work was supported, in part, by grants from Russian Ministry of Science and Education (No: 2010-1.2.1-102-022-033), and from the Molecular and Cellular Biology Programs of the Russian Academy of Sciences, and by a Grant-in-Aid for Creative Scientific Research (No. 17GS0314) from the Japanese Society for the Promotion of Science (JSPS) and by Scientific Research on Priority Areas "Comparative Genomics" (Nos: 17018022 and 18017016) from the Ministry of Education, Sports, Culture, Science, and Technology, Japan, to MM. PM acknowledges the support of INSA, JNU and DST/RAS. RC was supported by NSERC.

REFERENCES

Adir N, Zer H, Shochat S, Ohad I (2003) Photoinhibition a historical perspective. Photosynth Res 76: 343-370

Al-Khatib K, Paulsen GM (1989) Enhancement of thermal injury to photosynthesis in wheat plants and thylakoids by high light intensity. Plant Physiol 90: 1041-1048

Allakhverdiev SI, Klimov VV, Carpentier R (1994) Variable thermal emission and chlorophyll fluorescence in photosystem II particles. Proc Natl Acad Sci USA 91: 281-285

Allakhverdiev SI, Feyziev YM, Ahmed A, Hayashi H, Aliev JA, Klimov VV, Murata N, Carpentier R (1996) Stabilization of oxygen evolution and primary electron transport reactions in photosystem II against heat stress with glycinebetaine and sucrose. J Photochem Photobiol 34: 149-157

Allakhverdiev SI, Yruela Y, Picorel R, Klimov VV (1997) Bicarbonate is an essential constituent of the water-oxidizing complex of photosystem II. Proc Natl Acad Sci USA 94: 5050-5054

Allakhverdiev SI, Nishiyama Y, Suzuki I, Tasaka Y, Murata N (1999) Genetic engineering of the unsaturation of fatty acids in membrane lipids alters the tolerance of *Synechocystis* to salt stress. Proc Natl Acad Sci USA 96: 5862-5867

Allakhverdiev SI, Kinoshita M, Inaba M, Suzuki I, Murata N (2001) Unsaturated fatty acids in membrane lipids protect the photosynthetic machinery against salt-induced damage in *Synechococcus*. Plant Physiol 125: 1842-1853

Allakhverdiev SI, Nishiyama Y, Miyairi S, Yamamoto H, Inagaki N, Kanesaki Y, Murata N (2002) Salt stress inhibits the repair of photodamaged photosystem II by suppressing the transcription and translation of *psbA* genes in *Synechocystis*. Plant Physiol 130: 1443-1453

Allakhverdiev SI, Hayashi H, Nishiyama Y, Ivanov AG, Aliev JaA, Klimov VV, Murata N, Carpentier R (2003) Glycine betaine protects the D1/D2/Cytb559 complex of photosystem II against photo-induced and heat-induced inactivation. J Plant Physiol 160: 41-49

Allakhverdiev SI, Murata N (2004) Environmental stress inhibits the synthesis de novo of proteins involved in the photodamage-repair cycle of photosystem II in *Synechocystis* sp. PCC 6803. Biochim Biophys Acta 1657: 23-32

Allakhverdiev SI, Mohanty P, Murata N (2004) Dissection of photodamage at low temperature and repair in darkness suggests the existence of an intermediate form of photodamaged photosystem II. Biochem 42: 14277-14283

Allakhverdiev SI, Nishiyama Y, Takahashi S, Miyairi S, Suzuki I, Murata N (2005a) Systematic analysis of the relation of electron transport and ATP synthesis to the photodamage and repair of photosystem II in *Synechocystis*. Plant Physiol 137: 263-273

Allakhverdiev SI, Tsvetkova N, Mohanty P, Szalontai B, Moon BY, Debreczeny M, Murata N (2005b) Irreversible photoinhibition of photosystem II is caused by exposure of *Synechocystis* cells to strong light for a prolonged period. Biochim Biophys Acta 1708: 342-351

Allakhverdiev SI, Los DA, Mohanty P, Nishiyama Y, Murata N (2007) Glycinebetaine alleviates the inhibitory effect of moderate heat stress on the repair of photosystem II during photoinhibition. Biochim Biophys Acta 1767: 1363-1371

Allakhverdiev SI, Murata N (2008) Salt stress inhibits photosystems II and I in cyanobacteria. Photosynth Res 98: 529-539

Allakhverdiev SI, Kreslavski VD, Klimov VV, Los DA, Carpentier R, Mohanty P (2008) Heat stress: an overview of molecular responses in photosynthesis. Photosynth Res 98: 541-550

Allen JF (2004) Cytochrome b_6f: structure for signalling and vectorial metabolism. Trends Plant Sci 9: 130-137

Allen JF, Alexciev K, Hakansson G (1995) Regulation of redox signaling. Curr Biol 5: 869-872

Allen R (1995) Dissection of oxidative stress tolerance using transgenic plants. Plant Physiol 107: 1049-1054

Aminaka R, Taira Y, Kashino Y, Koike H, Satoh K (2006) Acclimation to the growth temperature and thermosensitivity of photosystem II in a mesophilic cyanobacterium, *Synechocystis* sp. PCC 6803. Plant Cell Physiol 47: 1612-1621

Andersson B, Aro E-M (2001) Photodamage and D1 protein turnover in photosystem II. In: Eds Aro E-M., Andersson B (eds) Regulation of Photosynthesis. Kluwer Acad Publ, Dordrecht, pp 377-393.

Aro E-M, Virgin I, Andersson B (1993) Photoinhibition of photosystem II: Inactivation, protein damage and turnover. Biochim Biophys Acta 1143: 113-134

Asada K (1999) The water-water cycle in chloroplasts: scavenging of active oxygen and dissipation of excess photons. Annu Rev Plant Physiol Plant Mol Biol 50: 601-639

Asada K (2006) Production and scavenging of reactive oxygen species in chloroplasts and their functions. Plant Physiol 141: 391-396

Balint I, Bhattacharya J, Perelman A, Schatz D, Moskovitz Y, Keren N, Schwarz R (2006) Inactivation of the extrinsic subunit of photosystem II, PsbU, in *Synechococcus* PCC 7942 results in elevated resistance to oxidative stress. FEBS Lett 580: 2117-2122

Baniwal SK, Bharti K, Chan KY, Fauth M, Ganguli A, Kotak S, Mishra SK, Nover L, Port M, Scharf KD (2004) Heat stress response in plants: a complex game with chaperones and more than twenty heat stress transcription factors. J Biosci 29:471–487

Barua D, Downs CA, Hechthorn SA (2003) Variation in chloroplast small heat-shock protein function is a major determinant of variation in thermotolerance of photosynthetic electron transport among ecotypes of *Chenopodium album*. Funct Plant Biol 30: 1071-1079

Berry JA, Björkman O (1980) Photosynthetic response and adaptation to temperature in higher plants. Annu Rev Plant Physiol 31: 491-543

Bondarava N, Beyer P, Krieger-Liszkay A (2005) Function of the 23 kDa extrinsic protein of photosystem II as a manganese binding protein and its role in photoactivation. Biochim Biophys Acta 1708: 63-70

Boyer JS (1982) Plant productivity and environment. Science 218: 443-448

Braun P, Greenberg BM, Scherz A (1990) D1-D2-cytochrome b559 complex from the aquatic plant *Spirodela oligorchiza*: correlation between complex integrity, spectroscopic properties, photochemical activity and pigment composition. Biochem 29: 10376-10387

Bukhov NG, Boucher N, Carpentier R (1998) Loss of the precise control of photosynthesis and increased yield of nonradioactive dissipation of excitation energy after mild heat treatment of barley leaves. Physiol Plant 104: 563-570

Bukhov NG, Mohanty P (1999) Elevated temperature stress effects on photosystems: Characterization and evaluation of the nature of heat induced impairments. In: Singhal GS, Renger G, Sopory SK, Irrgang K-D, Govindjee (eds) Concepts in Photobiology: Photosynthesis and Photomorphogenesis. Narosa Publishing House, New Delhi, pp 617-648

Bukhov NG, Carpentier R (2000) Heterogeneity of photosystem II reaction centers as influenced by heat treatment of barley leaves. Physiol Plant 110: 279-285

Burke JJ, Chen J (2006) Changes in cellular and molecular processes in plant adaptation to heat stress. In: Huang B (ed) Plant-Environment Interactions, 3rd edn. CRC Press, Boca Raton, pp 27-46

Caldas T, Demont-Caulet N, Ghasi A, Richarme G (1999) Thermoprotection by glycinebetaine and choline. Microbiol 145: 2543-2548

Carpentier R (1999) Effect of high-temperature stress on the photosynthetic apparatus. In: Pessarakli M (ed) Handbook of Plant and Crop Stress. Marcel Dekker Inc, New York, pp 337-348

Carratu L, Franceschelli S, Pardini CL, Kobayashi GS, Horvath I, Vigh L, Maresca B (1996) Membrane lipid perturbation modifies the set point of the temperature of heat shock response in yeast. Proc Natl Acad Sci USA 93: 3870-3875

Chen TH, Murata N (2002) Enhancement of tolerance of abiotic stress by metabolic engineering of betaines and other compatible solutes. Curr Opin Plant Biol 5: 250-257

Chen HX, Li WJ, An SZ, Gao HY (2004) Characterization of PS II photochemistry and thermostability in salt-treated *Rumex* leaves. J Plant Physiol 161: 257-264

Chen LS, Li P, Cheng L (2008) Effects of high temperature coupled with high light on the balance between photooxidation and photoprotection in the sun-exposed peel of apple. Planta 228: 745-756

Crafts-Brandner SJ, Salvucci ME (2000) Rubisco activase constrains the photosynthetic potential of leaves at high temperature and CO_2. Proc Natl Acad Sci USA 97: 13430-13435

Dat J, Yandenbeele S, Vranova E, Van Montagu M, Inze D, Van Breusegm F (2000) Dual action of the active oxygen species during plant stress responses. Cell Mol Life Sci 57: 779-795

Dat JF, Lopez-Delgado H, Foyer CH, Scott IM (1998) Parallel changes in H_2O_2 and catalase during thermotolerance induced by salicylic acid or heat acclimation in mustard seedlings. Plant Physiol 116: 1351-1357

Dau H, Iuzzolino L, Dittmer J (2001) The tetramanganese complex of photosystem II during its redox cycle. X-ray absorption results and mechanistic implications. Biochim Biophys Acta 1503: 24-39

Davletova S, Rizhsky L, Liang H, Shengqiang Z, Oliver DJ, Coutu J, Shulaev V, Schlauch K, Mittler R (2005) Cytosolic ascorbate peroxidase 1 is a central component of the reactive oxygen gene network of *Arabidopsis*. Plant Cell 17: 268-281

De Las Rivas J, Heredia P (1999) Structural predictions on the 33 kDa extrinsic protein associated with the oxygen evolving complex of photosynthetic organisms. Photosynth Res 61: 11-21

Demmig-Adams B, Adams WW (2002) Antioxidants in photosynthesis and human nutrition. Sci 298: 2149-53

Dietzel L, Bräutigam K, Pfannschmidt T (2008) Photosynthetic acclimation: state transitions and adjustment of photosystem stoichiometry--functional relationships between short-term and long-term light quality acclimation in plants. FEBS J 275: 1080-1088

Downs CA, Coleman JS, Heckathorn SA (1999) The chloroplast 22 kDa heat-shock protein: a lumenal protein that associates with the oxygen evolving complex and protects photosystem II during heat stress. J Plant Physiol 155: 477-487

El-Shitinawy F, Ebrahim MKH, Sewelam N, El-Shourbagy MN (2004) Activity of photosystem 2, lipid peroxidation, and the enzymatic antioxidant protective system in heat shocked barley seedlings. Photosynthetica 42: 15-21

Enami I, Kitamura M, Tomo T, Isokawa Y, Ohta H, Katoh S (1994) Is the primary cause of thermal inactivation of oxygen evolution in spinach PS II membranes release of the extrinsic 33 kDa protein or of Mn? Biochim Biophys Acta 1186: 52-58

Enami I, Kamo M, Ohta H, Takahashi S, Miura Taro, Kusayanagi M, Tanabe S, Kamei A, Motoki A, Hirano M, Tomo T, Satoh K (1998) Intramolecular cross-linking of the extrinsic 33 kDa protein leads to loss of oxygen evolution but not its ability of binding to photosystem II and stabilization of the manganese cluster. J Biol Chem 273: 4629-4634

Feller U, Crafts-Brandner SJ, Salvucci ME (1998) Moderately high temperatures inhibit ribulose-1,5-biphosphate carboxylase/oxygenase (Rubisco) activase-mediated activation of Rubisco. Plant Physiol 116: 539-546

Ferreira KN, Iverson TM, Maghlaoui K, Barber J, Iwata S (2004) Architecture of the photosynthetic oxygen-evolving center. Sci 303: 1831-1838

Fey V, Wagner R, Bräutigam K, Pfannschmidt T (2005) Photosynthetic redox control of nuclear gene expression. J Exp Bot 56: 1491-1498

Foyer CH, Lopez-Delgado H, Dat JF, Scott IM (1997) Hydrogen peroxide- and glutathione-associated mechanisms of acclimatory stress tolerance and signaling. Physiol Plant 100: 241-254

Foyer C, Noctor G (2005) Oxidant and antioxidant signaling in plants: a reevaluation of the concept of oxidative stress in a physiological context. Plant Cell Environ 28: 1056-1071

Georgieva K (1999) Some mechanisms of damage and acclimation of the photosynthetic apparatus due to high temperature. Bulg J Plant Physiol 25: 89-99

Gombos Z, Wada H, Murata N (1991) Direct evaluation of effects of fatty-acid unsaturation on the thermal properties of photosynthetic activities, as studied by mutation and transformation of *Synechocystis* PCC 6803. Plant Cell Physiol 32: 205-211

Gombos Z, Wada H, Hideg E, Murata N (1994) The unsaturation of membrane lipids stabilizes photosynthesis against heat stress. Plant Physiol 104: 563-567

Gong M, Chen B, Li ZG, Guo LH (2001) Heat shock-induced cross adaptation to heat, chilling, drought and salt stresses in maize seedlings and involvement of H_2O_2. J Plant Physiol 158: 1125-1130

Gorham J (1995) Betaines in higher plants—biosynthesis and role in stress metabolism. In: Wallsgrove RM (ed) Amino Acids and Their Derivatives in Higher Plants. Cambridge University Press, Cambridge, pp 171-203

Gounaris K, Brain ARR, Quinn PJ, Williams WP (1983) Structural and functional changes associated with heat-induced phase separation of non-bilayer lipids in chloroplast thylakoid membranes. FEBS Lett 153: 47-53

Gounaris K, Brain ARR, Quinn PJ, Williams WP (1984) Structural reorganization of chloroplast thylakoid membranes in response to heat stress. Biochim Biophys Acta 766: 198-208

Grover A (2002) Molecular biology of stress responses. Cell Stress Chap 7: 1-5

Hall AE (2001) Crop Responses to Environment. CRS Press LLC, Boca Raton

Havaux M (1993) Rapid photosynthetic adaptation to heat stress triggered in potato leaves by moderately elevated temperatures. Plant Cell Environ 16: 461-467

Havaux M (1996) Short-term responses of photosystem I to heat stress – induction of a PS II-independent electron transport through PS I fed by stromal components. Photosynthesis through PS I fed by stromal components. Photosynth Res 47: 85-97

Havaux M, Greppin H, Strasser RJ (1991) Functioning of photosystems I and II in pea leaves exposed to heat stress in the presence or absence of light. Planta 186: 88-98

Havaux M, Tardy F (1996) Temperature-dependent adjustment of the thermal stability of photosystem II *in vivo*: possible involvement of xanthophyll-cycle pigments. Planta 198: 324-333

Heckathorn S, Downs SA, Sharkey TD, Soleman JS (1998) The small, methionine-rich chloroplast heat-shock protein protects photosystem II electron transport during heat stress. Plant Physiol 116: 439-444

Heckathorn SA, Ryan SL, Baylis JA, Wang D, Hamilton EW, Cundiff L, Luthe DS (2002) *In vivo* evidence from an *Agrostis stolonifera* selection genotype that chloroplast small heat-shock proteins can protect photosystem II during heat stress. Funct Plant Biol 29: 933-944

Hong SW, Vierling E (2001) Hsp101 is necessary for heat tolerance but dispensable for development and germination in the absence of stress. Plant J 27: 25-35

Horton P, Ruban A, Walters R (1994) Regulation of light harvesting in green plants. Indication by non-photochemical quenching of chlorophyll fluorescence. Plant Physiol 106: 415-20

Horvath I, Glatz A, Varvasovszki V, Török Z, Pali T, Balogh G, Kovacs E, Nadasdi L, Benko S, Joo F, Vigh L (1998) Membrane physical state controls the signaling mechanism of the heat shock response in *Synechocystis* PCC 6803: identification of *hsp17* as a fluidity gene. Proc Natl Acad Sci USA 95: 3513-3518

Hung S-H, Yu C-W, Lin CH (2005) Hydrogen peroxide functions as a stress signal in plants. Bot Bull Acad Sin 46: 1-10

Inaba M, Grandall P (1988) Electrolyte leakage as an indicator of high-temperature injury to harvested mature green tomatoes. J Am Soc Hortic Sci 113: 96-99

Inaba M, Suzuki I, Szalontai B, Kanesaki Y, Los DA, Hayashi H, Murata N (2003) Gene-engineered rigidification of membrane lipids enhances the cold inducibility of gene expression in *Synechocystis*. J Biol Chem 278: 12191-12198

Joët T, Genty B, Josse EM, Kuntz M, Cournac L, Peltier G (2002) Involvement of a plastid terminal oxidase in plastoquinone oxidation as enhanced by expression of the *Arabidopsis thaliana* enzyme in tobacco. J Biol Chem 277: 31623-31630

Kalitulo LN, Pshybutko NL, Kabashnikova LF, Jahns P (2003) Photosynthetic apparatus and high temperature: role of light. Bulg J Plant Physiol 32: 281-289

Katoh S, San Pietro A (1967) Photooxidation and reduction of cytochrome-552 and NADP photoreduction by *Euglena* chloroplast. Arch Biochem Biophys 121: 211-219

Khan MS, Yu X, Kikuchi A, Asahina M, Kazuo N. Watanabe (2009) Genetic engineering of glycinebetaine biosynthesis to enhance abiotic stress tolerance in plants. Plant Biotechnol 26: 125-134

Kim K, Portis ARJr (2004) Oxygen-dependent H_2O_2 production by Rubisco. FEBS Lett 571: 124-128

Kimura A, Eaton-Rye JJ, Morita EH, Nishiyama Y, Hayashi H (2002) Protection of the oxygen-evolving machinery by the extrinsic proteins of photosystem II is also essential for development of cellular thermotolerance in *Synechocystis* sp. PCC 6803. Plant Cell Physiol 43: 932-938

Klimov VV, Baranov SV, Allakhverdiev SI (1997) Bicarbonate protects the donor side of photosystem II against photoinhibition and thermoinactivation. FEBS Lett 418: 243-246

Klimov VV, Allakhverdiev SI, Nishiyama Y, Khorobrykh AA, Murata N (2003) Stabilization of the oxygen-evolving complex of photosystem II by bicarbonate and glycinebetaine in thylakoid and subthylakoid preparations. Fun Plant Biol 30: 797-803

Komayama K, Khatoon M, Takenaka D, Horie J, Yamashita A, Yoshioka M, Nakayama Y, Yoshida M, Ohira S, Morita N, Velitchkova M, Enami I, Yamamoto Y (2007) Quality control of photosystem II: cleavage and aggregation of heat-damaged D1 protein in spinach thylakoids. Biochim Biophys Acta 1767: 838-46

Koussevitzky S, Suzuki N, Huntington S, Armijo L, Sha W, Cortes D, Shulaev V, Mittler R (2008) Ascorbate peroxidase 1 plays a key role in the response of *Arabidopsis thaliana* to stress combination J Biol Chem 283: 34197-34203

Kreslavski VD, Balakhnina TI, Khristin MS, Bukhov NG (2001) Pretreatment of bean seedlings by choline compounds increases the resistance of photosynthetic apparatus to UV radiation and elevated temperatures. Photosynthetica 39: 353-358

Kreslavski VD, Khristin MS (2003) Aftereffect of heat shock on fluorescence induction and low-temperature fluorescence spectra of wheat leaves. Russian J Biophys 48: 865-872

Kreslavski VD, Carpentier R, Klimov VV, Murata N, Allakhverdiev SI (2007) Molecular mechanisms of stress resistance of the photosynthetic apparatus. Membr Cell Biol 1: 185-205

Kreslavski V, Tatarinzev N, Shabnova N, Semenova G, Kosobrukhov A (2008) Characterization of the nature of photosynthetic recovery of wheat seedlings from short-time dark heat exposures and analysis of the mode of acclimation to different light intensities. J Plant Physiol 165: 1592-1600

Kreslavski VD, Lyubimov VYu, Shabnova NI, Balakhnina TI, Kosobryukhov AA (2009) Heat-induced impairments and recovery of photosynthetic machinery in wheat seedlings. Role of light and prooxidant-antioxidant balance. Physiol Mol Biol Plant 15: 115-122

Kreslavski VD, Balakhnina TI, Zharmukhamedov SK, Shabnova NI, Khristin MS (2010) Mechanism of enhancing PS II thermoresistance in wheat plants by chlorocholine chloride. Russian Agricultural Sciences 36(3): 156-159

Krieger-Liszkay A (2005) Singlet oxygen production in photosynthesis. J Exp Bot 56: 337-346

Law R, Crafts-Brandner SJ (1999) Inhibition and acclimation of photosynthesis to heat stress is closely correlated with activation of ribulose-1,5-bisphosphate carboxylase/oxygenase. Plant Physiol 120: 173-182

Li P, Cheng L, Gao H, Jiang C, Peng T (2009) Heterogeneous behavior of PS II in soybean (*Glycinemax*) leaves with identical PS II photochemistry efficiency under different high temperature treatments. J Plant Physiol 166: 1607-1615.

Los DA, Murata N (2004) Membrane fluidity and its role in the perception of environmental signals. Biochim Biophys Acta 1666: 142-157

Los DA, Suzuki I, Zinchenko VV, Murata N (2008) Stress responses in *Synechocystis*: regulated genes and regulatory systems. In: Herrero A and Flores E (eds) The Cyanobacteria: Molecular Biology, Genomics and Evolution. Caister Academic Press, London, pp 117-157.

Los DA, Zorina A, Sinetova M, Kryazhov S, Mironov K, Zinchenko VV (2010) Stress sensors and signal transducers in cyanobacteria. Sensors 10: 2386-2415

Low D, Brandle K, Nover L, Forreiter C (2000) Cytosolic heat-stress proteins Hsp17.7 class I and Hsp17.3 class II of tomato act as molecular chaperones *in vivo.* Planta 211: 575-582

Lu C, Qiu N, Wang B, Zhang J (2003) Salinity treatment shows no effects on photosystem II photochemistry, but increases the resistance of photosystem II to heat stress in halophyte *Suaeda salsa.* J Exp Bot 54: 851-60

Lubben TH, Keegstra K (1986) Efficient *in vitro* import of a cytosolic heat shock protein into pea chloroplasts. Proc Natl Acad Sci USA 83: 5502-5506

Maestri E, Klueva N, Perrotta C, Gullil M, Nguyen HT, Marmiroli N (2002) Molecular genetics of heat tolerance and heat shock proteins in cereals. Plant Mol Biol 48: 667-681

Mamedov MD, Hayashi H, Murata N (1993) Effects of glycinebetaine and unsaturation of membrane lipids on heat stability of photosynthetic electron transport and phosphorylation reactions in *Synechocystis* PCC 6803. Biochim Biophys Acta 1142: 1-5

Maxwell DP, Laudenbach DE, Huner NPA (1995) Redox regulation of light-harvesting complex II and cab mRNA abundance in *Dunaliella salina.* Plant Physiol 109: 787-795

McEvoy JP, Brudvig GW (2006) Water-splitting chemistry of photosystem II. Chem Rev 106: 4455-4483

Miller G, Mittler R (2006) Could heat shock transcription factors function as hydrogen peroxide sensors in plants? Annu Bot (Lond) 98: 279-288

Miller G, Suzuki N, Rizhsky L, Hegie A, Koussevitsky S, Mittler R (2007) Double mutants deficient in cytosolic and thylakoid ascorbate peroxidase reveal a complex mode of interaction between reactive oxygen species, plant development, and response to abiotic stresses. Plant Physiol 144: 1777-1785

Mittler R (2002) Oxidative stress, antioxidants and stress tolerance. Trends Plant Sci 7: 405-410

Mittler R, Vanderauwera S, Gollery M, Van Breusegem F (2004) Reactive oxygen gene network of plants. Trends Plant Sci 9: 490-498

Miyake C, Okamura M (2003) Cyclic electron flow within PS II protects PS II from its photoinhibition in thylakoid membranes from spinach chloroplasts. Plant Cell Physiol 44: 457-462

Mohanty P, Vani B, Prakash S (2002) Elevated temperature treatment induced alteration in thylakoid membrane organization and energy distribution between the two photosystems in *Pisum sativum.* Z Naturforsch C 57: 836-842

Mohanty P, Allakhverdiev SI, Murata N (2007) Application of low temperature during photoinhibition allows characterization of individual steps in photo damage and repair of photosystem II. Photosynth Res 94: 217-234

Murakami Y, Tsuyama M, Kobayashi Y, Kodama H, Iba K. (2000) Trienoic fatty acids and plant tolerance of high temperature. Science 287: 476-479

Murata N, Los DA (1997) Membrane fluidity and temperature perception. Plant Physiol 115: 875-879

Murata N, Takahashi S, Nishiyama Y, Allakhverdiev SI (2007) Photoinhibition of photosystem II under environmental stress. Biochim Biophys Acta 1767: 414-421

Myouga F, Motohashi R, Kuromori T, Nagata N, Shinozaki K (2006) An *Arabidopsis* chloroplast-targeted Hsp101 homologue, APG6, has an essential role in chloroplast development as well as heat-stress response. Plant J 48: 249-260

Nash D, Miyao M, Murata N (1985) Heat inactivation of oxygen evolution in photosystem II particles and its acceleration by chloride depletion and exogenous manganese. Biochim Biophys Acta 807: 127-133

Neta-Sharir I, Isaacson T, Lurie S, Weiss D (2005) Dual role for tomato heat shock protein 21: protecting photosystem II from oxidative stress and promoting color changes during fruit maturation. Plant Cell 17: 1829-1838

Nishiyama Y, Los DA, Hayashi H, Murata N (1997) Thermal protection of the oxygen-evolving machinery by PsbU, an extrinsic protein of photosystem II, in *Synechococcus* species PCC 7002. Plant Physiol 115: 1473-1480.

Nishiyama Y, Los DA, Murata N (1999) PsbU, a protein associated with photosystem II, is required for the acquisition of cellular thermotolerance in *Synechococcus* species PCC 7002. Plant Physiol 120: 301-308

Nishiyama Y, Yamamoto H, Allakhverdiev SI, Inaba M, Yokota A, Murata N (2001) Oxidative stress inhibits the repair of photodamage to the photosynthetic machinery. EMBO J 20: 5587-5594

Nishiyama Y, Allakhverdiev SI, Murata N (2005) Inhibition of the repair of photosystem II by oxidative stress in cyanobacteria. Photosynth Res 84: 1-7

Nishiyama Y, Allakhverdiev SI, Murata N (2006) A new paradigm for the action of reactive oxygen species in the photoinhibition of photosystem II. Biochim Biophys Acta 1757: 742-749

Nitta K, Suzuki N, Honma D, Kaneko Y, Nakamoto H (2005) Ultrastructural stability under high temperature or intensive light stress conferred by a small heat shock protein in cyanobacteria. FEBS Lett 579: 1235-1242

Oh JI, Kaplan S (2001) Generalized approach to the regulation and integration of gene expression. Mol Microbiol 39: 1116-1123

Ohnishi N, Murata N (2006) Glycinebetaine counteracts the inhibitory effects of salt stress on the degradation and synthesis of D1 protein during photoinhibition in *Synechococcus* sp. PCC 7942. Plant Physiol 141: 758-765

Papageorgiou GC, Murata N (1995) The unusually strong stabilizing effects of glycinebetaine on the structure and function of the oxygen-evolving photosystem II complex. Photosynth Res 44: 243-252

Pastori GM, Foyer CH (2002) Common components, networks and pathways of cross-tolerance to stress. The central role of "redox" and abscisic-acid-mediated controls. Plant Physiol 129: 460-468

Pfannschmidt T (2003) Chloroplast redox signals: how photosynthesis controls its own genes. Trends Plant Sci 8: 33-41

Pfannschmidt T, Liere K (2005) Redox regulation and modification of proteins controlling chloroplast gene expression. Antioxid Redox Signal 7: 607-18

Pospisil P, Haumann M, Dittmer J, Sole VA, Dau H (2003) Stepwise transition of the tetra-manganese complex of photosystem II to a binuclear $Mn_2(micro\text{-}O)_2$ complex in response to a temperature jump: a time-resolved structural investigation employing X-ray absorption spectroscopy. Biophys J 84: 1370-1386

Preczewski P, Heckathorn SA, Downs CA, Coleman JS (2000) Photosynthetic thermotolerance is positively and quantitatively correlated with production of specific heat-shock proteins among nine genotypes of tomato. Photosynthetica 38: 127-134

Pshybytko NL, Kruk J, Kabashnikova LF, Strzalka K (2008) Function of plastoquinone in heat stress reactions of plants. Biochim Biophys Acta 1777: 1393-1470

Pueyo JJ, Alfonso M, Andres C, Picorel R (2002) Increased tolerance to thermal inactivation of oxygen evolution in spinach photosystem II membranes by substitution of the extrinsic 33 kDa protein by its homologue from a thermophilic cyanobacterium. Biochim Biophys Acta 1554: 22-35

Roose JL, Wegener KM, Pakrasi HB (2007) The extrinsic proteins of photosystem II. Photosynth Res 92: 369-387

Sairam RK, Svastava GC, Saxena DC (2000) Increased antioxidant activity under elevated temperatures: A mechanism of heat stress tolerance in wheat genotypes. Biol Plant 43: 245-251

Sakamoto A, Murata N (2002) The role of glycinebetaine in the protection of plants from stress: clues from transgenic plants. Plant Cell Environ 25: 163-171

Salvucci ME (2008) Association of Rubisco activase with chaperonin-60b: a possible mechanism for protecting photosynthesis during heat stress. J Exp Bot 59: 1923-1933

Salvucci ME, Crafts-Brandner SJ (2004) Relationship between the heat tolerance of photosynthesis and the thermal stability of Rubisco activase in plants from contrasting thermal environments. Plant Physiol 134: 1460-1470

Sato N, Sonoike K, Kawaguchi A, Tsuzuki M (1996) Contribution of lowered unsaturation levels of chloroplast lipids to high temperature tolerance of photosynthesis in *Chlamydomonas reinhardtii*. J Photochem Photobiol 36: 333-337

Sazanov LA, Borrows PA, Nixon PJ (1998) The chloroplast Ndh complex mediates the dark reduction of the plastoquinone pool in response to heat stress in tobacco leaves. FEBS Lett 429: 115-18.

Schroda M, Vallon O, Wollman F-A, Beck CF (1999) A chloroplast-targeted heat shock protein 70 (HSP70) contributes to the photoprotection and repair of photosystem II during and after photoinhibition. Plant Cell 11: 1165-178

Seidler A (1996) The extrinsic polypeptides of photosystem II. Biochim Biophys Acta 1277: 35-60

Semenova GA (2004) Structural reorganization of thylakoid systems in response to heat treatment. Photosynthetica 42: 521-527

Sharkey TD (2005) Effects of moderate heat stress on photosynthesis: importance of thylakoid reactions, rubisco deactivation, reactive oxygen species, and thermotolerance provided by isoprene. Plant Cell Environ 28: 269-277

Sheng RY, Li PM, Xue GX, Zhao XX, Gao HY (2006) Choline chloride protects cell membrane and the photosynthetic apparatus in cucumber seedling leaves at low temperature and weak light. J Plant Physiol Mol Biol 32: 87-93

Shoumskaya MA, Paithoonrangsarid K, Kanesaki Y, Los DA., Zinchenko VV, Tanticharoen M, Suzuki I, Murata N (2005) Identical Hik-Rre systems are involved in perception and transduction of salt signals and hyperosmotic signals but regulate the expression of individual genes to different extents in *Synechocystis*. J Biol Chem 80: 21531-21538

Shutilova N, Semenova G, Klimov VV, Shnyrov V (1995) Temperature induced functional and structural transformations of the photosystem II oxygen-evolving complex in spinach subchloroplast preparations. Biochem Mol Biol Int 35: 1233-43

Shutova T, Kenneweg H, Buchta J, Nikitina J, Terentyev V, Chernyshov S, Andersson B, Allakhverdiev SI, Klimov VV, Dau H, Junge W, Samuelsson G (2008) The photosystem II-associated Cah3 in *Chlamydomonas* enhances the O_2 evolution rate by proton removal. EMBO J 27: 782-791

Singh A, Grover A (2008) Genetic engineering for heat tolerance in plants. Physiol Mol Biol Plant 14: 155-166

Slabas AR, Suzuki I, Murata N, Simon WJ, Hall JJ (2006) Proteomic analysis of the heat shock response in *Synechocystis* PCC 6803 and a thermally tolerant knockout strain lacking the histidine kinase 34 gene. Proteomics 6: 845-864

Srivastava A, Guissre B, Greppin H, Strasser RJ (1997) Regulation of antenna structure and electron transport in photosystem II of *Pisum sativum* under elevated temperature probed by the fast polyphasic chlorophyll *a* fluorescence transient: OKJIP. Biochim Biophys Acta 1320: 95-106

Strand A (2004) Plastid-to-nucleus signaling. Curr Opin Plant Biol 7: 621-625

Strasser RBJ (1997) Donor side capacity of PS II probed by chlorophyll a fluorescence transients. Photosynth Res 52: 147-155

Strasser RJ, Srivastava A, Tsimilli-Michael M (2000) The fluorescence transient as a tool to characterize and screen photosynthetic samples. In: Yunus M, Pathre U, Mohanty P (eds) Probing photosynthesis: mechanisms, regulation and adaptation. Taylor and Francis, UK, pp 445-483

Sung DY, Kaplan F, Lee KJ, Guy CL (2003) Acquired tolerance to temperature extremes. Trends Plant Sci 8: 179-187

Suzuki I, Kanesaki Y, Hayashi H, Hall JJ, Simon WJ, Slabas AR, Murata N (2005) The histidine kinase Hik34 is involved in thermotolerance by regulating the expression of heat shock genes in *Synechocystis*. Plant Physiol 138: 1409-1421

Suzuki N, Mittler R (2006) Reactive oxygen species and temperature stresses: a delicate balance between signaling and destruction. Physiol Plant 126: 45-51

Takahashi S, Nakamura T, Sakamizu M, van Woesik R, Yamasaki H (2004) Repair machinery of symbiotic photosynthesis as the primary target of heat stress for reef-building corals. Plant Cell Physiol 45: 251-255

Takahashi S, Murata N (2008) How do environmental stresses accelerate photoinhibition? Trends Plant Sci 13: 178-182

Tanaka Y, Nishiyama Y, Murata N (2000) Acclimation of the photosynthetic machinery to high temperature in *Chlamydomonas reinhardtii* requires synthesis *de novo* of proteins encoded by the nuclear and chloroplast genomes. Plant Physiol 124: 441-450

Tang Y, Wen X, Lu O, Yang Z, Cheng Z, Lu C (2007) Heat stress induces an aggregation of the light-harvesting complex of photosystem II in spinach plants. Plant Physiol 143: 629-638

Tefler A (2005) Too much light? How – carotin protects the photosystem II reaction center. Photochem Photobiol Science 4: 950-956

Tefler A, Barber J (1995) Role of carotenoid bound to the photosystem II reaction center. In: Mathis P (ed) Photosynthesis: From light to the Biosphere, Vol 3. Kluver Acad Publ, Dordrecht, pp 15-20

Thompson LK, Blaylock R, Sturtevant JM, Brudvig GW (1989) Molecular basis of the heat denaturation of photosystem II. Biochem 28: 6686-6695

Török Z, Goloubinoff P, Horvath I, Tsvetkova NM, Glatz A, Balogh G, Varvasovszki V, Los DA, Vierling E, Crowe JH, Vigh L (2001) *Synechocystis* HSP17 is an amphitropic protein that stabilizes heat stressed membranes and binds denatured proteins for subsequent chaperone-mediated refolding. Proc Natl Acad Sci USA 98: 3098-3103

Tóth SZ, Schansker G, Garab G, Strasser RJ (2007) Photosynthetic electron transport activity in heat-treated barley leaves: The role of internal alternative electron donors to photosystem II. Biochim Biophys Acta 1767: 295-305

Vani B, Saradhi PP, Mohanty P (2001) Characterization of high temperature induced stress impairments in thylakoids of rice seedlings. Indian J Biochem Biophys 38: 220-229

Vierling E (1991) The roles of heat shock proteins in plants. Annu Rev Plant Physiol Plant Mol Biol 42: 579-620

Vierling E, Mishkind ML, Schmidt GW, Key JL (1986) Specific heat shock proteins are transported into chloroplasts. Proc Natl Acad Sci USA 83: 361-365

Vigh L, Maresca B, Harwood JL (1998) Does the membrane's physical state control the expression of heat shock and other genes? Trends Biochem Sci 23: 369-374

Vigh L, Horvath I, Maresca B, Harwood JL (2007) Can the stress proteins response be controlled by 'membrane-lipid therapy'? Trends Biochem Sci 32: 357-364

Villarejo A, Shutova T, Moskvin O, Forssen M, Klimov VV, Samuelsson G (2002) A photosystem II-associated carbonic anhydrase regulates the efficiency of photosynthetic oxygen evolution. EMBO J 21: 1930-1938

Volkov R, Panchuk I, Mullineaux P, Schoffl F (2006) Heat stress-induced H_2O_2 is required for effective expression of heat shock genes in *Arabidopsis*. Plant Mol Biol 61: 733-746

Vranova E, Inzé D, Van Breusegem F (2002) Signal transduction during oxidative stress. J Exp Bot 53: 1227-1236

Wada H, Gombos Z, Murata N (1994) Contribution of membrane lipids to the ability of the photosynthetic machinery to tolerate temperature stress. Proc Natl Acad Sci USA 91: 4273-4277

Wahid A, Shabbir A (2005) Induction of heat stress tolerance in barley seedlings by pre-sowing seed treatment with glycinebetaine. Plant Growth Reg 46: 133-141

Wahid A, Gelani S, Ashraf M, Foolad MR (2007) Heat tolerance in plants: an overview. Environ Exp Bot 61: 199-223

Weis E (1981) The temperature-sensitivity of dark inactivation and light activation of the ribulose-1,5-bisphosphate carboxylase in spinach chloroplasts. FEBS Lett 129: 197-200

Wen X, Qiu N, Lu Q, Lu C (2005) Enhanced thermotolerance of photosystem II in salt-adapted plants of the halophyte *Artemisia anethifolia*. Planta 220: 486-97

Wilson KE, Ivanov AG, Öquist G, Grodzinski B, Sarhan F, Huner NPA (2006) Energy balance, organellar redox status, and acclimation to environmental stress. Can J Bot 84: 1355-1370

Yamamoto H, Miyake C, Dietz K-J, Tomizawa K, Murata N, Yokota A (1999) Thioredoxin peroxidase in the cyanobacterium *Synechocystis* sp. PCC 6803. FEBS Lett 447: 269-273

Yamane Y, Shikanai T, Koike H, Satoh K (2000) Reduction of Q_A in the dark: another cause of fluorescence F_o increases by high temperatures in higher plants. Photosynth Res 63: 23-34

Yang X, Liang Z, Lu C (2005) Genetic engineering of the biosynthesis of glycinebetaine enhances photosynthesis against high temperature stress in transgenic tobacco plants. Plant Physiol 138: 2299-2309

Yang X, Wen X, Gong H, Lu Q, Yang Z, Tang Y, Liang Z, Lu C (2007) Genetic engineering of the biosynthesis of glycinebetaine enhances thermotolerance of photosystem II in tobacco plants. Planta 225: 719-733

Yocum CF (1992) The calcium and chloride requirements for photosynthetic water oxidation. In: Pecoraro VL (ed) Manganese Redox Enzymes. VCH, New York, pp 71-83

Yordanov I (1993) Response of photosynthetic apparatus to temperature stress and molecular mechanisms of its adaptation. Photosynthetica 26: 517-531

Yordanov I, Dilova R, Petkova T, Pangelova V, Goltsev V, Süss K-H (1986) Mechanisms of the temperature damage and acclimation of the photosynthetic apparatus. Photobiochem Photobiophys 12: 147-155

Yordanov I, Goltsev V, Stefanov D, Chernev P, Zaharieva I, Kirova M, Gecheva V, Strasser RJ (2008) Preservation of photosynthetic electron transport from senescence-induced inactivation in primary leaves after decapitation and defoliation of bean plants. J Plant Physiol 165: 1954-1963

Yoshioka M, Uchida S, Mori H, Komayama K, Ohira S, Morita N, Nakanishi T, Yamamoto Y (2006) Quality control of photosystem II. Cleavage of reaction center D1 protein in spinach thylakoids by FtsH protease under moderate heat stress. J Biol Chem 281: 21660-21669

Zaharieva I, Taneva SG, Golstev V (2001) Effect of temperature on the luminescent characteristics in leaves of *Arabidopsis* mutants with decreased unsaturation of the membrane lipids. Bulg J Plant Physiol 27(3-4): 3-19

Zharmukhamedov SK, Shirshikova GN, Maevskaya ZV, Antropova TM, Klimov VV (2007) Bicarbonate protects the water-oxidizing complex of photosystem II against thermoinactivation in intact *Chlamydomonas reinhardtii* cells. Russian J Plant Physiol 54: 302-308

CHAPTER 15

Salinity Tolerance Strategies in Mangroves: Salt Shock Experiments for Assessing Stress-Triggered Effects During Salt Stress and Post-Stress Recovery in a True Mangrove *Bruguiera gymnorrhiza*

Bishnupriya Behera[1], Prasanna Mohanty[2] and Anath Bandhu Das[1*]

[1]*Cytogenetics and Mangrove Research Laboratory, Regional Plant Resource Centre, Nayapalli, Bhubaneswar-751015, India*

**Department of Agriculture Biotechnology, UAT, Bhubaneswar-751003*

[2]*INSA Honorary Scientist, Regional Plant Resource Centre, Nayapalli, Bhubaneswar-751015 and Professor in courtesy at DAVV, Indore-452001, India.*

ABSTRACT

Salinity is one of the major environmental stresses that affect the growth and productivity of plants by affecting photosynthesis and other metabolic processes. Halophytes are capable of thriving and growing under high concentration of NaCl. The understanding of salt tolerance mechanism, particularly in producing antioxidative enzymes and special salt stress marker proteins, is discussed in adapting adverse high salinity environment. A detailed antioxidative defense system against the production of reactive oxygen species (ROS)—an early indicator for adverse condition in both mangroves and mangroves associates is discussed. This aspect has been studied using a true mangrove *B. gymnorrhiza* in our laboratory with salt shock experiment. Two-month-old seedlings of *Bruguiera gymnorrhiza* were exposed to high salt shock (500 mM NaCl) for 6d under hydroponic culture to detect the changes in antioxidative enzyme system as well as leaf and thylakoid protein profile. Antioxidative enzymes, CAT, APX, GPX found increased 2.7, 2.4, 1.3-folds in salt treated samples as compared to control respectively. The total activity of SOD was increased 2-folds, whereas in gel study, a 2.5 times enhancement was noted in Fe-SOD and Cu/Zn SOD upon salinization as compared to control. Leaf protein profile analysis indicated that the intensity of several protein bands (10, 16, 23, 33, 37, 42, 44, 49 and 86 kDa) decreased with an increase in salt treatment. SDS-PAGE of thylakoid proteins showed that the intensity of many polypeptide bands ranging from 10 to 23 kDa and 33 to 90 kDa decreased as a result of NaCl exposure. The numbers of major bands found in SDS-PAGE were reduced to minor bands in urea-SDS-PAGE (23 kDa, 33 kDa, 37 kDa and 49 kDa). The present investigation showed that *B. gymnorrhiza* is able to tolerate such high salt shock by altering leaf soluble protein as well as enhanced level of antioxidative enzymes.

Keywords: Antioxidative enzymes, Mangrove, Salt shock, Sodium chloride, Thylakoid, *Bruguiera gymnorrhiza,* Polypeptides.

Abbreviations

ROS: Reactive Oxygen Species, *SOD:* Superoxide dismutase, *CAT:* Catalase, *APX:* Ascorbate peroxidase, *GR:* Glutathione reductase, *GPX:* Gluiacol peroxidase.

INTRODUCTION

Mangroves grow in inter-tidal zones at the interface between land and sea along the tropical and subtropical coastlines (Tomlinson, 1986). They are considered to be more salt adapted than any other halophytes, especially being capable of tolerating fluctuating salinities from hyposaline to hypersaline conditions (McMillan, 1986). Mangroves at their native habitats get periodically inundated by low and high tides leading to permanent water logging at places and fluctuations in salinity at other sites (Parida *et al.,* 2004). These conditions become worse in tropical mangroves. Overheating and desiccation is high at low tide condition, and evaporation causes the remaining water to be highly saline. At high tide, the warmth of water depletes the soluble oxygen, so the mangrove plants face the anaerobic or anoxic conditions in their habitat. These situations repeat seasons after seasons. Mangroves do cope with these adverse environmental conditions by a variety of adaptable features like salt secretion, production of pneumatophores, formation of prop-roots and viviparous-germination of seeds or propagules. Besides these morphological adaptive features, the biochemical strategies include regulated ion uptake and transport of ions from and to different plant parts, selective accumulation and exclusion of ions, compartmentalization of ions, synthesis and accumulation of osmolytes, changes in metabolic pathways, alternations in membrane transport and induction of hormone (Das *et al.,* 2006).

These salt management strategies of mangroves are thus special. Depending on their salt adaptation/ tolerance mechanisms, the mangroves are divided into two distinct groups: The 'secretors'—those having salt glands like in *Aegiceras corniculatum* and "non-secretors"—those lacking such morphological features for excretion of excess salt. Non-secretor mangroves like *Bruguiera* species are shown to exclude 99% of the salt in surrounding sea water by ultrafiltration (Scholander, 1968).

SALT STRESS IMPOSES OXIDATIVE STRESSES

Most stresses cause oxidative stress; however, the effect of salt stress on plant is a complex phenomenon. It decreases the osmotic potential of soil solution, creating water stress in plants. This water deficit in plants results in the accumulation of abscisic acid, which further triggers a signaling cascade in guard cells that ultimately results in the closure of stomata. All plants parts can produce reactive oxygen species (ROS), but it is mostly generated by mitochondria and chloroplasts. This stomatal closure limits CO_2 availability in leaf cells. As a whole, a disruption of coordination between carbon assimilation and chloroplast photofunction in electron transport chains leads to the transfer of electrons in alternating electron acceptor like O_2 (Tanaka *et al.,* 1999; Foyer and Noctor, 2000; Tausz *et al.*, 2004). A simplified pathway of ROS production is shown in Fig. 15.1.

The excess electron reduces molecular oxygen to produce reactive oxygen species (ROS) such as singlet oxygen ($^1O_2$), superoxide radical (O_2^-), hydroxyl radicals (OH^-) and hydrogen peroxide (H_2O_2). Superoxide radicals are synthesized in the chloroplasts and mitochondria and a little quantity is also reported to be produced in microbodies. Superoxide in combination with H_2O_2 produce highly toxic hydroxyl radical (OH). Hydroxyl radical can damage chlorophyll, protein, DNA, lipids and other important macromolecules, which affects plant metabolism and ultimately growth and yields (Figs. 15.1 and 15.2).

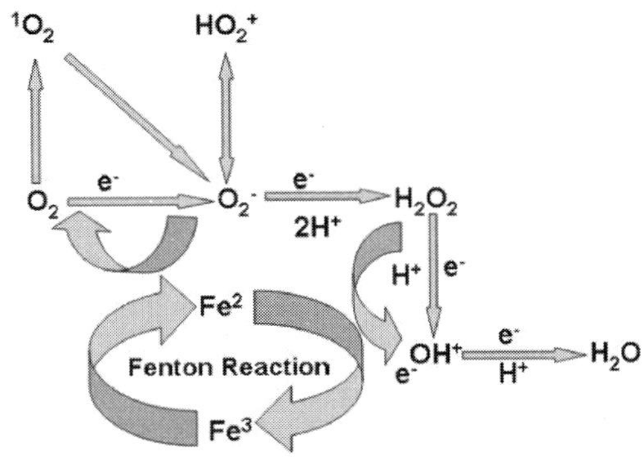

Fig. 15.1 Pathway of ROS production from oxygen.

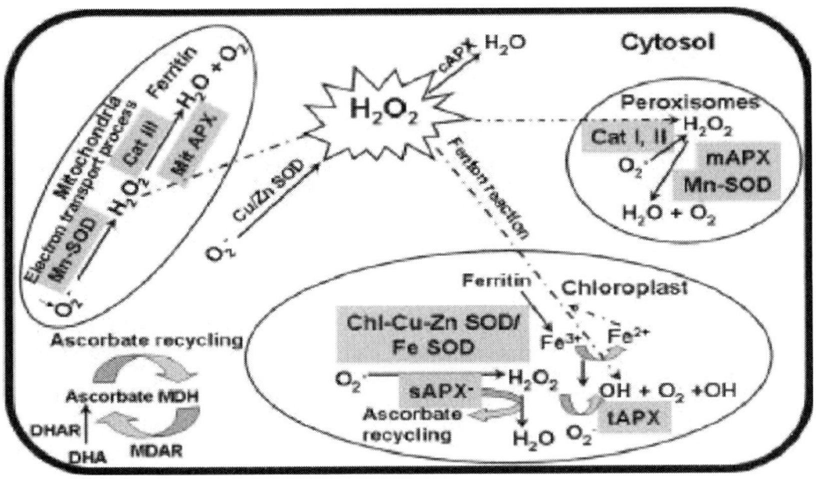

Fig. 15.2 Schematic representation of antioxidant enzyme activity in different organelles of the plant cell.

ANTIOXIDATIVE DEFENSE SYSTEM—A TRUE SCAVENGER OF ROS

Major bioenergetics organelles like chloroplasts, mitochondria and peroxisomes are the important intracellular generators of ROS. The formation of ROS is thus initiated by the univalent reduction of O_2 or the transfer of excess excitation energy to O_2. Plants possess both enzymatic and non-enzymatic mechanism to scavenge these reactive oxygen species (ROS). The major enzymes induced in response to ROS are superoxide dismutase (SOD; EC 1.15.1.1), catalase (CAT; EC 1.11.1.6), ascorbate peroxidase (APX; EC1.1 1.1.1 1), glutathione reductase (GR; EC1.6.4.2) and gluiacol peroxidase (GPX; EC 1.11.17). These enzymes either up-regulate or down-regulate to minimize the concentration of ROS to maintain cellular homeostasis. In addition to antioxidative enzyme systems, nonenzymatic antioxidants like ascorbic acid, glutathione, α-tocopherol and carotenoid as well as osmolytes like proline and glycine betaine also play a key role in scavenging free radicals in plants during salinity stress. Superoxide radical is regularly synthesized in the chloroplasts and mitochondria (Elstner, 1991; Rich and Bonner, 1987). superoxide dismutase (SOD) scavenges this superoxide radicals, which results in the production of H_2O_2. H_2O_2 is removed by APX or catalase. The combination of O_2^- and H_2O_2 in the presence of trace amounts of Fe^{2+} and Fe^{3+} produce highly toxic hydroxyl radicals (Figs. 15.1 and 15.2). Table 15.1 catalogues

Table 15.1 List of studies carried out on antioxidative enzymes during salt stress in mangroves and other halophytes.

Enzyme/ gene	*Plant species*	*Method used*	*Enzyme expression*	*Reference*
APX	*Bruguiera parviflora*	Activity staining on gel	Enhancement of APX activity in presence of NaCl stress in hydroponic culture	Parida *et al.* 2004, 2004a, Wang *et al.* 2004 Parida and Das 2005
APX	*Thellungiella halophila*	EST analysis	Presence of three APX in cDNA library from salt-treated plants	Wang *et al.* 2004
APX	*Aegiceras corniculatum*	Spectrophotometric assay	Activity decreased 58% on 4th day in root	Mishra and Dar 2003 Parida *et al.* 2004, 2004a
CAT	*Avicennia marina*	m RNA detection	Induction of AmCat I in salinity till 12h and subsequent decrease	Jithesh *et al.* 2006
CAT	*Bruguiera parviflora*	Spectrophotometric assay and activity staining on gel	Loss of total catalase activity with NaCl stress, CAT-2, and CAT-3 isoforms decreased	Parida *et al.* 2004, 2004a
CAT	*Bruguiera gymnorrhiza*	Spectrophotometric assay	Increase in total catalase upto sea water concentration	Takemura *et al.* 2000
CAT	*Bruguiera gymnorrhiza*	m RNA detection	A particular catalase transcript	Takemura *et al.* 2002
CAT	*Crithmum maritimum*	Spectrophotometric and isozyme analysis	Decrease in catalase activity with increased NaCl	Ben Amor *et al.* 2005
Ferritin crystallinum	*Mesembryanthemum*	Electron microscopic studies	Ferritin deposits present with NaCl stress in chloroplasts	Paramonova *et al.* 2004
Ferritin	*Avicennia marina*	mRNA analysis	Early induction of AmFer1 in salinity	Jithesh *et al.* 2006
CAT	*Aegiceras corniculatum*	Spectrophotometric assay	Activity decreased 72% in root	Mishra *et al.* 2003
CAT	*Aegiceras corniculatum*	Activity staining on gel	Two isoforms CAT 1, 2 decreased	Mishra *et al.* 2003
CAT	*Suaeda nudiflora*	Spectrophotometric analysis	Decrease in catalase activity in callus cultures with salinity stress	Cherian and Reddy 2003
POX	*Avicennia marina*	Spectrophotometric analysis	Enhanced peroxidase activity in root and shoot and no significant changes in leaf tissue	Cherian *et al.* 1999
POX	*Crithmum maritimum*	Spectrophotometric analysis	Transient increase in peroxidase activity	Ben Amor *et al.* 2005
POX	*Suaeda nudiflora*	Spectrophotometric analysis	Enhanced activity with NaCl stress	Cherian and Reddy 2003
GPX	*Aegiceras corniculatum*	Spectrophotometric assay	Activity decreases 80% on 4th day	Mishra *et al.* 2003

Contd...

Contd...

GPX	*Aegiceras corniculatum*	Activity staining on gel	Four isoforms of GPX- 1, 2, 3, 4 decreased upon salt treatment of 6d	Mishra *et al.* 2003
GPX	*Bruguiera parviflora*	Spectrophotometric assay	169% of GPX enhancement at 45d exposure of NaCl at 400 mM	Parida *et al.* 2004, 2004a, Parida and Das 2005
GPX	*Bruguiera parviflora*	Activity staining on gel	GPX-6 show maximum 73% increment in activity without any change in GPX-4, 5 whereas GPX- 1 increase by 61%,GPX-2 and 3 changed 34-38%	Parida *et al.* 2004, 2004a
SOD	*Avicennia marina*	Spectrophotometric analysis	Decrease in total activity in roots and shoots but increase in leaves with salinity stress	Cherian *et al.* 1999
SOD	*Avicennia marina*	mRNA analysis	Constitutive expression of Cu/Zn SOD in salinity	Jithesh *et al.* 2006
SOD	*Bruguiera parviflora*	Activity staining on gel	Increase in activity of Mn SOD and Fe SOD and no increase in activity of Cu/Zn SOD isoform with salinity stress	Parida *et al.* 2004, 2004a, Parida and Das 2005
SOD	Bruguiera gymnorrhiza	Spectrophotometric assay	Steep increase in SOD activity in salinity stress	Takemura *et al.* 2000
SOD	Bruguiera gymnorrhiza	mRNA detection	Induction of Cu/Zn SOD activity in salinity stress	Takemura *et al.* 2002
SOD	Crithmum maritimum	Spectrophotometric assay and isozyme analysis	Increase in total SOD activity in shoots and decrease in roots	Ben Amor *et al.* 2005
SOD	Suaeda nudiflora	Spectrophotometric assay	Decrease in total SOD activity in callus culture with salinity stress	Cherian and Reddy 2003
SOD	Suaeda salsa	Isozyme assay and spectrophotometric assay	Increase in activity of thylakoid bound SOD	Fang *et al.* 2005
SOD	Suaeda salsa	Staining on gel	Increase in Mn SOD and Fe SOD isoforms, Cu/Zn SOD I isoform activity decreased in presence of salinity stress, no change Cu/Zn SOD II isoform	Wang *et al.* 2004

Contd...

Contd...

SOD	*Thellungiella* halophila	mRNA detection	Higher mRNA transcript levels of Fe SOD	Taji et al. 2004
SOD	*Mesembryanthemum crystallinum*	mRNA detection	Increase in Fe SOD and Cu/Zn SOD during salt stress	Slesak et al. 2002, Hurst *et al.* 2004
SOD	*Mesembryanthemum crystallinum*	mRNA detection	Induction of all three SOD isoforms in salt stress	Slesak *et al.* 2002, Paramonova *et al.* 2004
Acid phosphatase	*Bruguiera parviflora*	Activity staining on gel	Out of the three isoforms of ACP, ACP-3 increased by 37% without much change in other isoforms	Parida *et al.* 2004, 2004a
Alkaline phosphatase	*Bruguiera parviflora*	Spectrophotometric assay	Activity increased to 10%, 20%, 39% dose dependent in100, 200, 400 mM NaCl	Parida *et al.* 2004, 2004a, 2005
Nitrate	*Bruguiera parviflora*	Spectrophotometric assay	Activity increased in 100 mM NaCl and gradually decline upto 63% in 400 mM NaCl	Parida *et al.* 2004, 2004a, Parida and Das 2005

various antioxidant enzymes that have been reported in mangroves and their associates as a salt adaptive strategy.

Earlier from our laboratory Parida *et al.* (2004, 2004a) reported that salt stress enhances the activity of superoxide dismutase, ascorbate peroxidase, glutathione reductase and guiacol peroxidase, whereas the activity of catalase decreased in leaves of hydroponically maintained, 250 mM NaCl treated *Bruguiera parviflora* seedlings. It was found that the total activity of SOD enhanced 128% as compared to untreated control, whereas activity staining data revealed that out of the five isoforms of SOD (one MnSOD, one Cu/Zn-SOD and three Fe-SOD), the expression of Mn-SOD and Fe-SOD-2 enhanced preferentially upon exposure to salt under hydroponic culture condition. A total 35% decrease in catalase activity was found from spectrophotometry analysis, whereas activity staining data indicated that out of the four isoforms (CAT-1, CAT-2, CAT-3 and CAT-4), CAT-2 isoform decreased to 45% as compared to control. It was suggested that the inactivation of catalase was possibly due to enhancement of H_2O_2. The activity of APX and GR also enhanced in response to salinity in this species. Among the high molecular weight isoforms, GPX- 1 increased by 61%, whereas GPX- 6 showed a ~70% increase in the concentration with regard to its control. It was further suggested that these two isoforms of GPX may have specific sensitivities to salinity.

When *Bruguiera gynmorrhiza* seedlings were treated with 500mM NaCl for a period of 9 days, it was found that the catalase activity increased to about nearly 4.9 times more as compared to untreated control plants (Takemura *et al.,* 2000). This suggested that catalase retained its full activity at least up to sea water salt concentration up to 9 days and again lowered till the end of the experimental treatment time of 16 days. This type of opposite behavior in catalase activity in both species of *Bruguiera* is difficult to explain at present. SOD activity also increased about 8.1 times upon exposure to 500 mM NaCl for 9 days as compared to untreated control. Takemura *et al.,* (2002) also reported that mRNA transcript of cytosolic Cu/Zn SOD increased in *B. gymnorrhiza* after the 5th and 6th day of salt treatment. However, there are only few reports about the mechanisms of generation of superoxide radicals in the cytosol and the role of cytosolic SOD in halophytes (Takemura *et al.,* 2002). In *Avicennia marina,* it was found that the total SOD activity increased in leaf tissues after six days of salt treatment, whereas surprisingly total SOD activity decreased in root and shoot tissue (Cherian *et al.,*1999). The m-RNA detection study by Jithesh *et al.* (2006) reported an enhancement in Cu/Zn SOD upon exposure to salinity. In another study, by the same group a class II enzyme, CAT I from *Avicennia marina,* was induced upon salt and other oxidative stress, such as exposure to H_2O_2 and light, in leaves (Jithesh *et al.,* 2006). Increased POX activity was observed in root and shoot tissues in *Avicennia marina* under NaCl stress conditions, whereas no such significant changes in POX were found in leaf tissues (Cherian *et al.,* 1999). However, in the roots of *Aegiceras corniculatum,* activity of antioxidative enzymes like CAT, APX and GPX decreased upon exposure to salinity for a short period for 4 days (Mishra and Das, 2003).

A decrease in total SOD activity was also observed in NaCl-treated callus cultures of *Suaeda nudiflora*—a halophyte. This result is not in agreement with the studies conducted on whole plants, where SOD activity was found to be increased under salt stress (Cherian *et al.,* 1999). The catalase activity was found to decrease by 60%, which showed a declining capacity of callus tissues to scavenge H_2O_2 with increasing salt stress. An enhancement in POX activity in salinized cell indicated that these cells had a higher capacity for decomposition of H_2O_2 and that POX performs a major role in this particular reaction (Cherian and Reddy, 2003). Both spectrophotometric and in-gel assay of *Suaeda salsa* showed an increase in the activity of thylakoid bound SOD upon exposure to salinization (Fang *et al.,* 2005). Wang *et al.* (2004) reported an increase in Mn-SOD and Fe-SOD isoforms and a decrease in Cu/Zn SOD I isoform, whereas no significant changes were found in Cu/Zn SOD II isoforms in *S. salsa* on salinization. Again, the total SOD activity was shown to increase in shoot tissues of another halophyte, *Crithmum maritimum,* in the presence of 50 mM NaCl stress. However, at a concentration of 200 mM NaCl, no significant difference in SOD activity was detected (Ben Amor *et al.,* 2005). On the contrary, SOD activity in root tissues of *C. maritimum* decreased upon treatment with 50 mM as well as 200 mM NaCl (Ben Amor *et al.,* 2005), whereas total catalase activity decreased with increasing NaCl concentration in *Crithmum maritimum* (Ben Amor *et al.,* 2005). A transient increase in POX activity was also observed in this study. These data reflect that each mangrove plant has its own regulatory control of inducing specific antioxidative enzymes. Table 15.1 depicts different antioxidative enzyme systems in mangroves and their associates.

GENERAL PHYSIOLOGICAL CHANGES INDUCED BY SALT STRESS IN MANGROVES WITH SPECIAL EMPHASIS ON A TRUE MANGROVE *Bruguiera gymnorrhiza*

Bruguiera gymnorrhiza, popularly known, as "Burma mangrove", is a prominent broadleaved mangrove species in Indian mangrove forests. It belongs to the family Rhizophoraceae and is considered a true non-salt secreting tree mangrove (Fig. 15.3). This mangrove vegetation is mainly seen at the deltaic regions of Mahanadi river and Bhiterkanika in the east coastal region of India (Das *et al.,* 1997). It is stated that *B. gymnorrhiza* is more salt tolerant than *B. parviflora,* but no detailed study has been done on this aspect. As it is the dominant species for zonation, we conducted a short term experiment of high NaCl treatment on *B. gymnorrhiza* in hydroponic culture.

Fig. 15.3 *Bruguiera gymnorrhiza* with vivipary seeds in natural condition at Bhitarkanika mangrove forests of Orissa.

Salt Shock Experiments

In this section we present a part of our findings on the response of antioxidative enzymes to high salt (500 mM NaCl) in the leaf of hydroponically maintained mangrove, *Bruguiera gymnorrhiza* (Fig. 15.4). When two-month-old seedlings of *B. gymnorrhiza* were exposed to 500 mM NaCl for a period of 9 days, specific changes in the activity of selective antioxidative enzymes like CAT, SOD, APX and GPX were observed in both spectrophotometry and activity staining assays.

The total activity of SOD increased 2-fold, whereas in gel study, a 2.5 times enhancement was noted in Fe-SOD and Cu/Zn SOD upon salinization as compared to control (Fig. 15.7). The catalase activity increased 2.1 times in 9 days salt stress leaf samples of *B. gymnorrhioza* as compared to untreated control samples. The activity staining study revealed that out of the four isoforms of catalase (CAT-1, CAT-2, CAT-3, CAT-4), CAT-1 and CAT-2 enhanced 1.2 and 1.4 times respectively upon 9 days salinization (Fig. 15.5). The ascorbate peroxidase activity increased 2.3 times as compared to untreated control (Fig. 15.6). Total activity of GPX increased 2.6 times as compared to control. Six isoforms of GPX (GPX-1, GPX-2, GPX-3, GPX-4, GPX-5 and GPX-6) were found from native electrophoresis and activity staining data. Out of these six isoforms, GPX-1 and GPX-3 enhanced 2 and 1.27 times respectively as compared to control, whereas GPX-2 increased 3.4 times in 3 days and 2.4 times in 9 days salt treatment (Fig. 15.8).

Fig. 15.4 *Bruguiera gymmorrhiza* seedlings treated with NaCl in hydroponic culture under growth chamber condition at 22±2°C temperature and 80% relative humidity.

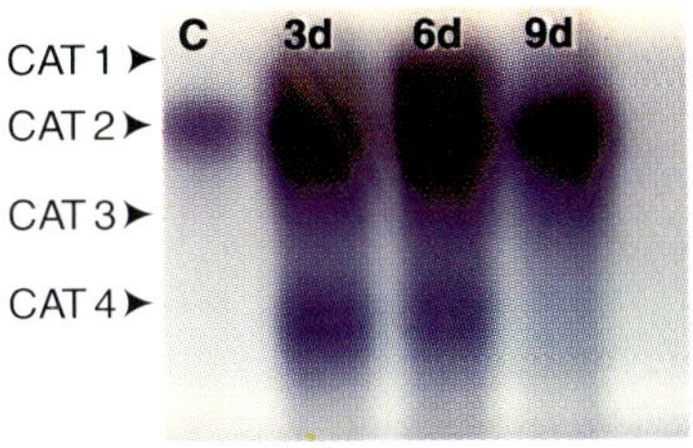

Fig. 15.5 Activity staining of CAT in leaves of *B. gymnorrhiza* during 500 mM NaCl treatment.

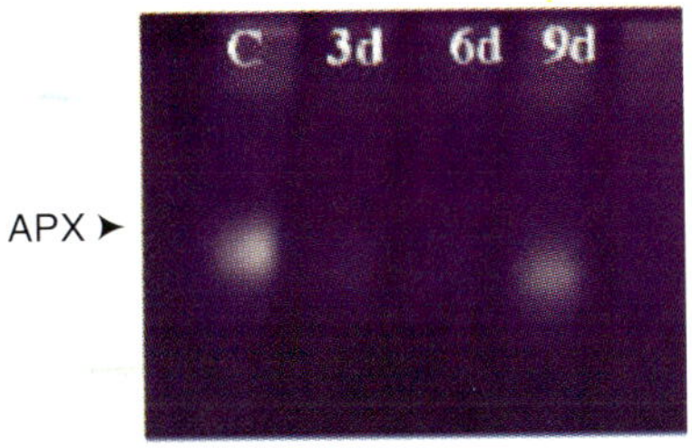

Fig. 15.6 Activity staining of APX in leaves of *B. gymnorrhiza* during 500 mM NaCl treatment.

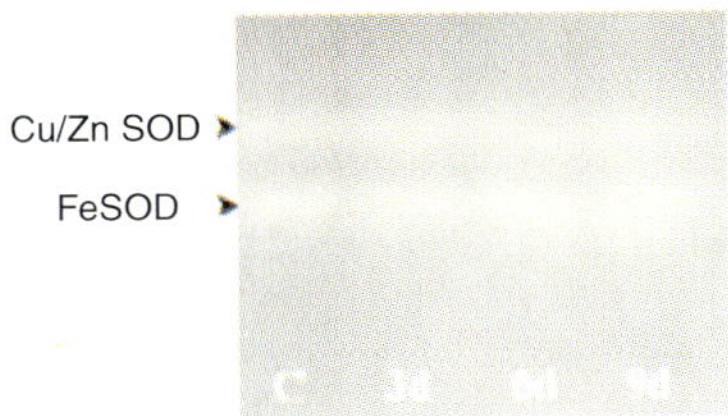

Fig. 15.7 Activity staining of SOD in leaves of *B. gymnorrhiza* during 500 mM NaCl treatment.

Changes of Soluble Leaf and Thylakoid Protein Profile

When one-year-old seedlings of *B. gymnorrhiza* were exposed to 500 mM NaCl for a short period of 6 days under hydroponic culture condition, the total soluble leaf protein increased abruptly after 2 days of salt exposure and again decreased on 6 days as compared to control (Behera *et al.*, 2009). From SDS-PAGE analysis of leaf protein, it was found that the intensity of several protein bands ranging from 10-86 kDa decreased on 6 days as compared to control (Fig. 15.9). The thylakoid membranes isolated from control and NaCl treated leaves were analyzed by SDS-PAGE and it was found that the intensity of many protein bands ranging from 10 to 86 kDa showed extensive decrease as compared to control (Fig. 15.10). The progressive decrease in protein bands due to an increase in duration of salt exposure suggested that

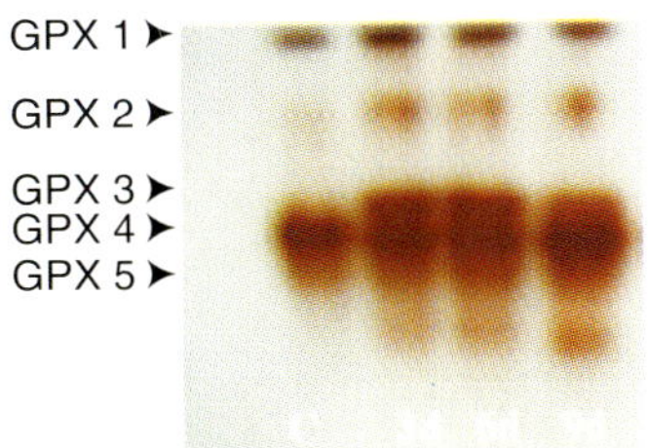

Fig. 15.8 Activity staining of GPX in leaves of *B. gymnorrhiza* during 500 mM NaCl treatment.

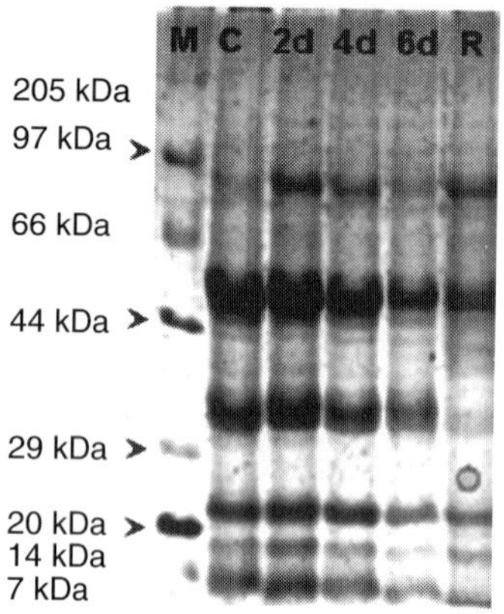

Fig. 15.9 Changes in leaf total soluble protein profile in leaf of *B. gymnorrhiza* upon exposure to 500 mM NaCl treatment and post-recovery.

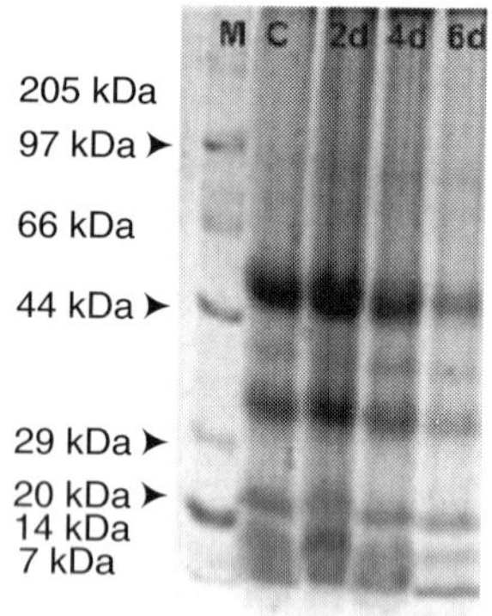

Fig. 15.10 Changes in leaf total thylakoid protein profile in leaf of *B. gymnorrhiza* upon exposure to 500 mM NaCl treatment and post-recovery.

the target site of salt stress could be the degradation of specific thylakoid proteins. Parida *et al.* (2005) reported a novel protein of molecular mass 23 kDa in *B. parviflora* whose intensity gradually decreases with increasing concentration of salinity. In another study, several proteins were induced in salt stressed *B. gymnorrhiza,* but the induction of 33 kDa Mn-stabilizing protein is most important to understand the underlying mechanism of salt stress to maintain the capacity of PSII functioning (Sugihara *et al.*, 2000). This 33 kDa protein is identified with oxygen evolving enhancer protein 1 (Sugihara *et al.*, 2000).

CONCLUSION AND PERSPECTIVES

Salinity is one of the major environmental stresses that affect the growth and productivity of plants by affecting photosynthesis and other metabolic processces. Halophytes are plants of salty environment capable of thriving and growing under high concentration of NaCl (Hellebust, 1976; Flowers *et al.*, 1986). So, the understanding of salt tolerance mechanism would be useful in transferring salt tolerance genes to crop plants. Among the halophytes, mangroves can be used as the best model for salt stress research because they are able to adapt not only in stress conditions ranging from high salinity at one extreme, but also in a complete lack of salinity due to freshswater conditions. Mangroves possess specialized adoptable features to cope with this adverse condition like high salinity. An early indicator of the adverse condition is the production of reactive oxygen species (ROS) within different cellular compartments of the plant cell. So, in this investigation, we discussed in detail antioxidative defense system against the reactive oxygen species in both mangroves and mangroves associates. This aspect has been studied in our laboratory using a true mangrove, *B. gymnorrhiza,* and salt shock experiment design was proposed.

ACKNOWLEDGEMENTS

The authors thank to the Director, Regional Plant Resource Centre, Bhubaneswar, for providing laboratory facilities. The financial support for this work from Council of Scientific and Industrial Research, New Delhi (38(1 123)/EMR-II) to Dr. A. B. Das is thankfully acknowledged. P. Mohanty thanks Indian National Science Academy, New Delhi, for Honorary Scientist affiliation at RPRC, Bhubaneswar and University of Indore for courtesy support as Honorary Professor.

REFERENCES

Behera B, Das AB, Mohanty P (2009) Changes of soluble proteins in leaf and thylakoid exposed in high saline condition of mangrove taxa *Bruguiera gymnorrhiza.* Physiol Mol Biol Plants 15(1): 55-59

Ben Amor N, Ben Hamed K, Dedez A, Grignon C, Abdelly C (2005) Physiological and antioxidative responses of the perennial halophyte *Crithmum maritimum* to salinity. Plant Sci 68: 889-899.

Cherian S, Reddy MP, Pandya JB (1999) Studies on salt tolerance in *Avecennia marina* (Forstk.) Vierh: effect of NaCl salinity on growth, ion accumulation and enzyme activity. Indian J Plant Physiol 4: 266-270

Cherian S, Reddy MP (2003) Evaluation of NaCl tolerance in callus cultures of *Suaeda nudiflora.* Moq Biol Plant 46: 193-198

Das AB, Mishra S, Mohanty P (2006) Physiology of mangroves: strategies on stress adaptation with special emphasis on tolerance to high salinity. *In:* (ed Hemantarajan, A) Advances in Plant Physiol, Scientific Publishers, Jodhpur 9: 101-154

Das P, Basak UC, Das AB (1997) Restoration of mangrove vegetation in Mahanadi delta of Orissa, India. Mangroves and Salt Marshes 1: 155-161

Elstner EF, (1991) Mechanisms of oxygen activation in different compartments of plant cell. In (eds) Pell, EJ, Stefen KL) Active Oxygen/Oxidative Stress and Plant Metabolism. American Society of Plant Physiology, Rockville, MD, pp 13-25

Fang ZQ, Yuan LY, Hong PC, Ming LC, Shan WB (2005) NaCl enhances thylakoid-bound SOD activity in the leaves of C_3 halophyte *Suaeda salsa* L. Plant Sci 168: 423-430

Flowers TJ, Hajibagheri MA, Clipson NJW (1986) Halophytes. Quart Rev Biol 61: 313-337

Foyer HC, Noctor G (2000) Oxygen processing in photosynthesis: regulation and signaling. New Phytol 146: 359-388

Hellebust JA (1976) Osmoregulation. Annu Rev Plant Physiol 27: 485-505

Hurst AC, Grams TEE, Ratajczak R (2004) Effects of salinity, high irradiance, ozone and ethylene on mode of photosynthesis, oxidative stress and oxidative damage in the C_3/CAM intermediate plant *Mesembryanthemum crystallinum* L. Plant Cell Environ 27: 187-197

Jithesh MN, Prashanth SR, Sivaprakash KR, Parida AK (2006) Monitoring expression profiles of antioxidant genes to salinity, iron, oxidative, light and hyperosmotic stresses in the highly salt tolerant grey mangrove, *Avicennia marina* (Forsk.) Vierh. by mRNA analysis. Plant Cell Rep 25: 865-876

McMillan C (1986) Isozyme patterns among populations of black mangrove *Avicennia germinans* L., from the gulf of Mexico-Caribean and Pacific Panama. Contrib Marine Sci 29: 130-162

Mishra S, Das AB (2003) Effect of NaCl on leaf salt secretion and antioxidative enzyme level in roots of mangrove, *Aegiceras corniculatum.* Indian J Exp Biol 41: 160-166

Paramonova NV, Shevyakova NI, Kuznetsov VV (2004) Ultrastructure of chloroplast and their storage inclusions in the primary leaves of *Mesembryanthemum crystallinum* affected by putrescine and NaCl. Russ J Plant Physiol 1: 86-96

Parida AK, Das AB, Mohanty P (2004) Defense potentials to NaCl in a mangrove, *Bruguiera parviflora:* differential changes of isoforms of some antioxidative enzymes. J Plant Physiol 161: 531-542

Parida AK, Das AB, Mohanty P (2004a) Investigation on the antioxidative defense responses to NaCl stress in a mangrove, *Bruguiera parviflora*: Differential regulations of isoforms of some antioxidative enzymes. Plant Growth Regulation 42(3): 213-226

Parida AK, Das AB (2005) Salt tolerance and salinity effects on plants: a review. Ecotoxicology and Environmental Safety 60: 324-349

Parida AK, Mittra B, Das AB, Das TK, Mohanty P (2005) High salinity reduces the content of a highly abundant 23 kDa protein of the mangrove, *Bruguiera parviflora.* Planta 221: 135-140

Rich PR, Bonner WD Jr (1987) The sites of superoxide anion generation in high plant mitochondria. Arch Biochem Biophys 188: 206-213

Scholander PF (1968) How mangrove desalinate seawater. Physiol Plant 21: 251-261

Slesak I, Miszalski Z, Karpinska B, Niewiadomska E, Ratajczak R, Karpinski S (2002) Redox control of oxidative stress responses in the C_3-CAM intermediate plant *Mesembryanthemum crystallinum.* Plant Physiol Biochem 40: 669-677

Sugihara K, Hanagata N, Dubinsky Z, Baba S, Karube J (2000) Molecular characterization of cDNA encoding oxygen evolving enhancer protein 1 increased by salt treatment in the mangrove, *Bruguiera gymnorrhiza.* Plant Cell Physiol 41: 1279-1285

Taji T, Seki M, Satox, Sakurai T, Kobayashi M, Ishiyama K (2004) Comparative genomics in salt tolerance between *Arabidopsis* and *Arabidopsis*-related halophyte salt stress using *Arabidopsis* microarray. Plant Physiol. 135: 1-13

Takemura T, Hanagata N, Sugihara K, Baba S, Karube I, Dudinsky Z (2000) Physiological and biochemical responses to salt stress in the mangrove *Bruguiera gymnorrhiza.* Aquat Bot 68: 15-28

Takemura T, Hanagata N, Dubinsky Z, Karube I (2002) Molecular characterization and response to salt stress of mRNAs encoding cytosolic Cu/Zn superoxide dismutase and catalase from *Bruguiera gymnorrhiza.* Trees 16: 94-99

Tanaka Y, Hibino T, Hayashi Y, Tanaka A, Kishitani S, Takabe T (1999) Salt tolerance of transgenic rice overexpressing yeast mitochondrial Mn-SOD in chloroplasts. Plant Sci 148: 131-138

Tausz M, Sircelj H, Grill D (2004) The glutathione system as a stress marker in plant ecophysiology: is a stress-response concept valid? J Exp Bot 55: 1955-1962

Tomlinson PB (1986) The Botany of Mangroves, 1st edition. Cambridge University Press, Cambridge.

Wang B, Luttge U, Ratajczak R (2004) Specific regulation of SOD isoforms by NaCl and osmotic stress in leaves of the C_3 halophyte *Suaeda salsa* L. J Plant Physiol 161: 285-293

CHAPTER 16

Metal Toxicity and Photosynthesis

Anjali Aggarwal, Iti Sharma, B.N. Tripathi*, A.K. Munjal, Mamta Baunthiyal and V. Sharma

Department of Bioscience and Biotechnology, Banasthali University, Banasthali-304022

ABSTRACT

The excess concentration of metals deleteriously affect the plant metabolism particularly photosynthesis. The toxicity of metals on photosynthesis is mainly associated with metal induced structural and functional modifications of photosynthetic machinery of plants. Metal ions can inhibit the activities of various photosynthetic enzymes as well as chlorophyll biosynthesis. Metals also damage the chloroplast membrane system and interfere with photosynthetic electron transport processes. This chapter summarizes various aspects of metal toxicity and photosynthesis.

Keywords: Heavy metals toxicity, plant metabolism, metal interaction, photosynthesis.

Abbreviations

Cd: Cadmium, *Cu:* Copper, *Zn:* Zinc, *Pb*: Lead, *Co:* Cobalt, *Hg:* Mercury, *Fe:* Iron, *Ni:* Nickel, *Mn:* Manganese, *PS-I, II:* Photosystem I and II, *ROS:* Reactive oxygen species.

INTRODUCTION

The concentration of heavy metals (Cd, Cu, Zn, Pb, Co, Hg) in the environment is increasing continuously as a consequence of the increased environmental pollution from industrial, agricultural, energetic and municipal sources (Adrino,1986). However few metals, e.g., Cu, Zn, Co, Fe in trace amounts are essential for various metabolic activities of plants. But excess of all kinds of metals (essential and non-essential both) adversely affect the plant metabolism (Hall, 2002). In plants, metals exert their toxic action mostly by damaging chloroplast and disturbing photosynthesis. The inhibition of photosynthesis is the consequence of interference of metal ions with photosynthetic enzymes and chloroplast membranes. A plethora of information is now available on interference of metals with photosynthesis. The present chapter deals with the various aspects of metal and plant interaction in relation to its effect on photosynthetic metabolism of plants and algae.

ROLE OF SOME HEAVY METALS IN PHOTOSYNTHESIS: DEFICIENCY AND TOXICITY

Although heavy metals inhibit plant metabolism, but some metals like Cu, Fe, Mn are essential for the photosynthesis. Metals are constituents of a great number of proteins and enzymes consequently essential for maintaining optimum plant metabolism. The deficiency of these metals has direct impact on the process of photosynthesis, but elevated concentration of these metals also becomes toxic to plant and affects photosynthesis in many ways. The roles of some of the heavy metals have been described in the following paragraphs.

Copper

In plants more than 50% of the Cu present is found associated with plastocyanin, an important component of the electron transport chain of PSI in the chloroplast. Therefore, copper deficiency reduces PSI electron transport due to decreased formation of plastocyanin (Baszynski *et al.,* 1988). A decrease in PSII is also observed in Cu deficient plants (Henriques, 1989). Cu deficient plants exhibit a disintegration of the thylakoid membrane of chloroplasts decrease in pigment content, reduced plastoquinone synthesis and lower unsaturated C18 fatty acid contents (Baszynski *et al.,* 1988; Henriques, 1989).

Copper can interfere with the biosynthesis of the photosynthetic machinery, changing the pigment and protein composition of photosynthetic membranes. A lower content of chlorophyll, inactivation of enzymes and proteins linked to photosynthesis process and modification of thylakoid membranes occurs under copper toxicity. Copper-induced chlorosis can result from the inhibition of pigment accumulation and retardation of chlorophyll integration into photosystems. Magnesium can be removed by copper from the chlorophyll, present in both antenna complexes and reaction centers, thus damaging the structure and function of the chlorophyll (Kupper *et al.,* 2003).

Iron

Iron is required for the synthesis of chlorophyll. Important redox enzyme involved in photosynthesis include the haem-containing cytochrome and non-haem iron-sulfur protein. During the course of electron transfer, the iron is reversibly reduced from the Fe^{3+} to Fe^{2+} state. Iron is not a constituent of the chlorophyll molecule itself and its precise role in the chlorophyll synthesis remains a mystery. There is no definite evidence that any of the enzymes involved in the chlorophyll synthesis are iron dependent. Instead, iron requirement may be related to more general need for the synthesis of the chloroplast constituents, especially e- transport proteins. Iron deficiency invariably leads to a simultaneous loss of chlorophyll and degeneration of chlorophyll structure.

Manganese

Manganese is required as a cofactor for a number of enzymes involved in photosynthesis, particularly decarboxylase and dehydrogenase enzymes. Its well known function is in the evolution of photosynthetic oxygen. Manganese, in the form of a manganoprotein, is part of the oxygen evolving complex (OEC) where it appears to be involved in the accumulation of changes during the oxidation of water. Mn deficiency may cause extreme chlorosis between the leaf veins as well as discoloration and deformities in seeds.

MECHANISM OF HEAVY METAL TOXICITY TOWARDS PHOTOSYNTHESIS: DIRECT AND INDIRECT EFFECT

Heavy metals affect both light and dark reactions of photosynthesis directly or indirectly. The extent to which the toxicity of these metals may influence photosynthesis depends on the way these events are investigated (*in vitro* or *in vivo*) as well as the age of plants. The direct effects of heavy metals on the light reactions are on the photosynthesis O_2 evolution, reduction of NADP and photophosphorylation. Indirectly it affects the process of chlorophyll synthesis and degradation, changes in Chl *a*/*b* ratio, competition with other essential metals, etc. Some information concerning the possible mechanisms of heavy metal toxicity on photosynthesis is given below.

Indirect Effects

In higher plants, photosynthesis is reduced indirectly by heavy metal accumulation in leaves which influences the functioning of stomata. Cd concomitantly decreased net photosynthesis and transpiration in excised *Acer sacchanium* L. leaves (Lamoreaux RJ and Chaney, 1978). Its primary effect is therefore proposed to be on stomatal function. Other heavy metals like Cd, Ni, and Pb also cause stomatal closure in epidermal peels on detached leaves of *Helienthus annus* L. floating on a solution containing either of these elements (Bazzaz FA *et al.*, 1974). This simultaneously led to inhibition of photosynthesis and transpiration as these are stomata related functions. This was also revealed by Ni in *Zea mays* L. In leaves of metal treated plants the net CO_2 fixation rate per unit chlorophyll increased, but this activity decreased when expressed on a leaf area basis (Baszynski *et al.*, 1980). Moreover Zn, Cd also decreased the water use efficiency (amount of CO_2 photosynthetically fixed per mol of water transpired) (Van Assche F and Clijsters 1980; Carlson RW *et al.*, 1975).

Reduction of photosynthetic pigments by the heavy metals also indirectly influences the photosynthesis. Photosynthetic pigments are easily measurable and frequently used to determine stress for regulatory purposes. Heavy metals decrease the chlorophyll content of algae, the chlorophyll *a*/*b* ratio, pheophytin levels but increase protochlorophyll levels and carotenoid/chlorophyll ratios in the majority of cases (De Filippis and Pallaghy, 1994). Photosynthetic pigments were found to be reduced under the excessive concentrations of Hg (Rai *et al.*, 1981), Cu (Gross *et al.*, 1970; Wong and Chang, 1991; Genter, 1996), Cr (Corradi *et al.*, 1995), Ni (Xylander and Braune, 1994; Genter, 1996), Cd (Conway, 1978; Rai *et al.*, 1981), and Zn (De Filippis and Pallaghy, 1976; Rai *et al.*, 1981), etc. There are few reports that showed the enhancement of pigments after the exposure to metals (Deviprasad and Deviprasad, 1982)—0.1 to 1.0 mg l^{-1} Ni caused significant increase in chlorophyll *a* content of *Chlorella pyrenoidosa* (Wong and Chang, 1991). In the case of *N. palea*, slight stimulation of Chl *a*, Chl *b* and carotenoids was noticed at 20 μgl^1 Cd (Sathya and Balakrishnan, 1988). Carotenoids: Chlorophyll ratio was found to increase or decrease (De Filippis and Pallaghy, 1976, Rai *et al.*, 1981) by heavy metals depending on the metals used for experiments. A marked decrease in Chl *a* and Chl *b* ratio was observed in Cr treated *Scenedesmus* in comparison to that of control (Corradi *et al.*, 1995). In view of the high redox potential of many heavy metals, the simplest explanation consistent with those observations is that metals inhibit the reductive steps in the biosynthetic pathway of photosynthetic pigments. In addition, the key enzyme, protochlorophyllide reductase, which is involved in the reduction of protochlorophyll to chlorophyll, is well known to be inhibited by heavy metals (De Filippis and Pallaghy, 1994).

Besides the above effects, heavy metals also induce the deficiency of other essential elements, which causes inhibition of metal-ion dependent reactions such as photolysis of water; the deficient essential metal can also be substituted by the abundantly present toxic one in metalloproteins. Heavy metals have also been shown to increase susceptibility to photoinhibiton *in vitro* using isolated thylakoids (Cedeno-Maldonado and Swader, 1972; Patsikka *et al.*, 2001) or PS II particles (Jegerschold *et al.*, 1995; Yruela *et al.*, 1996), but *in vivo* susceptibility to photoinhibition is particularly severe. Patsikka *et al.* (2002) have shown *in vivo* that excess Cu predisposes photosystems II to photoinhibition in bean plants; the reason behind photoinhibition is not only due to toxicity of Cu, but at higher concentrations of Cu, it competes with Fe and deficiency of Fe causes decrease in leaf chlorophyll.

The exact information on the concentration of polluting heavy metals on the leaf and in respective cell compartments is of major importance for assessing their possible mechanisms of action on photosynthesis.

Direct Effects

Heavy metals directly affect the photosynthetic machinery by binding to the various sensitive sites of the photosynthetic apparatus. In chloroplast, heavy metals disturb the architecture of thylakoid membranes, which in turn, change some light reaction processes, directly especially those associated with PSII. It is evident that excess Cu has a strong effect on chloroplast fine structure, resulting in degradation of grana stacks and stroma lamellae and an increase in the number and size of plastoglobuli and intrathylakoidal inclusions (Baszynski *et al.*, 1988). Sandmann and Boger (1980) correlated these disturbances with processes of enhanced lipid peroxidation. Various authors gave evidences that the toxic effect of metal on phototropic organisms strongly appears to be related to the increase in the levels of lipid peroxidation and protein carbonylation as well as the production of antioxidant defense systems (Gallego *et al.*, 1996; Devi and Prasad, 2005; Tripathi *et al.*, 2006). Proteins and lipids found embedded in the thylakoid membranes are directly involved in photosynthesis. Metal induced oxidative damage to these essential protein and lipids might be primarily responsible for the inhibition of electron flow. In contrast, Halliwell and Gutteridge (1999) suggest that oxidative damage is merely a consequence of metal toxicity, and a direct interaction of metal either with the donor or with the acceptor site of PSII leads to the interruption of electron transport (Patsikka *et al.*, 2001). Defense systems of organisms against ROS also play an important role in photosynthetic activity. Knauert and Knauer (2008) concluded that differences in species-specific sensitivities, measured as photosynthetic activity, were not caused by differences in the cellular ROS content of the algae, but probably by different species-specific ROS defense systems.

Metals are also known for their strong binding affinity to sulfhydryl groups, which are essential for enzymatic activity and protein structure. (De Fillippis and Pallaghy,1994). S-H interaction with metals is suggested to explain the inhibitory effect of 10 μM Pb (Hampp *et al.*, 1973) and 10 μM Cd (Ernst, 1980) on the activity of the chloroplast enzymes ribulose 1, 5-bisphosphate-carboxylase (RuBPC; E.C.4.1.1.39) and phosphoribulokinase (E.C.2.7.1.18) *in vitro*.

TARGET SITES OF HEAVY METAL INTERACTION WITH PHOTOSYNTHESIS

Photosynthetic Pigments

The inhibition of chlorophyll biosynthesis by metals has been well described by De Filippis and Pallaghy (1976) and De Filippis *et al.* (1981). The sensitivity of metals is found for two key enzymes of the chlorophyll biosynthetic pathway: δ-aminolaevulinic acid (ALA)-dehydratase (EC 4.2.1.24) and protochlorophyllide reductase (Vangrosveld and Clijsters, 1994). δ- ALA-dehydratase, the chloroplastic enzyme, which is not membrane bound, catalyses the conversion of δ- ALA in porphobilinogen in the synthesis of chlorophyll. This enzyme is found to be sensitive to different metals (Vallee and Ulmer, 1972; Vangrosveld and Clijsters, 1994). In the green alga *Chlorella vulgaris* grown in media containing sublethal mercury concentrations, chlorophyll biosynthesis was strongly inhibited and protochlorophyll accumulated (De Filippis and Pallaghy, 1976). A more metal sensitive alga, *Euglena gracilis* showed a decrease in chlorophyll and a marked enhancement in protochlorophyllide levels in the presence of sublethal concentrations of Zn, Cd and Hg (De Filippis *et al.*, 1981). In these reports it was speculated that metals inhibit protochlorophyllide reductase, which is responsible for the final reductive step of chlorophyll biosynthesis.

Photosynthetic Enzymes

From *in vitro* experiments, at least two potential metal sensitive sites can be derived in the photosynthetic electron transport chain: the water splitting enzymes at the oxidising side of PSII and NADPH-oxidoreductase at the reducing side of PSI (Vangrosveld and Clijsters, 1994). De Filippis *et al.* (1981) proposed that in *Euglena gracilis* the reduction of photosynthetic capacity under metal stress can be attributed to inhibition of the both, the water splitting enzyme and NADPH-oxidoreductase. The *in vitro* studies plants treated with Zn, Cd, and Hg clearly suggest that some enzymes of the carbon fixation pathway, especially energy dependent enzymes, are also affected by metals (De Filippis and Pallaghy, 1994). A majority of metals affect the key enzymes of Calvin cycle, ribulose-1, 5-bisphosphate carboxylase/oxygenase (Vangrosveld and Clijsters, 1994) and phosphofructokinase. Interaction of heavy metals with functional SH- groups is generally proposed as the mechanism of inhibition of enzymes.

Photosystems

In the photosynthetic electron transport chain, photosystem (PS) is the site most sensitive to metal ions (Prasad and Strazalka, 1999). All the metals studied are potential inhibitors of PSII, whereas PSI was reported to be less sensitive. The inhibitory effect of metals on reducing side of PSI electron transport system has been reported in algae and plants by various authors (Shioi *et al.,* 1978, Baszynski *et al.* 1988). The cyt b_6/f complex has also been suggested as an inhibitory site (Singh and Singh,1987).

PSII is a large protein complex that consists of about 20 protein subunits and is found embedded in the thylakoid membrane of the chloroplasts. It catalyses the light-driven reduction of plastoquinone by electrons from water which is oxidized to molecular oxygen (Andersson and Styring, 1991, Vermaas *et al.,* 1993). The most apparent effect of the toxic action of metal is inhibition of oxygen evolution accompanied by quenching of variable fluorescence (Hsu and Lee, 1988; Arellano *et al.,* 1995). It is found that metal ions inhibit both the donor and the acceptor sides of PSII, but the most sensitive site of metal inhibitor is located on the oxidation (donor) side of PSII (Cedeno and Swader, 1972; Vierke and Stukmeier, 1977), where a reversible inhibition of Tyrz (the redox active tyrosine residue of the D1 protein) takes place. At higher metal concentration, primary quinine acceptor Q_A (Jegerschold *et al.,* 1995), pheophytin-Q_A-Fe region (Prasad and Strzalka K, 1999), non-haem iron (Singh and Singh, 1987, Jagerschold *et al.,* 1995) and secondary quinine acceptor Q_B (Mohanty *et al.,* 1989) are identified as the target sites of metal inhibitory action on the acceptor site of PSII. Apart from the oxygen-evolving complex, cytochrome b_{559} (cyt b_{559}) in its high potential form is the most sensitive component. Yruela *et al.* (1996) found that Cu decreased the level of the photoreduced cytb_{559} and slowed down its rate of photoreduction, whereas Jegerschold *et al.* (1995) observed the conversion of low potential form at high Cu concentration. Burda *et al.* (2003) suggest that the primary target sites in photosystems II for Cu are tyrosine$_z$, high and low potential cytb_{559}, and chlorophyll$_z$, and that these sites are the source of Cu-induced fluorescence quenching and oxygen evolution inhibition in PSII.

Besides these interactions of heavy metal with photosynthesis, the process of photosynthesis also plays an important role in the recovery of photosynthetic organisms from previous exposure to heavy metals. This was shown by Tripathi *et al.* (2004) in *Scenedesmus* sp. that photosynthesis was quickly recovered after the transfer of algal cells to basal medium. This is due to the test algae used photosynthesis as acclimatory responses as they prepare a background for the recovery of other parameters, including growth, by generating energy, forming photosynthate and establishing the usual catabolism to attain normal conditions.

ACKNOWLEDGEMENTS

Authors pay hearty tribute to Prof. Govindjee for his work on photosynthesis. BNT thank to Professor P Mohanty for his support and encouragement and Professor K N Guruprasad for kindly offering the invitation to write this chapter. The support from Professor Aditya Shastri, Vice-Chancellor, Banasthali University is also gratefully acknowledged.

REFERENCES

Adrino DC (1986) Trace Elements in the Terrestrial Environment. Springer-Verlag, New York.,

Andersson B, Styring S (1991) Photosystem II: molecular organization, function, and acclimation. Current Topic in Bioenergetics 16: 1-81

Arellano JB, Lazaro JJ, Lopez-Gorge J, Baron M (1995) The donor side of Photosystem II as the copper-inhibitory binding site. Photosynthetic Research 50: 698-701

Baszynski T, Wajda L, Krol M, Wolinska D, Krupa Z, Tukendorf A (1980) Photosynthetic activities of cadmium-treated tomato plants. Physiologia Plantarum 48: 365-370

Baszynski T, Tukendorf A, Ruszkowska M, Skorzynska E, Maksymiec W (1988): Characteristics of the photosynthetic apparatus of copper non-tolerant spinach exposed to excess copper. Journal of Plant Physiology 132: 708-713

Bazzaz FA, Carlson RW, Rolfe GL (1974) The effect of heavy metals on plants I. Inhibition of gas exchange in sunflower by Pd, Cd, Ni and TI. Environmental Pollution 7: 241-246

Burda K, Kruk J, Strzalkak, Schmid GH (2002) Stimulation of oxygen evolution in photosystem II by copper (II) ions. Z Naturforsch 57c: 853-857

Carlson RW, Bazzaz EA, and Rolfe GL (1975) The effects of heavy metals on plants. II. Net Photosynthesis and transpiration of whole corn and sun flower plants treated with Pb, Cd, Ni, and TI. Enviormental Research 10: 113-120

Cedeno-Maldonado A, Swader JA (1972) The cupric ions as an inhibitor of photosynthetic electron transport in isolated chloroplast. Plant Physiology 50: 698-701

Conway HL (1978) Sorption of arsenic and cadmium and their effect on growth micronutrient utilization and photosynthesis pigment composite of *Asterionella formosa.* J Fish Res Board Can 35: 286-294

Corradi GM, Gorbi G, Rieci A, Torelli A, Bassi M (1995) Chromium–induced sexual reproduction gives rise to a Cr tolerant progeny in *Scenedesmus actus.* Ecotoxicol Environ Safety 32: 12-19

De Filippis LF, Pallaghy CK (1976) The effect of sublethal concentration of mercury and zinc on *Chlorella.* I Growth characteristic and uptake of metals. Z Pflanzenphysiol 78: 197-207

De Filippis LF, Hampp R, Ziegler H (1981) The effect of sub-lethal concentration of zinc, cadmium and mercury on *Euglena* II. Respiration, photosynthesis and photochemical activities. Arch Microbiol 128: 407-411

De Filippis LF, Pallaghy CK (1994) Heavy metals: Sources and biological effects. In: LC Rai, JP Gaur, CJ Soeder (eds), Algae and Water Pollution. E. Schweizerbart'sche, Verlagsbuchhandlung, Stuttgart 31-77

Devi Prasad PV, Devi Prasad PS (1982) Effect of cadmium, lead and nickel on three freshwater green algae. Water, Air Soil Pollut 17: 263-268

Devi SR, Prasad MNV (2005) Antioxidant capacity of *Brassica juncea* plants exposed to elevated levels of copper. Russian Journal of Plant Physiology 52: 246-73

Ernst WHO (1980) Biochemical aspects of cadmium in plants. In: Nriagu JO (ed) Cadmium in the Environment Part 1, New York: Wiley & Sons, pp 639-653

Gallego SM, Benavides MP, Tomaro ML (1996) Effect of heavy metal ion excess on sunflower leaves: evidence for involvment of oxidative stress. Plant Science 121: 151-9

Genter RB (1996) Ecotoxicology of inorganic stresses. In: RJ Stevenson, ML Bothwell, RL Lowe (eds), Algal Ecology: Freshwater Benthic Ecosystems. Academic Press, San Diego, pp 403-468

Gross RE, Pungo P, Dugger WM (1970) Observation of the mechanism of copper damage in *Chlorella*. Plant Physiol 46: 183-185

Hall JL. (2002) Cellular mechanisms for heavy metal detoxification and tolerance. Journal of Experimental Botany 53: 1-11

Halliwell B, Gutteridge JMC (1999) Free radicals in Biology and Medicine. 3rd ed. Oxford University Press Inc., New York, p 936

Hampp R, Ziegler H and Ziegler I (1973) Der Einfluss von Bleiionen auf Enzyme des reduktiven Pentosephosphatcyclus. Biochem Physiol Pflanzen 164: 588 595

Henriques F (1989) Effects of copper deficiency on the photosynthetic apparatus of suger beet (*Beta vulgaris* L.). Journal of Plant Physiology 135: 453-458.

Hsu BD, Lee JY (1988) Toxic effect of copper on photosystem II of spinach chloroplasts. Plant Physiology 87: 116-119

Jegerschold C, Arellano JB, Schroder WP, van Kan PJM, Baron M, Styring S (1995) Copper (II) inhibition of electron transfer through photosystem II studied by EPR spectroscopy. Biochemistry 34: 12747-12754

Knauert S, Knauer K (2008) The role of reactive oxygen species in copper toxicity to two fresh water green algae. Journal of Phycology 44: 311-319

Kupper H, Setlik I, Setlikova E, Ferimazova N, Spiller M, Kupper FC (2003) Copper-induced inhibition of photosynthesis: Limiting steps of *in vivo* copper chlorophyll formation in *Scenedesmus quadricauda*. Functional Plant Biology 30: 1187-1196

Lamoreaux RJ and Chaney WR (1978) The effect of cadmium on net photosynthesis, transpiration, and dark respiration of excised silver maple leaves. Physiologia Plantarum 43: 231-236

Mohanty N, Vass I, Demeter S (1989) Copper toxicity affects photosystems II electron transport at the secondary quinone acceptor (Q_B). Plant Physiology 90 175-179

Patsikka E, Aro E-M, Tyystjarvi E (2001) Mechanism of copper-enhanced photoinhibition in thylakoid membranes. Physiologia Plantarum 113: 142-150

Patsikka E, Kairavuo M, Sersen F, Aro Eva-Mari, Tyystjarvi E (2002) Excess Copper Predisposes photosystem II to photoinhibition *in vivo* by outcompeting iron and causing decrease in leaf chlorophyll. Plant Physiology 129: 1359-1367

Prasad MNV, Strzalka K (1999) Impact of heavy metals on photosynthesis. In: (Prasad MNV, Hagemeyer J eds) Heavy Metal Stress in Plants, Springer Verlag, Berlin, pp 117-138

Rai LC, Gau JP, Kumar HD. (1981) Phycology and heavy-metal pollution. Biol Rev Cambridge Phil Soc 56: 99-151

Sandmann G Boger P (1980) Copper mediated lipid peroxidation processes in photosynthetic membranes. Plant Physiology 66: 797-800

Sathya KS, Balakrishnan KP. (1988) Physiology of phytoplankton in relation to metal concentration: Effect of cadmium on *Scenedesmus bijugatus* and *Nitzschia palea*. Water Air Soil Pollut 38: 283-29

Shioi Y, Tamai H, Sasa T (1978) Effect of copper on photosynthetic electron transport systems in spinach chloroplasts. Plant & Cell Physiol 19 (2): 203-209

Singh DP, Singh SP (1987) Action of heavy metals on Hill activity and O_2 evolution in *Anacystis nidulans*. Plant Physiology 83: 12-14.

Tripathi BN, Mehta SK, Guar JP (2004) Recovery of uptake and assimilation of nitrate in *Scenedesmus* sp. previously exposed to elevated levels of Cu^{2+} and Zn^{2+}. Journal of Plant Physiology 161: 543-547

Tripathi BN, Guar JP (2006): Physiological behaviour of *Scenedesmus* sp. during exposure to elevated levels of Cu and Zn and after withdrawal of metal stress. Protoplasma 229 1-9

Tripathi BN, Mehta SK, Amar A, Guar JP (2006) Oxidative stress in *Scenedesmus* sp. during short- and long-term exposure to Cu^{2+} and Zn^{2+}. Chemosphere 62: 538-44

Vallee BL, Ulmer DD. (1972) Biochemical effects of mercury, cadmium and lead. Annu Rev Biochem 41: 91-128

Van Assche F, Clijsters H (1980) Zinc mediated effects on leaf CO_2 diffusion conductances and net photosynthesis in *Phaseolus vulgaris* L. Photosynthesis Research 1: 171-180

Vangronsveld J, Clijsters H (1994) Toxic effects of metals. In: ME Farago (ed.), Plant and the Chemical Elements: Biochemistry, Uptake, Tolerance and Toxicity. VCH Verlagsgesellschaft, Weinheim, pp 150-177.

Vermaas WFJ, Styring S, Schroder WP, Andersson B (1993) Photosynthetic water oxidation: the protein framework. Photsynthetic Reseach 38: 249-63

Vierke G, Stuckmeier P (1977) Binding of Copper (II) to proteins of the photosynthetic membrane and its correlation with inhibition of electron transport in class II chloroplasts of spinach. Z Naturforsch 32: 605-610

Wong PK, Chang L (1991) Effects of copper, chromium and nickel on growth, photosynthesis, and chlorophyll *a* synthesis of *Chlorella pyrenoidosa* 251. Environ Pollut 72: 127-140

Xyländer M, Braune W (1994) Influence of nickel on the green alga *Haematococcus lacustris* Rostafinski in various phases of its life cycle. J Plant Physiol 144: 86-93

Yruela I, Pueyo JJ, Alonso PJ, Picorel R (1996) Photoinhibition of photosystems II from higher plants: effect of copper inhibition. Journal of Biological Chemistry 271: 27408-27415

CHAPTER 17

Effects of Heavy Metals on Photosynthetic Carbon Assimilation and Thermoluminescence in Unicellular Alga, *Chlorella pyrenoidosa*

Gururaj Maralihalli and Jayashree K. Sainis

Molecular Biology Division, Bhabha Atomic Research Centre, Mumbai-400085

ABSTRACT

Chlorella pyrenoidosa, when grown in the presence of 100 μM of cadmium nitrate, cupric chloride, lead nitrate, uranium nitrate, and cobalt chloride showed decline in growth rate by 30 to 74% depending on the salt present in the medium. The rate of CO_2 fixation increased after incubation with 100 μM of lead nitrate and cupric chloride and decreased after incubation with cadmium nitrate and cobalt chloride. Uranium nitrate did not affect the rate of CO_2 fixation. Thermoluminescence (TL) studies showed two characteristic peaks at ~6°C (Q) and ~34°C (B). TL peak temperature (Tm) of both Q and B peak showed changes indicating that redox potential of Q_A^-, Q_B^- and S states was affected after incubation with these heavy metal salts.

Keywords: Thermoluminescence, Peak temperature, Photophoshorylation, Carbon assimilation.

Abbreviations

DCMU: 3-{3′, 4′-dichlorophenyl}-1, 1-dimethylurea, *PSII:* Photosystem II, *QA:* primary quinone acceptor, *QB:* secondary quinone acceptor, *TL:* thermoluminescence.

INTRODUCTION

Algae are known to grow in most inhospitable environments (Kalin *et al.,* 2005). Heavy metal polluted and chromium contaminated sites can support the growth of algae (Cervantes *et al.,* 2001). Heavy metals are known to affect both light and dark reactions of photosynthesis. Among photosynthetic activities, PSII has been shown to be more vulnerable to Cd^{2+}, Hg^{2+} and Zn^{2+}. The disorganization of chloroplast, inactivation of oxygen evolving centers, impairment of electron transport and increase in susceptibility of PSII to photo inhibition was observed in treated samples. Photosynthetic carbon assimilation and photophosphorylation were shown to be affected by heavy metals (Betrand and Poirer, 2005). El-sheek *et al.* (2003) had shown the effects of mercury and cobalt on photosynthetic electron transport in *Chlorella kessleri, Monoraphidium minitum* and diatom *Nitzschia perminuta.* PSII was reported to be a major site of action of chromate and other heavy metals in *Scenedesmus obliquus* (Mallick and Mohn, 2003). There was a concentration dependent increase in lipid peroxidation, carotenoid content and increase in SOD activity by copper in *Chlorella vulgaris* (Mallick, 2004). Chromate inhibited photosynthesis by the oxidative stress resulting in photo destruction of D1 as indicated by high temperature thermoluminescence band as was shown in the case of *Chlorella pyrenoidosa* (Horcsik *et al.,* 2007).

We report here the effect of salts of copper, lead, uranium, cadmium and cobalt on growth, CO_2 fixation and thermoluminescence of Chlorella *pyrenoidosa.*

MATERIAL AND METHODS

Algal Culture and Growth Conditions

The culture of *Chlorella pyrenoidosa* was from culture collection at National Chemical Laboratory, Pune, India. The cells (1×10^6 cells ml^{-1}) were inoculated in Tris-Acetate-Glycerophosphate (TAgP) (Collard and Matagne, 1990) medium containing 100 μM of metal salts, namely, cupric chloride, lead nitrate, uranium nitrate, cadmium nitrate, or cobalt chloride and were incubated at 25°C under continuous illumination with fluorescent lamps of 24 Wm^{-2} of light intensity. After keeping for 24 h in light, OD at 655 nm was measured. The growth rate of algal culture was determined as described by Wong and Chang (1991). $k = (\ln X_1 - \ln X_0)/(T_1 - T_0)$ where k is the specific growth rate, X_1 is absorbance at 655 nm at 24 hours (T_1) and X_0 is the absorbance at 655 nm at the time of inoculation (T_0).

CO_2 Fixation

The log phase *Chlorella* culture obtained was treated with 100 μM of metal salts, namely, uranium nitrate, lead nitrate, cadmium nitrate, cupric chloride and cobalt chloride each separately in light (24 Wm^{-2}) at 25°C for 1 or 24 h. 20 mM $NaH^{14}CO_3$ (specific activity 0.5 mCi/mole) as possible detectable carbon source was added to 100 μl of treated and incubated mixture, which was further incubated in light for a period of 10 min. The reaction was stopped by the addition of 100 μl of 6N acetic acid. The resulted acid stable ^{14}C labeled product was measured using scintillation counter.

Thermoluminescence

Log phase culture was treated with 100 μM of metal salts as described above, the treated culture was centrifuged and the pellet was resuspended in medium to a final concentration of 250 μg chlorophyll /ml and was kept in dark for 10 minutes, and subsequently exposed to light for 4 minutes (210 W m^{-2}) at 25°C. The cells were immediately frozen in liquid nitrogen under the same illumination. The thermoluminescence glow peaks were recorded using the TL set up having a microcontroller-based data recording system, acquiring data at half a second interval (Desai, 1992). The rate of heating was 16-18°C min^{-1} during TL measurement. The glow curves were deconvoluted and fitted into Guassian peaks using Microsoft MS origin. Statistical analysis was done using student's't' test for independent samples at 5% level of significance.

RESULTS

Chlorella pyrenoidosa, when grown in the presence of 100 μM of cupric chloride, lead nitrate, uranium nitrate, cadmium nitrate, or cobalt chloride showed decline in growth rate by 30-74% depending on the salt (Fig. 17.1).

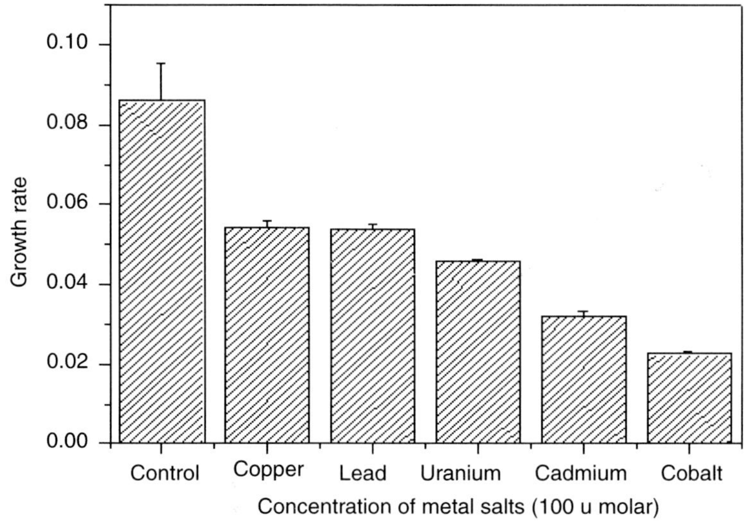

Fig. 17.1 Effect of uranium nitrate, lead nitrate, cadmium nitrate, cupric chloride and cobalt chloride on growth rate of *Chlorella pyrenoidosa*. Equal number of cells of *Chlorella pyrenoidosa* were inoculated in TAgP medium containing 100 μM of metal salts each separately and OD at 655 nm was measured at the time of inoculation and 24 hours after growth to calculate growth rate as described in material and methods.

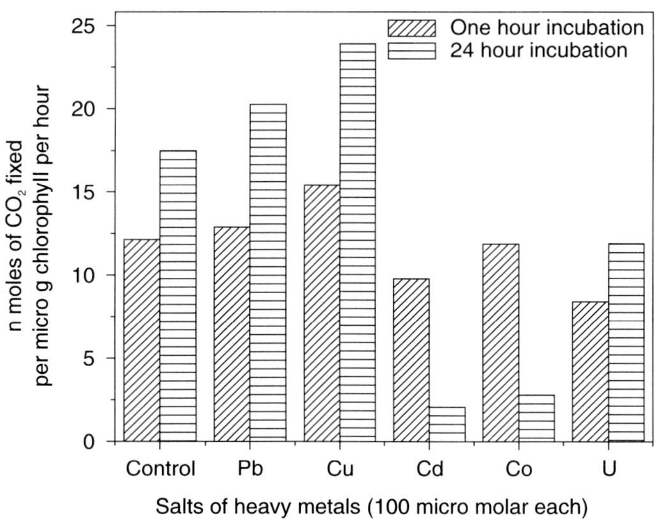

Fig. 17.2 Effect of uranium nitrate, lead nitrate, cadmium nitrate, cupric chloride and cobalt chloride on CO_2 fixation by **Chlorella** *pyrenoidosa. Chlorella pyrenoidosa* cells were incubated with 100 μM of uranium nitrate, lead nitrate, cadmium nitrate, cupric chloride or cobalt chloride as indicated in figure, in light (24 Wm^{-2}) at 25°C for 1 or 24 h. CO_2 fixation was monitored by measuring incorporation of $NaH^{14}CO_3$ in acid stable ^{14}C labeled compounds. The rates are plotted against the heavy metal used during incubation.

To investigate the effect of heavy metals on photosynthesis, the log phase *Chlorella* culture was incubated with 100 μM of each of the heavy metal salts for 1 h and 24 h. Figure 2 shows that there was an increase in rate of CO_2 fixation in untreated cells after 24 h incubation in light. In case of cells

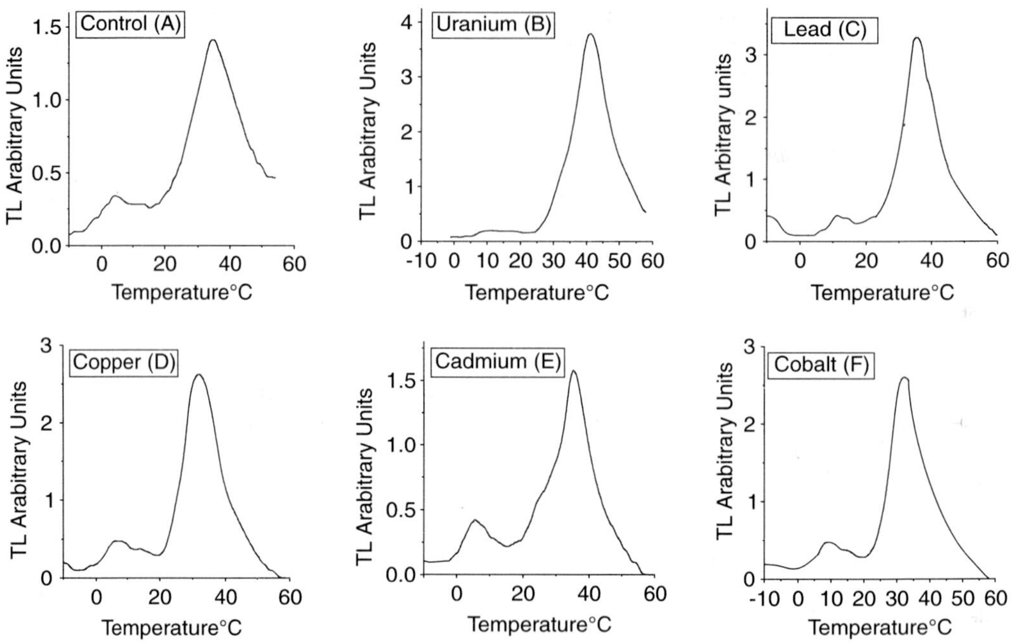

Fig. 17.3 Effect of uranium nitrate, lead nitrate, cadmium nitrate, cupric chloride and cobalt chloride on Thermoluminescence by *Chlorella pyrenoidosa*. TL glow curves of *Chlorella pyrenoidosa* cells incubated with 100 μM of metal salts namely uranium nitrate, lead nitrate, cadmium nitrate, cupric chloride and cobalt chloride for 24 hours. **A:** Untreated cells, **B:** Cells treated with uranium nitrate **C:** Cells treated with lead nitrate, **D:** Cells treated with cupric chloride, **E:** Cells treated with cadmium nitrate and **F:** Cells treated with cobalt chloride.

treated with lead nitrate and copper chloride there was 10-15% and 31-37% increase in rate of CO_2 fixation respectively, especially after 24 hours of treatment. In case of Cd^{+2} and Co^{+2} the CO_2 fixation rates were not affected after treatment for one hour, but decreased by 85-90 % after incubation for 24 hours. Uranium showed 25-30% decrease as compared to untreated controls both after 1 and 24 hours of incubation.

Figure 17.3 A-F shows the TL glow peaks obtained after treatment of *Chlorella* cells for 24 hours with different heavy metal salts. Two peaks were obtained one at 5.65 ± 0.48°C, regarded as Q peak and the other at 33.8 ± 1.4°C which is a B peak in case of cells not treated with heavy metal salts (Fig. 17.3A). The cells treated with DCMU showed enhancement of Q peak and suppression of B peak (data not shown). Figure 17.3 B-F shows the TL glow curves of cells treated with heavy metal salts for 24 hours. Table 17.1 shows the Tm for Q and B peaks, the area under the peaks and the ratio of area under B and Q peaks as well as ΔTm values for control as well as treated cells. There was an increase in Tm of Q peak for cells treated with uranium and lead nitrate (Table 17.1). The Tm of Q peak was in the range of 5-8°C for the cells treated with salts of cadmium, cobalt and copper. The area under Q peak was not significantly affected by treatment with heavy metal salts (Table 17.1) indicating that treated cells did not show any significant differences in the yield of TL as compared to untreated cells. However, the cells treated with uranium nitrate showed significant increase in Tm of B peak to around 42°C whereas for all other metal salts, the B peak was not significantly affected. There was an apparent increase in the area of B peak as well as in the ratio of area of B peak to Q peak in cells treated with uranium and lead nitrate, whereas for others there was no discernible change (Table 17.1). The ΔTm showed significant decrease in case of cells treated with lead nitrate.

Table 17.1 Data on TL glow curves obtained from *Chlorella pyrenoidosa* treated with various heavy metal salts.

	Tm Q peak	*Area Q peak*	*Tm B peak*	*Area B peak*	*Ratio of Area of B peak/Q peak*	*ΔTm*
			Control			
Mean	5.65	4.93	33.81	38.78	8.26	28.16
± SE	0.48	0.60	1.38	5.06	1.87	1.44
			Uranium nitrate			
Mean	15.43*	4.26	41.88*	59.67	14.56	26.45
± SE	3.17	0.88	0.60	6.70	1.36	2.61
			Lead nitrate			
Mean	13.57*	3.73	35.90	46.80	18.48	22.33*
± SE	2.43	1.28	2.25	1.01	8.79	0.76
			Cupric chloride			
Mean	7.99	5.62	33.88	42.23	8.28	25.89
± SE	1.55	1.24	1.33	1.75	1.75	0.31
			Cadmium nitrate			
Mean	5.51	11.94	32.61	34.07	4.46	27.10
± SE	2.41	5.31	2.64	8.72	1.83	2.13
			Cobalt chloride			
Mean	7.00	4.88	31.55	33.10	6.87	24.55
± SE	1.28	0.96	1.53	6.16	0.88	1.20

Log phase culture was treated with 100 μM of metal salts, namely, uranium nitrate, lead nitrate, cadmium nitrate, cupric chloride and cobalt chloride each separately for 24 hours in light at room temperature. The treated culture was centrifuged and the pellet was resuspended in medium to a final concentration of 250μg chlorophyll /ml and was kept in dark for 10 minutes, and subsequently exposed to light for 4 minutes (210W m^{-2}) at 25°C. The cells were immediately frozen in liquid nitrogen under the same illumination. The thermoluminescence glow peaks were recorded using the TL setup. The glow curves were deconvoluted and fitted into Guassian peaks using Microsoft MS origin. Tm and area of Q and B peaks was calculated using MS origin. Δ Tm and ratio of area of B /Q peak was computed using the data obtained. * indicates values which were significantly different as compared to controls at 5% level.

DISCUSSION

The *Chlorella pyrenoidosa* culture examined here was found to be more sensitive to cobalt as compared to copper, cadmium and other heavy metal salts, indicating that the algae may be using different mechanisms to endure the heavy metal stress. It has been shown that *Chlamydomonas acidophila* was resistant to Co and Cu but was sensitive to Cd (Nishikawa and Tominaga, 2001).

We investigated the effect of heavy metals on photosynthesis by monitoring CO_2 fixation and thermoluminescence. Lead and copper treatments enhanced the rate of CO_2 fixation whereas cadmium and cobalt treatments resulted in decrease in the rates. Uranium showed only slight decrease as compared to untreated controls both after one and 24 h of incubation. Treatment with cadmium chloride and cobalt chloride drastically reduced the rate of CO_2 fixation indicating that the photosynthesis may be an important target in the case of cadmium and cobalt.

Thermoluminescence is considered to be an effective tool to monitor PSII *in vivo* (Sane, 2004). TL arises from thermally activated recombination between the positively charged states of oxygen evolution complex and negatively charged acceptors of PSII viz Q_A^- and Q_B^-. Charge recombination between S_2/S_3 states of the water splitting system and Q_B^- gives rise to B band with Tm at 32-35°C and charge recombination between $S_2Q_A^-$ gives rise to Q band with Tm at 5-12°C. The increase in Tm for a recombining species is interpreted as an increase in activation energy and redox potential difference of the charged pairs.

The themoluminescence glow curves of *Chlorella* cells showed two peaks one at 5.6°C (Q peak) and other at 33.8°C (B peak) for untreated controls. *Chlorella* cells treated with uranium and lead showed increase in Tm of Q peak suggesting an increase in activation energy of recombination for S_2 & Q_A^-. Similar change in redox potential of Q_A was reported in Arabidopsis, acclimatized to the cold temperatures (Sane *et al.,* 2003). The Tm for B peak was found to have increased considerably in uranium nitrate treated cells whereas cells treated with lead, cadmium, cobalt and copper showed no discernible change in Tm of B peak. The narrowing of the redox potential gap between Q_A and Q_B resulted in decreased rate constant for electron transfer between Q_A and Q_B which would lead to accumulation of Q_A (Sane *et al.,* 2003) and inhibit the formation of radical charge pair $P680^+$ and $Pheo^-$. The gap existing in the activation energy between the two charge pairs was reduced significantly in cells treated with lead nitrate whereas in others there was slight decrease (Table 17.1). Since, the activation energies are directly related to the redox potentials of the participating species it is apparent that the redox potentials of the participating species must have undergone a change (DeVault and Govindjee, 1990). In the production of Q and B peaks, the participating species involved are the S_2 state from donor side and the two-quinone molecules Q_A and Q_B on the acceptor side of PS II. Thus, the redox potential of either the S_2 state or the Q_A and Q_B or both must have changed. The increase in Tm of Q peak in the case of lead treatment may be attributed to increase in activation energy of $Q_{A.}$ Uranium appears to act differently on PSII. In case of uranium, both B and Q peaks have shifted to higher temperatures by 8-9 degrees indicating increase in redox potential of S_2 and S_3 states, on the donor side of PSII. Thus, uranium probably leads to stabilization of $S_2Q_A^-$ and $S_2Q_B^-$ charge pairs. A number of workers have noted such changes in the redox potentials of Q_A and Q_B under various stress and acclimation conditions (Ivanov *et al.,* 2006). Copper and mercury have been observed to downshift $S_2Q_B^-$ (B_2) band besides decreasing its intensity (Horvath *et al.,* 1999; El-sheek and Mustafa, 1999). Recently, it was shown that in *Chlorella pyrenoidosa.* cells, treated with 20g m^{-3} - 40g m^{-3} of $K_2Cr_2O_7$ for 8-72 hours. PSII function was inhibited due to enhanced photo destruction of the reaction centres (Horscik *et al.,* 2007). However, although no change in redox potential of Q_A and Q_B was observed, a decrease in intensity of TL was reported.

Previous studies on *Synechocystis* using site directed mutagenesis have shown that a narrowing of the gap in the Tm of S_2Q_A and S_2Q_B by 6-8°C in a temperature range of 15 to 30°C resulting in redox potential difference of 22 mV between these two acceptors (Minagawa *et al.,* 1999). In our studies, the temperature gap between S_2Q_A and S_2Q_B was around 28°C for controls. The ΔTm was reduced by 6°C in case of lead treated cells indicating that lead treatment may also lead to similar change in redox potential.

Interestingly, stimulation of CO_2 fixation was seen in case of lead and copper treated cells, however decrease in ΔTm was observed only after lead treatment. Thus, the decrease in ΔTm did not appear to correlate with stimulation of CO_2 fixation. There was drastic reduction in rate of CO_2 fixation in cells treated with cobalt and cadmium. However, in these cases, there was no appreciable change in Tm of Q and B peaks as well as in the yield of TL. Therefore, in the cells treated with cobalt and cadmium, the site of inhibition of photosynthesis may not be located in PSII.

Normally fluorescence studies are done to examine the effects of stress on photosynthetic apparatus since it monitors the quenching of excitation energy in the antenna system prior to charge separation. TL provides a differential insight into charge separation and stabilization of PSII thereby providing detailed information about the changes that occur in donor and acceptor side. Our results showed that TL studies should be accompanied by measurements of CO_2 assimilation rates as we did not find correlation between TL and CO_2 assimilation.

ACKNOWLEDGEMENTS

This paper is dedicated to the memory of Late Shri S. S. Rane, Control and Instrumentation Group, Bhabha Atomic Research Centre, Mumbai-400085, who had done a lot of work for obtaining the data related to thermoluminescence. We also thank Dr. P. V. Sane, former director NBRI for useful suggestions. We thank Mr. Shrikant Iyer for technical help.

REFERENCES

Bertrand M, Poirer I (2005) Photosynthetic organisms and excess metals. Photosynthetica 43: 345-353

Cervantes C, Campos-Garcia J, Devars S, Gutierrez-Corona F, Loza-Tavera H, Torres-Guzman JC, Moreno Sanchez R (2001) Interactions of chromium with microorganisms and plants. FEMS Microbiol Rev 25: 335-347

Collard JM., Matagne RF (1990) Isolation and genetic analysis of *Chlamydomonas reinhardtii* strains resistant to cadmium. Appl Environ Microbiol 56: 2051-2055

Desai TS (1992) Flash induced glow peaks from green plant leaves. Indian J Experimen Biol 30: 135-139

DeVault D, Govindjee (1990) Photosynthetic glow peaks and their relationship with the free energy changes. Photosynth. Res 24: 175-181

El-Sheek M M., Mustafa M (1999) Inhibition of the photosynthetic electron transport in the unicellular green alga *Chlorella kessleri* by mercury at multiple sites. Cytobios 98: 25-37

El-Sheek M M, EL-Naggar AH, Osman MEH, El-Mazaly E (2003) Effect of cobalt on growth, pigments and photosynthetic electron transport in *Monoraphidium minutum* and *Nitzchia perminuta*. Braz J Plant Physiol 15: 159-166

Horcsik Z, Kovács L, Láposi R, Mészáros I, Lakatos G, Garab G (2007) Effect of chromium on photosystem II in the unicellular green alga, *Chlorella pyrenoidosa*. Photosynthetica 45: 65-69

Horvath G, Arellano JB, Droppa M, Baron M (1999) Alterations in photosystem II electron transport as revealed by thermoluminescence of Cu-poisoned chloroplasts. Photosynth Res 57: 175-181

Ivanov AG, Sane PV, Krol M, Gray GR, Balseris A, Savitch LV, Öquist G, Huner N PA (2006) Acclimation to temperature and irradiance modulates PS II charge recombination. FEBS Lett 58: 2797-2802

Kalin M, Wheeler W N, Meinrath G (2005) The removal of uranium from mining waste water using algal/microbial biomass. J Environ Radioactivity 78: 151-177

Mallick N (2004) Copper induced oxidative stress in the chlorophyaceae microalgae *Chlorella vulgaris*: response of the antioxidative system. J Plant Physiol 161: 591-597

Mallick N, Mohn FH (2003) Use of chlorophyll fluorescence in metal stress research: a case study with the green microalgae *Scenedesmus*. Ecotoxicol Environ Saf 55: 64-69

Minagawa J, Narusaka Y, Inoue Y, Satoh K (1999) Electron transfer between Q_A and Q_B in photosystem II is thermodynamically perturbed in photo tolerant mutants of *Synechocystis* sp. *PCC* 6803. Biochemistry 38: 770-775

Nishikawa K, Tominaga N (2001) Isolation, growth, ultrastructure and metal tolerance of the green alga *Chlamydomonas acidophila* (Chlorophyta). Biosci Biotech Biochem 65: 2650-2656

Sane PV (2004) Thermoluminescence A technique for probing photosystem II. In: (ed R. Carpentier) Methods in Molecular Biology. Humana Press Inc., Totowa, pp 229-248

Sane PV, Ivanov AG, Hurry V, Huner NPA, Oquist G (2003) Changes in redox potential of primary and secondary electron accepting quinones in photosystem II confer increased resistant to photoinhibition in low temperature acclimatized *arabidopsis*. Plant Physiol 132: 2144-2151

Wong P K, Chang L (1991) Effect of copper, chromium and nickel on the growth, photosynthesis and chlorophyll *a* synthesis of *Chlorella pyrenoidosa*. Environmental Pollution 72: 127-139

CHAPTER 18

Photosynthetic Regulation of Senescence in Green Leaves: Involvement of Sugar Signaling

Basanti Biswal[1,*], Pranab K. Mohapatra[1], Mukesh K. Raval[2] and Udaya C. Biswal[3]

[1]*Laboratory of Biochemistry and Molecular Biology, School of Life Sciences, Sambalpur University, Jyotivihar-768019, Orissa, India*

[2]*Department of Chemistry, Gangadhar Meher College, Sambalpur-768004, Orissa*

[3]*Biology Enclave, Basant Vihar, Jyotivihar-768019, Orissa*

ABSTRACT

Several lines of evidence suggest photosynthetic modulation of leaf developmental program through a network of sugar signaling. This review evaluates the possibility of involvement of the signaling in induction of senescence in green leaves. The results on expression of the genes associated with senescence indicate that the genes associated with photosynthesis are down-regulated with up-regulation of the genes associated with senescence suggesting a link between the two processes. Although photosynthetic control of leaf senescence is believed to be mediated by sugar signaling, the exactnature of the signaling and its downstream regulation of senescence related gene expression yet remains unclear primarily because of the emergence of two opposite schools of thought. On one hand, sugar in excess is shown to induce senescence and on the other hand, induction of the process is demonstrated by its starvation. Recently we have shown a decline in photosynthesis, consequently loss in sugar as a signal for induction of cell wall bound hydrolases including β-glucanase and β-glucosidase which cause breakdown of cell wall polysaccharides to provide respiratory sugar to execute energy dependent senescence program in the absence of photosynthesis. We thus conclude, sugar starvation is a signal for induction of senescence related catabolic processes. Nevertheless, the two opposite hypotheses, induction of senescence by excess or starvation of sugar, are critically discussed and attempt is made to rationalize the contradictory experimental findings through the action of hexokinase, a sensor of sugar. The review concludes with remarks on some of the challenging areas of research for future and experimental limitations to work with sugar signaling in senescing system.

Keywords: Carbon nitrogen ratio, Developmental age, Hormone, Leaf senescence, Nutrient mobilization, Photosynthesis, Sugar signaling.

Abbreviations

ABA: Abscisic acid, *Chl:* Chlorophyll, *DCMU*: 3-(3,4-Dichlorophenyl)-1,1-dimethyl urea, *GUS:* b-Glucuronidase, *HXK:* Hexokinase, *LHC:* Light harvesting complex, *NADPH:* Nicotinamide adenine dinucleotide phosphate (reduced) *PAG:* Photosynthesis associated gene, *PAR:* Photosynthetically active radiation, *PS I:* Photosystem I *PS II:* Photosystem II, Q_A: Plastoquinone bound to D2 protein of PS II, Q_B: Plastoquinone bound to D1 protein of PS II, *RC II,* Photosynthetic reaction center II, *SAG:* Senescence associated gene, *SDG:* Senescence down regulated gene, *SSU:* Small subunit of Rubisco.

INTRODUCTION

Senescence in green leaves is regulated both by intrinsic and various environmental factors. The intrinsic factors include phytohormones, age, and metabolites. Currently, sugar, one of the important metabolites, has been recognized as an intrinsic regulator of senescence process. Sugar not only acts as a source of energy but also as a signal which, through several complex signal transduction pathways, modulates plant development, specifically the development of green leaf including senescence. Since the photosynthetic production of sugar in green leaves is the primary source of metabolite, it is considered a modulator of senescence. The review attempts to summarize the basics of photosynthesis, its role in leaf developmental programme and finally elaborates its role in regulating senescence through sugar signaling.

Photosynthesis is a photochemical process in which chlorophyll (Chl) and other pigments absorb light energy, which is subsequently utilized in oxidation of H_2O to O_2 and reduction of $NADP^+$ to NADPH. The photoredox reactions occur by electron transport through a chain of protein units embedded in thylakoid membrane of photosynthetic apparatus. The electron transport results in the production of a proton gradient across the thylakoid membrane. The proton gradient drives the synthesis of adenosine triphosphate (ATP). The chemical potentials stored in NADPH and ATP are used for the reduction of CO_2 to glucose through Calvin-Benson-Basham enzyme cycle. Besides carbohydrates, other nutrients namely, amino acids and lipids are also synthesized in chloroplasts utilizing the chemical potential stored by photosynthesis. In higher plants, major part of photosynthetic process occurs in green leaves. The primary function of a leaf is to harvest solar energy and convert it into carbohydrate, which provides building blocks for cell components and is the source of energy for sustaining life in diverse biosystems on the earth. The photosynthetic efficiency of green leaves varies during the development, maturation and senescence. It significantly declines during senescence when the leaves turn yellow (Biswal *et al.* 2003).

CHLOROPLAST IS THE MASTER REGULATOR OF LEAF DEVELOPMENT, MATURITY AND SENESCENCE

Leaf, being the major source of nutrition, enjoys a special status in higher plants. It has its own developmental pace, which may not be synchronous with the plant development as a whole. Leaf, like any other living system, has three major phases of development: growth, maturity and senescence.

Since photosynthesis is the major function of a leaf, all its other activities center around the activity of photosynthetic apparatus. Practically a leaf is for photosynthesis, by photosynthesis and of photosynthesis. Therefore, the photosynthetic apparatus that contains more than 70% of leaf protein largely determines the different developmental phases of the leaf. The link between photosynthesis and developmental program of the leaf has been critically discussed earlier (Biswal *et al.*, 2003). The location of proplastids in juvenile cells of a very young leaf, differentiation of chloroplasts in developing leaves, presence of mature chloroplasts in fully expanded leaves and gerontoplasts in yellow senescing leaves are indicative of strong coupling between the development of the organelle and the organ. The factors associated with the developmental transition from leaf maturity to senescence are not properly understood. There must be a device that suppresses senescence till leaf maturity and/or activates the switch for induction of the process once the leaves are mature.

During leaf senescence, chloroplast is the first organelle to exhibit the degradative symptoms and the last one to survive when all other organelles are completely dismantled during the last phase of senescence followed by necrosis and leaf fall indicating perfect step-wise coordination between leaf senescence and chloroplast demolition. The interdependence of chloroplast demolition and leaf senescence, however, is not properly understood. Photosynthetic production of sugar, an important metabolic signal for plant development and production of reactive oxygen species (ROS), a known regulator of senescence may modulate senescence process in green leaves (Biswal and Biswal, 1988; 1990; Dangl *et al.*, 2000; Jing *et al.*, 2008).

Apart from photosynthetic modulation of leaf senescence, the plastid gene expression and synthesis of specific proteins in the chloroplast may be associated with the initial stage of leaf senescence (Garcia *et al.*, 1983; Matile, 1992). It is, however, not possible yet to precisely identify these proteins. The contribution of plastid genes for leaf senescence program is further supported by mutational studies (Guiamet *et al.*, 1991; Nooden and Guiamet, 1996). The plastid to nuclear signaling and the role of chloroplast in modulation of leaf senescence have, in recent days, become one of the fascinating areas of senescence study in green plants (Krause *et al.*, 2005; Zapata *et al.*, 2005; Grabowski *et al.*, 2008).

WHAT SIGNALS THE ONSET OF SENESCENCE?

The signals from developing sink, growth regulators, metabolites and various environmental factors including shading/dark regulating the expression of senescence associated genes have been suggested (Buchanan-Wollaston, 1997; Gan and Amasino, 1997; Biswal, 1999; He and Gan, 2001; He *et al.*, 2002; Buchanan-Wollaston *et al.*, 2003, 2005; Lim *et al.*, 2007). However, studies on the mechanism of signal perception, its processing, and final expression of genes encoding the proteins for initiation, progression and completion of senescence are limited (Biswal and Biswal, 1999; Biswal *et al.*, 2003). The process of senescence may involve multi-channeled regulatory paths (Gan and Amasino, 1997; Bleecker, 1998; Lim *et al.*, 2007). The possibility of crosstalks between these channels also cannot be ruled out.

DECLINE IN PHOTOSYNTHESIS AS A POSSIBLE SIGNAL FOR INDUCTION OF SENESCENCE PROGRAM IN GREEN LEAVES

Chloroplast is the major target of senescence induced catabolic process. The degradation process is associated with the loss of photosynthetic pigments, changes in the ultrastructure of chloroplasts, loss in primary photochemical reactions associated with both photosystem I (PS I) and photosystem II (PS II) of thylakoid membranes and decline in ATP synthesis. Macromolecular degradation and loss in the efficiency of Calvin-Benson-Basham cycle for the fixation of CO_2 are the major symptoms of chloroplast dismantling during leaf senescence (see Biswal *et al.*, 2003). The question whether these photosynthetic changes are the cause or consequence of senescence still remains controversial. But the recent literature indicates that the catabolic process of senescence is induced with decrease in photosynthesis after the leaves are fully expanded (van Doorn, 2008).

Association of Photosynthesis with the Onset of Senescence in Plant Canopy

Fig. 18.1 Schematic illustration of induction of leaf senescence (yellowing) in the bottom leaf by the light transmitted through the upper leaf. The nutrients (N) released during breakdown of macromolecules from senescing bottom leaves are transported to developing seeds and young organs of the plant (modified from Quirino *et al.* 2000).

The link of photosynthesis with the onset of leaf senescence is better understood during progressive leaf senescence in plant canopy. During progressive leaf senescence, the lower leaves of many plants in the canopy receive less light compared to the upper ones (Fig. 18.1). The light environment of leaf changes from full sunlight during its emergence at the top to extremely low intensity received by it when it is mature. Its location is changed from top to the bottom of the canopy of a plant. The intensity of light received may vary at different canopy levels and may even be reduced at the bottom to the extent of 10% of the light received at the top of the canopy. Decrease in the intensity of light transmitted through and reflected from the upper leaves may also be associated with significant alterations in spectral energy distribution. Since Chls absorb red light preferentially to far red components of the solar spectrum, the transmitted light is enriched with far red and the ratio of photon fluence rates at red to far red is reported to be significantly reduced. Under low photosynthetically active radiation (PAR) condition, photosynthesis declines progressively, reaching the compensation point when the leaves do not contribute to the assimilatory needs of the plant. Since they are no longer in a position for the net carbon gain, they start senescing, mobilize essential nutrients to other growing parts of the plant before they turn complete yellow and fall. Therefore, the change in the canopy structure of the plant acts as a signal for senescence induction in older leaves at the bottom (Fig. 18.1). The appearance of senescence syndrome, therefore, is a response of the older leaves below the canopy to the loss of the photosynthetically active radiation. These leaves below the canopy shift their function from photosynthesis to mobilization of nutrients.

Induction of Age Dependent Senescence may be Mediated through a Loss in Photosynthesis

The developmental age is known to govern leaf senescence under natural condition (Hensel *et al.,* 1993; Lim *et al.,* 2007). In fact, age as one of the important factors for induction and progress of senescence was suggested much earlier in the leaves of *Hibiscus* (Misra and Biswal, 1973) and *Ervatamia* (Misra and Biswal, 1974). However, various intrinsic and extrinsic (environmental) factors that are integrated into the age information influence senescence (Wingler *et al.,* 2006). This integrated senescence response provides plants the capability to fine-tune the initiation timing, progression rate, and nature of leaf senescence. Although the exact mechanism of aging induced senescence is not understood clearly, it has been speculated to be associated with age dependent decline in photosynthesis (Hensel *et al.,* 1993; Guo and Gan, 2005).

The well executed experiments of Hensel *et al.* (1993) in *Arabidopsis* indicate that an age dependent loss in photosynthesis is the signal responsible for induction of senescence program. This proposition is supported further by the observation of a significant decline in the level of the transcripts of LHC and SSU of Rubisco just after full expansion of *Arabidopsis* leaves (Hensel *et al.,* 1993; see also the review by Weaver *et al.*, 1997). Although the link between decline in photosynthesis and initiation of senescence is suggested, the precise nature of the signaling molecules that interplay between these two processes is yet to be clearly understood. The loss in photosynthesis, however, has been demonstrated to be linked to the expression of genes associated with senescence process (Biswal *et al.* (2003).

Down-regulation of Photosynthesis is Accompanied by Up-regulation of Genes Associated with Senescence Program

The decline in photosynthetic activity is associated with the induction of several genes associated with senescence (Smart, 1994; Bleecker and Patterson, 1997; Buchanan-Wollastan, 1997; Nooden *et al.*, 1997; Biswal and Biswal, 1999; Biswal, 1999, 2005; Quirino *et al.*, 2000; Chandlee, 2001; Biswal *et al.*, 2003; Buchanan-Wollastan *et al.*, 2003, 2005; Lim *et al.*, 2003, 2007; Yoshida, 2003; Lin and Wu, 2004; Biswal *et al.*, 2010). The genes that are up-regulated during senescence are termed senescence associated genes (*SAGs*), and those down-regulated are termed senescence down-regulated genes (*SDGs*) (Biswal, 1999). Most of the *SDGs* are also known to be photosynthesis-associated genes (*PAGs*). *SAGs* are associated with a broad range of metabolic activities, namely, macromolecular degradation, nutrient mobilization and recycling, defense and cell rescue mechanisms, transcriptional regulation and signal transduction (Gepstein, 2004; Guo *et al.*, 2004; Lim *et al.*, 2007; Balazadeh *et al.*, 2008; Biswal *et al.*, 2010). Isolation, cloning and characterization of some of the *SAGs* and *PAGs* have been reported. But most of the *SAGs* expressed during senescence are yet to be functionally identified. Up-regulation of *SAGs* and down-regulation of *PAGs* are correlative phenomena and have been reviewed in Biswal *et al.* (2003) (Fig. 18.2). The next question is how does the loss in photosynthesis act as a signal for induction of senescence program?

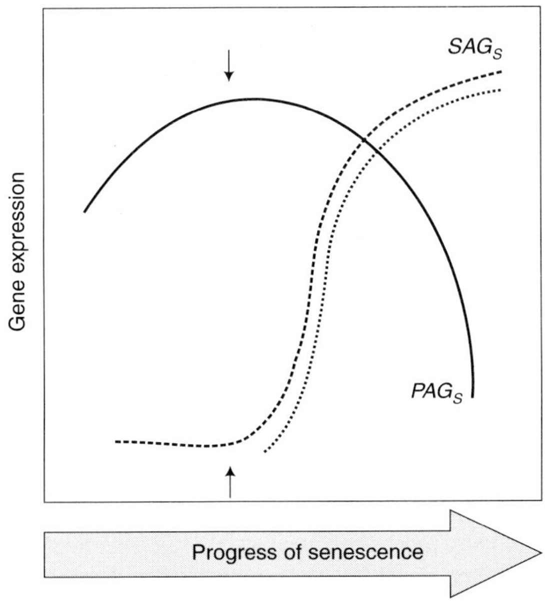

Fig. 18.2 A model represents correlation between down-regulation (↓) of photosynthesis-associated genes (PAGs) and up-regulation (↑) of senescence-associated genes (SAGs). Two types of SAGs are known to express during senescence. Some SAGs exhibit increased expression only and therefore, referred to as the SAGs nonspecific to senescence, whereas the other SAGs are expressed only during senescence and, therefore, are specific to the process. ; ——, photosynthesis associated genes; -----, senescence associated genes nonspecific to the process; ......, senescence associated genes specific to the process (modified from Gan and Amasino,1997).

PHOTOSYNTHESIS LINKED SUGAR SIGNALING AND LEAF SENESCENCE

Sugar is an energy carrier and carbon backbone provider molecule in metabolic processes. It also maintains osmotic potential in plants. Besides being a major nutrient, its role in the complex signaling process of plant development and senescence is well known (Price *et al.,* 2004; Baena-Gonzalez and Sheen, 2008; van Doorn, 2008; Wingler *et al.,* 2009). There is an increasing evidence that sugar sensing system includes changes in gene expression. The sensing system may act as a gene expression switch. Although sugars control cellular activity at different levels, starting from transcription and translation to stability of proteins, the precise nature of their role in regulating senescence of green leaves is yet to be known.

The literatures indicate that cellular sugars can dramatically affect plant senescence (Fujiki *et al.,* 2001; Pourtau *et al.,* 2006; van Doorn, 2008; Wingler and Roitsch, 2008). However, the results available are contradictory. Both sugar starvation and excess sugar are reported to induce leaf senescence (see the review by Wingler *et al.,* 2006; Wingler and Roitsch, 2008).

Sugar Starvation Induces Senescence

Low sugar level is extensively reported to induce senescence (van Doorn, 2008). In a classic experiment, Thimann *et al.* (1977) showed that glucose treatment inhibited yellowing of floating leaf segments of oats in dark. Mannitol was found to have no effect, which eliminates the effect of osmosis. Similarly, increase in transcripts of a *SAG sen1* is inhibited by sugar treatment in cut leaves of *Arabidopsis* placed in darkness (Oh *et al.,* 1996). Repression of dark induced expression of many *SAGs* in the presence of sucrose provides support to low sugar induced senescence (Chung *et al.,* 1997; Fujiki *et al.,* 2001). In the past, senescence of green leaves has been extensively examined under dark or unfavorable light resulting sugar starvation conditions—the condition that results in induction of several senescence associated genes including the genes that code for the catabolic enzymes for degradation of lipids, proteins and other cellular constituents to meet the requirement of respiratory substrate (Biswal *et al.,* 2003). The induction of this kind of senescence program is basically an adaptational mechanism of energy requiring senescing leaves in the absence of photosynthesis.

In this context, the work of Chung *et al.* (1997) is remarkable. They reported the regulation of *sen 1* by various senescence inducing factors including darkness, and examined the promoter activity of *sen 1* by employing transgenic tobacco plants harboring the promoter sequence of the gene fused with β-glucuronidase (*GUS*) reporter gene. Darkness or treatment of the transgenic plants with DCMU in light exhibits induction of activity of the promoter with concomitant decrease in the level of sugar that suggests sugar repression of *sen 1* expression. Since darkness is extensively reported to induce *SAGs* (Azumi and Watanabe, 1991; Kleber-Janke and Kripinska, 1997; Nakabayashi *et al.,* 1999; Fujiki *et al.,* 2001, 2005; Lee *et al.,* 2001; Yoshida *et al.,* 2001; Lin and Wu, 2004; Buchanan and Wollasten, 2005), photosynthetic decline and consequently sugar starvation may regulate the senescence. This proposition was supported by the observation of sugar induced repression of *SAG 12* (Noh and Amasino, 1999a). These authors have critically discussed the possible regulation of *SAG 12* promoter modulated by sugars (Noh and Amasino, 1999a, b). The gene encodes a cysteine protease and is expressed only in age dependent senescent tissues with a decline in photosynthesis.

Recently we have demonstrated senescence induced decline in photosynthesis as a signal for breakdown of cell wall polysaccharides Mohapatra *et al.* (2010). As discussed earlier, loss of photosynthetic production of sugar causes expression of several senescence related genes, most of which code for the enzymes for the degradation of macromolecules and their transport Biswal *et al.* (2003). The loss is also reported to induce the expression of several glycosyl hydrolases of cell wall in *Arabidopsis* in different experimental conditions (Lee *et al.*, 2007). Leaf senescence induced decline in photosynthesis and enhancement of β-glucosidase activity observed in the leaves of *Arabidopsis* are shown in Fig. 18.3. The activity of cell wall bound β-glucosidase, from the leaves of *Arabidopsis thaliana* at different stages of senescence and corresponding photosynthetic status of the leaves indicate photosynthetic modulation of the enzyme activity. Statistical analysis suggests a distinct correlation between the loss in photosynthetic parameters and concomitant enhancement in the activity of the enzyme (Mohapatra *et al.*, 2010). In addition to β-glucosidase, leaf senescence also causes enhancement in the activity of another enzyme, namely β-glucanase, that breaks down cell wall polysaccharides to oligosaccharides, the substrate for β-glucosidase. Both the enzymes act in a sequence to break down the polysaccharides to respiratory and remobilizable sugars during senescence (Biswal *et al.*, 2003). The execution of senescence program requires energy. As discussed earlier, the leaves in the background of senescence-induced decline in photosynthesis, the major source of energy for green plants, use stored organic molecules like polysaccharides, lipids and proteins for respiratory substrates (Biswal, 2005). The degradation of macromolecules, however, follows a pattern. The sequence of degradation of starch, lipids, proteins and polysaccharides bound to the cell wall is shown in Fig. 18.4. The *SAGs* coding for the degradative enzymes for catabolism of these macromolecules are expressed at different phases of senescence process. This is a kind of adaptational response of senescing leaves to loss in photosynthetic production of sugar. In senescing leaves, the cells retain the cell wall till the last when most of the cellular materials are dismantled and the breakdown products are remobilized to different growing parts of the plant. Therefore, cell wall polysaccharides in addition to physical support may be the last target for production of sugars for transport to other growing organs and/or for the use as respiratory substrate to

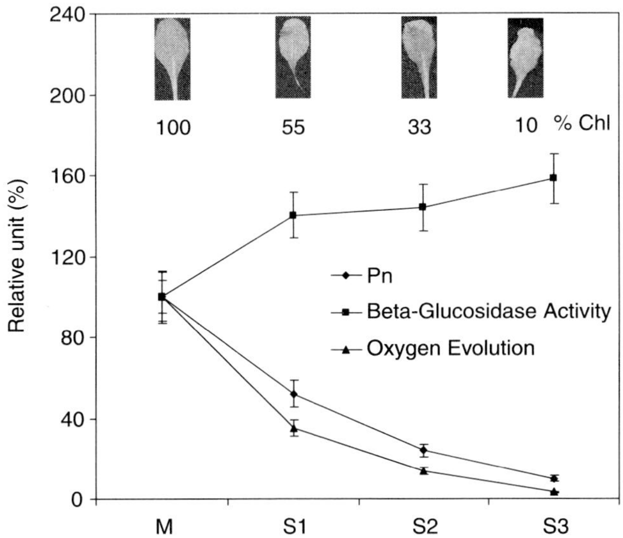

Fig. 18.3 Changes in the rate of oxygen evolution, net photosynthetic rate (Pn) and β-glucosidase enzyme activity at mature and different senescence phases of *Arabidopsis thaliana* (modified from Mohapatra *et al.*, 2010).

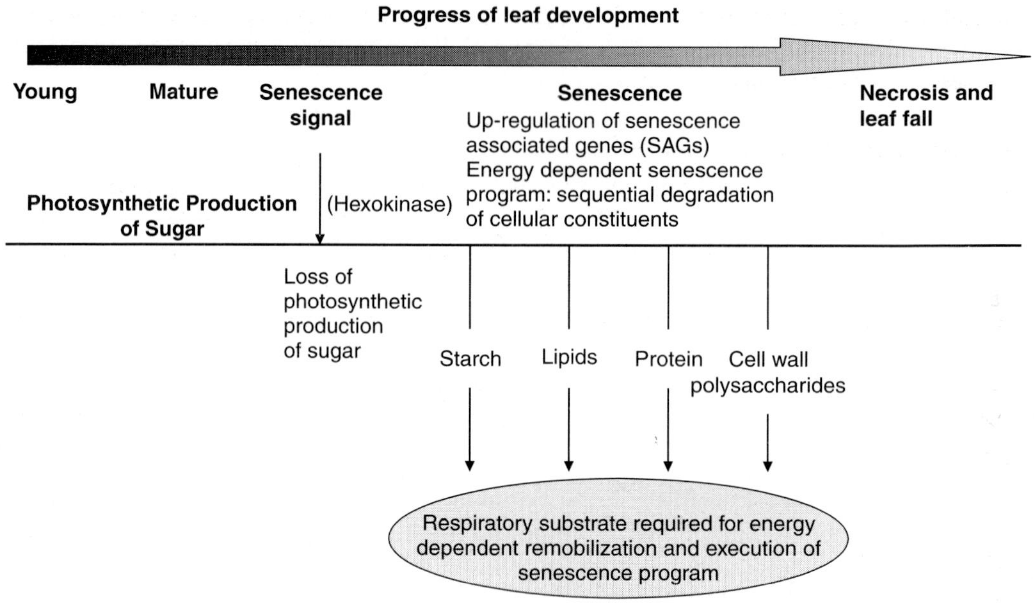

Fig. 18.4 In the background of senescence induced loss in photosynthetic production of sugar, degradation of starch, lipids, proteins and cell wall polysaccharides provide sugar and other transportable materials for remobilization and production of energy through respiration for completion of the process.

supply metabolic energy. The proposition that the cell wall polysaccharides could be the last target of the senescence program is supported by earlier observations of late expression of *din2* gene coding for the enzyme β-glucosidase during dark induced or natural senescence (Fujiki *et al.*, 2001, 2005). Several genes associated with senescence are induced in dark. Some of these genes are expressed early and some at late phase of dark incubation. The proteins synthesized at different phases of senescence carry out different functions for execution of senescence program. The late expression of *din2* matches with maximum activity of β-glucosidase encoded by the gene at late phase of senescence when the activity of photosynthesis was almost negligible.

A decline in photosynthetic production of sugar in the leaves of *Arabidopsis* as a signal for enhancement in the activity of the enzyme is confirmed by the results shown in Table 18.1. The activity of β-glucosidase is relatively high in the leaves incubated in dark and light with DCMU, the conditions that create an environment of sugar starvation. The enzyme activity is, however, suppressed when the leaves are treated with light as well as incubated with exogenous sugar (Table 18.1). These experiments

Table 18.1 Alteration in the status of sugar and variation in the activity of β-glucosidase of fully mature excised leaves treated with continuous white light (LC), Continuous dark (DC), Continuous light with 3% (w/v) sucrose (L+S), Continuous light with 10 μM DCMU (L+DCMU) (modified from Mohapatra *et al.*, 2010).

Treatment	*β-glucosidase activity in %*
LC	100
DC	129
L + S	60
L+DCMU	167

indicate that loss in photosynthesis and enhancement in the activity of β-glucosidase during senescence are not two parallel events. The link of photosynthetic decline with enhancement of the enzyme activity during senescence involves sugar signaling.

Sugar Accumulation Induces Senescence

Induction of senescence by excess sugar is opposite to the hypothesis of induction of the process by sugar starvation as described earlier. Sugar induced stimulation of senescence was first demonstrated in light incubated leaves by Khudairi (1970). Sugar accumulation by prevention of phloem export by girdling treatment is known to induce early senescence in wheat and barley (Feller and Fischer, 1994; Parott *et al.*, 2005, 2007). Feeding of sugar to whole plant is observed to cause leaf yellowing (Wingler *et al.*, 2004). The concentration of the soluble sugar shoots up at the onset of leaf senescence in a wide range of species (Nooden *et al.*, 1997; Keskitalo *et al.*, 2005; Wingler *et al.*, 2006). Pourtau *et al.* (2006) with well designed experiments confirmed that *Arabidopsis* leaf senescence is induced and not repressed by sugars. The hypothesis of senescence induction by sugar accumulation is further supported by the experiments relating to initiation of the process at various concentrations of cellular nitrogen. Low nitrogen induced leaf senescence is known to accumulate sugar, which is considered the factor for induction of the process. Similarly, the induction of senescence has been demonstrated in *Arabidopsis* plants grown in 2% glucose with low nitrogen supply (Pourtau *et al.*, 2004; Wingler *et al.*, 2004).

A Possible Explanation for the Controversy of Sugar Modulation of Senescence

As discussed, the role of sugar in the induction of leaf senescence under natural conditions is still unclear and appears to be controversial. However, nature is not controversial. The apparent controversial results must be analyzed in the light of source-sink balances that may affect the partitioning of sugar within a plant and can consequently induce leaf senescence. In a young growing leaf, there is demand for sugar, hence it acts as a sink organ. If a cut young leaf is placed in dark, thereby depriving it from photosynthates, it may senesce due to low sugar or starvation. On the other hand, a mature fully expanded leaf acts as a source organ. Its demand for sugar decreases, rather it provides sugar to the rest of the plant and especially to the expanding leaves. Accumulation of carbohydrates in the mature leaf may signal for decline in photosynthesis and senescence (Wingler *et al.*, 1998, 2004).

Although the sugar sensing mechanism of *SAG* expression is not clearly known, the hexokinase (HXK) that catalyzes phosphorylation of hexose has been considered a possible candidate involved in the mechanism. Overexpression of *Arabidopsis* hexokinase encoding gene (*AtHXK1*) in transgenic tomato plants resulting in reduction in photosynthesis, consequently a rapid induction of senescence has been demonstrated by Dai *et al.*, (1999). The variation in the hexokinase expression in transgenic tomato plants like HK37, HK4 and HK38 and the changes in the activity of the enzyme in inducing senescence are shown in the transgenic plants. Also overexpression of HXK produces hypersensitivity to glucose (Dai *et al.*, 1999; Xiao *et al.*, 2000). An *Arabidopsis gin2* mutant having lesion in HXK1 is glucose insensitive and exhibits delayed senescence (Moore *et al.*, 2003). These experiments indicate that hexokinase acts as a sugar sensor during senescence in leaves.

Here we have tried to explain the contradictory results on sugar signaling through the action by hexokinase. Low sugar levels are known to enhance photosynthesis, and the accumulation of high levels of glucose and sucrose represses the transcription of *PAGs* in green leaves under normal light conditions (Rolland *et al.*, 2002). High level of sugar may signal the reduction of source activity, i.e., decline in photosynthesis activity by feedback mechanism (Jongebloed *et al.,* 2004). It is likely that hexokinase plays a role at this point of leaf development. When green leaves are fully mature with maximum level of sugar through photosynthesis, hexokinase senses it and that signals down-regulation of photosynthesis, the initial event of senescence process (Fig. 18.4). The decline in photosynthesis causes gradual depletion of sugar resulting in induction of next phase of senescence. This model can accommodate the role of sugar in regulation of senescence, both in excess and under starvation condition. The model is supported by the experimental findings, which suggests that the *SAGs* expressed late in the process of senescence are repressed by sugar, but the *SAGs* expressed early are sugar induced (Paul and Pellny, 2003; Wingler *et al.,* 2006).

CROSSTALK BETWEEN SUGAR SIGNALING, NITROGEN METABOLISM AND PHYTOHORMONE NETWORK

The availability of nitrogen significantly modulates developmental processes in plants. Masclaux *et al.* (2000) and Diaz *et al.* (2008) have demonstrated that the changes in C/N ratio during sink/source transition of leaves are the important factors for initiation of senescence. The proposition is supported by the observations made by Wingler *et al.* (2004) and Pourtau *et al.* (2004). These authors have observed induction of senescence in the leaves with sugars in combination with low nitrogen supply. Nitrogen depletion induced senescence in leaves is demonstrated to be effectively reversed by the addition of exogenous nitrogen (Schildhauer *et al.,* 2008).

Similarly leaf senescence characterized by decrease in fluorescence parameter Fv/Fm is observed to set early in *Arabidopsis* when the plant is grown with 111 mM glucose and 4.7 mM nitrogen. However, the early onset of senescence is eliminated when the concentration of nitrogen is raised to 30 mM along with the same concentration of glucose (111 mM) (Diaz *et al.,* 2005). It seems that the relative concentration of sugar and nitrogen actually signals senescence. There exists a crosstalk between C and N metabolic pathway generating signal for senescence. Sugars are reported to be involved in the regulation of nitrogen metabolizing enzymes (Masclaux-Daubresse *et al.,* 2005). Nitrogen remobilization is marked by the activity of cytosolic glutamine synthetase, glutamate dehydrogenase and CND41-like proteases (Diaz *et al.,* 2008). Both sugar and inorganic nitrogen act as metabolites as well as signaling molecules (Price *et al.,* 2004).

Phytohormones are extensively reported to regulate leaf senescence. Some of them delay the process and others accelerate it. In most of the cases, molecular mechanism of their action is known. The possible interaction between the hormones and sugar cannot be ruled out. The crosstalk between sugar signaling and signal transduction through cytokinins, auxins, ABA and ethylene is suggested (van Doorn, 2008). Hexokinase (HXK1), a sugar sensor in high molecular weight complexes in nucleus, controls at the level of transcription which ultimately results in counteracting the effect of ethylene, a well known senescence accelerator (Rolland *et al.,* 2006). The interaction between sugar signaling and phytohormones, namely ethylene and ABA in regulating senescence has been critically reviewed by van Doorn (2004). Exogenous sugar is likely to decrease the concentration and sensitivity of ethylene, the promoter of senescence (van Doorn, 2004). On the other hand, interaction between cytokinins, well known senescence retardant with sugar signaling has been demonstrated by Wingler *et al.* (1998) and Moore *et al.* (2003). The participation of cytokinins in sugar signaling in modulating senescence is clearly shown by Balibrea-Lara *et al.* (2004). In their elegant experiments, they have shown cytokinin dependent induction of extracellular invertase that results in the activation of sugar utilization, and consequently a delay in senescence.

CONCLUSION

The study of involvement of sugar signaling in modulating plant development is rather recent. In spite of some data available on participation of sugar signaling with the onset of senescence of green leaves, the mechanism of signal transduction regulating the process remains unclear. The available literature rather raises several questions that have to be addressed in future in order to have a clear understanding of this fascinating area of senescence regulation. Some of the fundamental questions and perspectives of the area are provided below.

1. Several regulating factors for the induction of senescence have been known recently using genetic and reverse genetic tools, but the picture of integration of the factors in a network to develop the mechanism of induction of the process remains hazy. The question whether photosynthetic production of sugar as a signal could play crucial role in the interconnecting network in green leaves remains unaddressed. The interplay between photosynthesis and induction of senescence and the molecular players of the network should be identified by recently available advanced genetic tools.
2. The concept whether sugar accumulation or starvation causes initiation of leaf senescence remains unresolved. With present literatures it is difficult to rationalize the contradictory results with these opposite hypotheses. We have to wait.
3. Hexokinase, as a major sensor of sugar for the expression of genes both in green plants and yeast, is known. But we do not know yet whether the downstream regulation of gene expression through hexokinase is same or different in photosynthetic and non-photosynthetic systems. Petal senescence like senescence in green leaves is sensitive to sugar signaling (van Doorn, 2004), but we do not have sufficient experimental data to evaluate whether the mechanism of sugar modulation is same or different in these systems.
4. It is long known that source-sink transition plays a key role in the induction of senescence primarily mediated by changes in C/N ratio, but the question of precise metabolic processes involved in changing the ratio is yet to be answered. The relation between source and sink is not a static one and the changes depend on several factors including the sink strength, number of sinks competing for same pool of sugars. Sugar signaling becomes more complex when source sink transition is explained in terms of its response to diverse metabolic signals, plant signaling network mediated by phytohormones and different environmental factors. This is a challenging area for future study.
5. Some pathways for induction of senescence in green leaves are suggested to be independent of sugar signaling (Lee *et al.,* 2004; van Doorn, 2008). For example, photoperiod effect on senescence is likely to be independent of senescence related sugar signaling (Wingler *et al.,* 2006). The senescence response in this case may be controlled by flowering and action of different photomorphogenic receptors like phytochrome and blue light receptors (Biswal and Biswal, 1988; Wingler *et al.,* 2006) and therefore, needs further critical evaluation.
6. Artificial experimental systems used to examine the involvement of sugar signaling for induction of senescence in green leaves is a major limitation to reach a conclusion. Future research in the area should concentrate on the use of intact plants grown in diurnal light dark cycle in field. The true picture of senescence regulated by sugar could be revealed by accurate measurement

of nutrients including carbon and nitrogen at the level of tissues and cells before the onset of senescence.

7. Finally a word for senescence program: How does a leaf know when to senesce?

Senescence is obviously an age dependent developmental program because the process does not occur in young leaves. However, apparent correlation with age may not provide the true picture of the program. The biggest question is how the leaf senses age for initiation of senescence process. Is time the only factor? If time is the factor, how is the early or delay in senescence explained in different environmental conditions? There must be a more sophisticated mechanism than the senescence alarm of the bio-clock in leaf ringing at a pre-determined set time. Rather it must have a dynamic alarm setting mechanism and several intrinsic and extrinsic factors regulating the process in an interdependent network system. It is possible that sugars could play a major role in integrating all internal and environmental signals contributing to the dynamics of senescence program. It is likely that these signals, through metabolic network, determine the level of sugar critical for initiation and progression of senescence. This proposition appears to be complex and should be supported by more critical experimentation in future.

REFERENCES

Azumi Y, Watanabe A (1991) Evidence for a senescence associated gene induced by darkness. Plant Physiol 95: 577-583

Baena-Gonzalez E, Sheen J (2008) Convergent energy and stress signaling Trends Plant Sci 13: 474-482

Balazadeh S, Parlitz S, Mueller-Roeber B, Meyer RC (2008) Natural developmental variations in leaf and plant senescence in *Arabidopsis thaliana* Plant Biol 10: 136-147

Balibrea-Lara ME, Garcia MC, Gonzalez FT, Ehness R, Lee TK, Proels R, Tanner W, Roitsch T (2004) Extracellular invertase is an essential component of cytokinin-mediated delay of senescence. Plant Cell 16: 1276-1287

Biswal B (1999) Senescence-associated genes of leaves. J Plant Biol 26: 43-50

Biswal B (2005) Formation and demolition of chloroplasts during leaf ontogeny. In: Pessarakli M (ed) Handbook of Photosynthesis. CRC Press, Florida, pp 109-122

Biswal B, Biswal UC (1999) Leaf senescence, physiology and molecular biology. Curr Sci 77: 775-782

Biswal B, Raval MK, Mohapatra PK, Biswal UC (2010) Leaf senescence and transformation of chloroplasts to gerontoplasts. In: Eaton-Rye J, Tripathy BC (eds) Photosynthesis: Plastid Biology, Energy Conversion and Carbon Assimilation. Springer Verlag, The Netherlands (In Press)

Biswal UC, Biswal B (1988) Ultrastructural modifications and biochemical changes during senescence of chloroplasts. Int Rev Cytol 113: 271-321

Biswal UC, Biswal B (1990) Plant senescence and changes in photosynthesis. Biol Edn 7: 56-72

Biswal UC, Biswal B, Raval MK (2003) Chloroplast biogenesis: from proplastid to gerontoplast. Kluwer Academic Publishers, Dordrecht, The Netherlands

Bleecker AB (1998) The evolutionary basis of leaf senescence: method to the madness? Curr Opin Plant Biol 1: 73-78

Bleecker AB, Patterson SE (1997) Last exit: senescence, abscission and meristem arrest in *Arabidopsis*. Plant Cell 9: 1169-1179

Buchanan-Wollaston V (1997) The molecular biology of leaf senescence. J Exp Bot 48: 181-199

Buchanan-Wollaston V, Earl S, Harrison E, Mathas E, Navabpour S, Page T, Pink D (2003) The molecular analysis of leaf senescence: a genomics approach. Plant Biotech J 1: 3-22

Buchanan-Wollaston V, Page T, Harrison E, Breeze E, Lim PO, Nam HG, Lin JF, Wu SH, Swidzinski J, Ishizaki K, Leaver CJ (2005) Comparative transcriptome analysis reveals significant differences in gene expression and signaling pathways between developmental and dark/starvation-induced senescence in *Arabidopsis*. Plant J 42: 67-585

Chandlee JM (2001) Current molecular understandings of the genetically programmed process of leaf senescence. Physiol Plant 113: 1-8

Chung BC, Lee SY, Oh SA, Rhew TH, Nam HG, Lee CH (1997) The promoter activity of *sen1*, a senescence-associated gene of *Arabidopsis*, is repressed by sugars. J Plant Physiol 151: 339-345

Dai N, Schaffer A, Petreikov M, Shahak Y, Giller Y, Ratner K, Levine A, Granot D (1999) Over expression of *Arabidopsis* hexokinase in tomato plants inhibits growth, reduces photosynthesis, and induces rapid senescence. Plant Cell 11: 1253-1266

Dangl JL, Dietrich RA, Thomas H (2000) Senescence and programmed cell death. In: Buchanan B, Gruissem W and Jones R (eds) Biochemistry and Molecular Biology of Plants, American Society of Plant Physiologists, USA. pp 1044-1099

Diaz C, Purdy S, Christ A, Morot-Gaudry J-F, Wingler A, Masclaux-Daubresse (2005) Characterization of markers to determine the extent and variability of leaf senescence in *Arabidopsis*, a metabolic profiling approach. Plant Physiol 138: 898-908

Diaz C, Lemaître T, Christ A, Azzopardi M, Kato Y, Sato F, J Morot-Gaudry J-F, Dily F-L, Masclaux-Daubresse C (2008) Nitrogen recycling and remobilization are differentially controlled by leaf senescence and development stage in *Arabidopsis* under low nitrogen nutrition. Plant Physiol 147: 1437-1449

Feller U, Fischer AM (1994) Nitrogen metabolism in senescing leaves. Critic Rev Plant Sci 13: 241-273

Fujiki Y, Yoshikawa Y, Sato T, Inada M, Nishida I, Watanabe A (2001) Dark inducible genes from *Arabidopsis thaliana* are associated with leaf senescence and repressed by sugars. Physiol Plant 111: 345-352

Fujiki Y, Nakagawa Y, Furumoto T, Yoshida S, Biswal B, Ito M, Watanabe A, Nishida I (2005) Response to darkness of late responsive dark inducible genes is positively regulated by leaf age and negatively regulated by calmodulin-antagonist sensitive signalling in *Arabidopsis thaliana.* Plant Cell Physiol 46: 1741-1746

Gan S, Amasino RM (1997) Making sense of senescence: molecular genetic regulation and manipulation of leaf senescence. Plant Physiol 113: 313-319

Garcia S, Martin M, Sabater B (1983) Protein synthesis by chloroplasts during senescence of barley leaves. Physiol Plant 57: 260-266

Gepstein S (2004) Leaf senescence- not just a 'wear and tear' phenomenon. Genome Biol 5: 212

Grabowski E, Miao Y, Mulisch M, Krupinska K (2008) Single-stranded DNA binding protein Whirly1 in barley leaves is located in plastids and nucleus of the same cell. Plant Physiol 147: 1-5

Guiamet JJ, Schwartz E, Pichersky E, Nooden LD (1991) Characterization of cytoplasmic and nuclear mutations affecting chlorophyll and chlorophyll binding proteins during senescence in soybean. Plant Physiol 96: 227-231

Guo Y, Cai Z, Gan S (2004) Transcriptome of *Arabidopsis* leaf senescence. Plant Cell Environ 27: 521-549

Guo Y, Gan S (2005) Leaf senescence: signals, execution, and regulation. Curr Top Dev Biol 71: 83-112

He Y, Gan S (2001) Identical promoter elements are involved in regulation of the OPR1 gene by senescence and jasmonic acid in *Arabidopsis*. Plant Mol Biol 47: 595-605

He Y, Fukushige H, Hildebrand DF, Gan S (2002) Evidence supporting a role of jasmonic acid in *Arabidopsis* leaf senescence. Plant Physiol 128: 876-884

Hensel LL, Grbic V, Baumgarten DA, Bleecker AB (1993) Developmental and age related processes that influence the longevity and senescence of photosynthetic tissues in *Arabidopsis*. Plant Cell 5: 553-564

Jing HC, Hebeler R, Oeljeklaus S, Sitek B, Stuhler K, Meyer HE, Sturre MJG, Hille J, Warscheid B, Dijkwel PP (2008) Early leaf senescence is associated with an altered cellular redox balance in *Arabidopsis cpr5lold1* mutants. Plant Biol 10: 85-98

Jongebloed U, Szederkenyi J, Hartig K, Schobert C, Komor E (2004) Sequence of morphological and physiological events during natural ageing and senescence of a castor bean leaf: sieve tube occlusion and carbohydrate back-up precede chlorophyll degradation. Physiol Plant 120: 338-346

Keskitalo J, Bergquist G, Gardestrom, Jansson S (2005) A cellular time table of autumn senescence. Plant Physiol 139: 1635-1648

Khudairi K (1970) Chlorophyll degradation by light in leaf discs in the presence of sugar. Physiol. Plant 23: 613-622

Kleber-Janke T, Krupinska K (1997) Isolation of cDNA clones for genes showing enhanced expression in barley leaves during dark induced senescence as well as during senescence under field conditions. Planta 203: 332-340

Krause K, Kilbienski I, Mulisch, M, Rödiger A, Schäfer A, Krupinska K (2005) DNA-binding proteins of the Whirly family in *Arabidopsis thaliana* are targeted to the organelles. FEBS Lett 579: 3707-3712

Lee E-J, Koizumi N, Sano H. (2004). Identification of genes that are up-regulated in concert during sugar depletion in *Arabidopsis*. Plant Cell Environ 27: 337-345

Lee EJ, Matsumura Y, Soga K, Hoson T, Koizumi N (2007) Glycosyl hydrolases of cell wall are induced by sugar starvation in *Arabidopsis*. Plant Cell Physiol 48: 405-413

Lee RH, Wang CH, Huang LT, Chen SC (2001) Leaf senescence in rice plants: cloning and characterization of senescence up-regulated genes. J Exp Bot 52: 1117-1121

Lim PO, Woo HR, Nam HG (2003) Molecular genetics of leaf senescence in *Arabidopsis*. Trends Plant Sci 8: 272-278

Lim PO, Kim HJ, Nam HG (2007) Leaf senescence. Annu Rev Plant Biol 58: 115-136

Lin JF, Wu SH (2004) Molecular events in senescing *Arabidopsis* leaves. Plant J 39: 12-28

Masclaux C, Valadier MH, Brugie`re N, Morot-Gaudry JF, Hirel B (2000) Characterization of the sink/source transition in tobacco (*Nicotiana tabacum L.*) shoots in relation to nitrogen management and leaf senescence. Planta 211: 510-518

Masclaux-Daubresse C, Carrayol E, Valadier M-H (2005) The two nitrogen mobilisation- and senescence-associated *GS1* and *GDH* genes are controlled by C and N metabolites. Planta 221: 580-588

Matile P (1992) Chloroplast senescence. In: Baker N, Thomas H (eds) Crop photosynthesis: spatial and temporal determinants, Elsevier, Amsterdam, pp 413-440

Misra G, Biswal UC (1973) Factors concerned in leaf senescence. Effect of age, chemicals,petiole and photoperiod on senescence in detached leaves of *Hibiscus rosa sinensis* L.. Bot Gaz 134: 5-11

Misra G, Biswal UC (1974) Factors concerned in leaf senescence. II. Senescence in detached leaves of *Ervatamia divaricata* Burkill. J Ind Bot Soc 53: 65-70

Mohapatra PK, Patro L, Raval MK, Ramaswamy NK, Biswal UC, Biswal B (2010) Senescence-induced loss in photosynthesis enhances cell wall β-glucosidase activity. Physiol Plant 138: 346-355

Moore B, Zhou L, Rolland F, Hall Q, Cheng WH, Liu YX, Hwang I, Jones T, Sheen J (2003) Role of the *Arabidopsis* glucose sensor HXK1 in nutrient, light, and hormonal signaling. Science 300: 332-336

Nakabayashi K, Ito M, Kiyosue T, Shinozaki K, Watanabe A (1999) Identification of *clp* genes expressed in senescing *Arabidopsis* leaves. Plant Cell Physiol 40: 504-514

Noh Y, Amasino RM (1999a) Identification of a promoter region responsible for the senescence-specific expression of *SAG 12*. Plant Mol Biol 41: 181-194

Noh Y, Amasino RM (1999b) Regulation of developmental senescence is conserved between *Arabidopsis* and *Brassica napus*. Plant Mol Biol 41: 195-206

Nooden LD, Guiamet JJ (1996) Genetic control of senescence and aging in plants. In Schneider EL, Rowe JW (eds) Handbook of the biology of aging, Academic Press, Orlando, pp 94 -118

Nooden LD, Guiamet JJ, Jhon I (1997) Senescence mechanisms. Physiol Plant 101: 746-753

Oh SA, Lee SY, Chung IK, Lee CH, Nam HG (1996) Senescence-associated gene of *Arabidopsis thaliana* is distinctively regulated during natural and artificially induced leaf senescence. Plant Mol Biol 30: 739-754

Parrott DL, Yang L, Shama L, Fischer AM (2005) Senescence is accelerated, and several proteases are induced by carbon "feast" conditions in barley (*Hordeum vulgare L.*) leaves. Planta 222: 989-1000

Parrot DL, Mcinnerney K, Feller U, Fischer AM (2007) Steam-girdling of barley (*Hordeum vulgare*) leaves leads to carbohydrate accumulation and accelerated leaf senescence, facilitating transcriptomic analysis of senescence associated genes. New Phytol 176: 56-69

Paul MJ, Pellny TK(2003) Carbon metabolite feedback regulation of leaf photosynthesis and development. J Exp Bot 54: 539-547

Pourtau N, Mare`s M, Purdy S, Quentin N, Rue¨l A, Wingler A (2004) Interactions of abscisic acid and sugar signalling in the regulation of leaf senescence. Planta 219: 765-772

Pourtau N, Jennings, Pelzer E, Pallas J, Wingler A (2006) Effect of sugar-induced senescence on gene expression and implications for the regulation of senescence in *Arabidopsis*. Planta 224: 556-568

Price J, Laxmi A, St Martin SK, Jang JC (2004) Global transcription profiling reveals multiple sugar signal transduction mechanisms in *Arabidopsis*. Plant Cell 16: 2128-2150

Quirino BF, Noh YS, Himelblau E, Amasino RM (2000) Molecular aspects of leaf senescence. Trends Plant Sci 5: 278-282

Rolland F, Moore B, Sheen J (2002) Sugar sensing and signaling in plants. Plant Cell 14: S185-205

Rolland F, Baena-Gonzalez E, Sheen J (2006) Sugar sensing and signalling in plants: conserved and novel mechanisms. Annu Rev Plant Biol 57: 675-709

Schildhauer J, Wiedemuth K, Humbeck K (2008) Supply of nitrogen can reverse senescence processes and affect expression of genes coding for plastidic glutamine synthetase and lysine-ketoglutarate reductase/saccharopine dehydrogenase. Plant Biol 10: 76-84

Smart CM (1994) Gene expression during leaf senescence. New Phytol 126: 419-448

Thimann KV, Tetley RM, Krivak BM (1977) Metabolism of oat leaves during senescence V. senescence in light. Plant Physiol 59: 448-454

van Doorn WG (2004) Is petal senescence due to sugar starvation? Plant Physiol 134: 35-42

van Doorn WG (2008) Is the onset of senescence in leaf cells of intact plants due to low or high sugar levels? J Exp Bot 59: 1963-1972

Weaver LM, Himelblau E, Amasino RM (1997) Leaf senescence: gene expression and regulation. In Setlow JK (ed) Genetic Engineering Vol 19. Plenum Press, New York, pp 215-234

Wingler A, von Schaewen, Leegood RC, Lea PJ, Quick WP (1998) Regulations of leaf senescence by cytokinin, sugars and light. Plant Physiol 116:329-335

Wingler A, Mare`s M, Pourtau N (2004) Spatial patterns and metabolic regulation of photosynthetic parameters during leaf senescence. New Phytol 161: 781-789

Wingler A, Purdy S, MacLean JA, Pourtau N (2006) The role of sugars in integrating environmental signals during the regulation of leaf senescence. J Exp Bot 57:391-399

Wingler A and Roitsch T (2008) Metabolic regulation of leaf senescence: interactions of sugar signalling with biotic and abiotic stress responses. Plant Biol 10: 50-62

Wingler A, Masclaux-Daubresse C, Fischer AM (2009) Sugars, senescence, and ageing in plants and heterotrophic organisms. J Exp Bot 60:1063-1066

Xiao W, Sheen J, Jang JC (2000) The role of hexokinase in plant sugar signal transduction and growth and development. Plant Mol Biol 44: 54-61

Yoshida S (2003) Molecular regulation of leaf senescence. Curr Opin Plant Biol 6: 79-84

Yoshida S, Ito M, Nishida I, Watanabe A (2001) Isolation and RNA gel blot analysis of genes that could serve as potential molecular markers for leaf senescence in *Arabidopsis thaliana*. Plant Cell Physiol 42: 170-178

Zapata J M, Guéra A , Esteban-Carrasco A , Martín M, Sabater B (2005) Chloroplasts regulate leaf senescence: delayed senescence in transgenic *ndhF*-defective tobacco. Cell Death Differ 12: 1277-1284

CHAPTER 19

UV Radiation and Photosynthesis

Guruprasad, K.N. and Priyanka Singh

School of Life Sciences, Devi Ahilya University, Khandwa Road, Indore-452001, India.

ABSTRACT

Plants are exposed to ambient UV-B (280-315 nm) and UV-A (315-400 nm) radiation while using sunlight for photosynthesis. Solar UV radiations are known to reduce photosynthetic efficiency of the plants by causing structural and functional changes in the chloroplast. Studies with artificial UV-B sources have shown that UV-B strongly affects PS II. Both oxidizing and reducing sides of PS II complex are targeted by UV-B. Mn cluster of water oxidation complex, D1 and D2 proteins, quinone molecules and cytochrome b are the primary targets of UV-B. UV-B also affects Calvin cycle enzymes, Rubisco and Sedoheptulose1,7-bisphosphatase. UV-B radiation affects several photosynthetic genes encoded by nucleus and chloroplast. Exclusion of ambient UV under field conditions results in the enhancement of the rate of photosynthesis and accumulation of biomass. Reduction in ambient UV radiation has several agricultural advantages.

Keywords: Photosynthesis, PS II, UV-B, UV-A, UV-exclusion, Rubisco, Sedoheptulose 1,7- bisphosphatase.

Abbreviations

UV-B: Ultraviolet-B, *UV-A:* Ultraviolet-A, *PS II:* Photosystem II, *CPD:* Cyclobutane pyrimidine dimer, *PAR:* Photosynthetically active radiation, *PPFD:* Photosynthetic photon flux density

INTRODUCTION

All living forms on this earth are dependent on the sunlight energy and photosynthesis is the only natural process by which the physical form of sunlight energy is converted to usable chemical form. Sunlight is trapped by the plant chloroplast and converted to biochemical compounds in the form of ATP, NADPH and organic matter. Composition of the sunlight received at the global surface differs from the original extraterrestrial sunlight, especially in the UV part of the spectrum. The extraterrestrial solar radiation has three UV components—UV-C (200-280 nm) UV-B (280-315 nm) and UV-A (315-400 nm)—out of which the bactericidal UV-C is totally eliminated by the stratospheric oxygen/ozone layer. Amongst the other two components, a small portion of UV-B (22 w/m^2 extraterrestrial and 5 w/m^2 global) and a large portion of UV-A (88 w/m^2 extraterrestrial and 63 w/m^2 global) reach the earth's surface (Schulze, 1977) (Fig. 19.1).

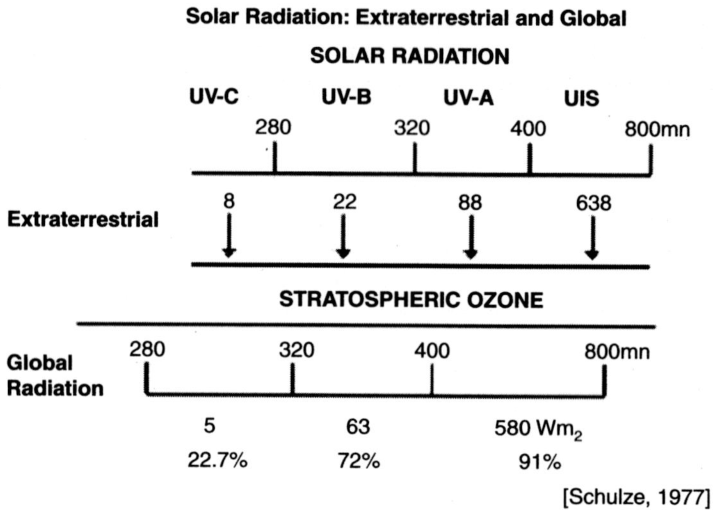

Fig. 19.1 Spectral distribution of the solar radiation.

The process of photosynthesis uses the photons of wavelength beyond 400 nm (PAR), but the plants are exposed to UV-B and UV-A part of the natural solar spectrum and these radiations are known to reduce the photosynthetic efficiency of the plants by causing structural and functional changes in the chloroplast.

There have been two major approaches to assess the suppression imposed by UV-B and UV-A on the process of photosynthesis. Several laboratories have used artificial UV light sources, especially UV-B lamps either used alone or by supplementing white light sources to record the damaging effects of this part of solar spectrum on photosynthesis (Caldwell and Flint, 1997). Most of these studies were prompted after the discovery of "ozone hole" (Farman *et al.*, 1985) and the impending environmental concern about leakage of higher amounts of UV-B radiation from stratosphere. In this assessment approach, often unrealistically high levels of UV-B radiation along with inadequate level of UV-A and low photosynthetic photon flux densities (PPFD) have been used (Edwards, 1992). These studies have, however, helped in analyzing the specific sites of damage by UV-B, since multiple target sites exist for UV-B in higher plants (Bornman, 1989). Much of the damaging effects of solar UV are focused on UV-B radiation, although the studies on phytoplanktons have demonstrated that at least half of the damage caused by solar radiation between 290 and 400 nm is caused by UV-A (Holm-Hansen *et al.*, 1993). Damages caused to photosynthesis by UV-B radiation are listed in Box 19.1.

Box 19.1 Damage of photosynthesis by UV radiation

- Damage to PS II electron transport
- Disruption of thylakoid membrane
- Reduction of chlorophyll content, possibly by reduced expression of genes encoding chlorophyll binding proteins
- Inhibition of water oxidizing complex and Mn catalytic activity
- Down-regulation of mRNA transcript for photosynthesis associated genes
- Damage to RUBP carboxylase (CO_2 binding enzyme)

Control measures to limit the UV-B leakage (and cure the 'ozone hole') have subsequently been initiated globally with perceptible improvements in preventing the damage. The second approach has been to assess the inhibition on photosynthesis imposed by ambient UV-B and UV-A radiation.

In these studies, photosynthetic performance of plants in natural sunlight has been compared with that of plants grown under UV-excluded or reduced sunlight. Specific UV-B and UV-A/B absorbing filters have been used in these studies which transmit only PAR wavelengths (Guruprasad *et al.,* 2007). These studies have not only observed the damaging effects of UV-B and UV-A on photosynthesis, but have also helped in focusing on the agricultural/environmental advantages of solar UV exclusion. The present chapter summarizes the status of researches on targets and mechanism of UV-B damage (using artificial UV-B radiation) and the imposition of inhibition on photosynthesis by ambient solar UV (using exclusion filters).

UV-B AND LIGHT REACTION OF PHOTOSYNTHESIS

Photosynthetic process comprises a two reactions: the light dependent reaction carried out in the thylakoid region of the chloroplast is the primary step for energy assimilation and the light independent reaction referred to as 'dark reaction' is responsible for the assimilation of carbon dioxide carried out in the stromal region of the chloroplast. Approximately half of the identified genes encoding the proteins of light reactions are chloroplast encoded and the other half are nuclear encoded. Genes encoding all the proteins for the dark reaction are nuclear encoded with the exception of the gene encoding for the large subunit of ribulose-1-5-phosphate carboxylase /oxygenase which is chloroplast encoded. UV-B radiation is absorbed by several biological molecules including nucleic acids, proteins, amino acid, lipids, phytoharmones, etc. (Berkelaar *et al.,* 1996). Being the energy harvesting organelle of the plants, chloroplasts become the primary targets of solar UV induced damages (Mpoloka *et al.,* 2007). PS II complex is most sensitive to UV-B (Vass 1996) and UV-B targets multiple sites of this complex. UV-B affects light harvesting complex, especially chlorophyll a (Renger *et al.,* 1986; Lingakumar and Kulandaivelu, 1993; Roh and Choi, 2003; Unal *et al.,* 2009) water oxidation complex and causes structural alterations in D1/D2 polypeptides, which affects charge separation (Renger *et al.,* 1989).

Studies with artificial UV-B sources have shown that UV-B may strongly affect PSII (Figueroa and Gomez, 2001; Hader *et al.,* 2001), whereas PS I appear to be relatively insensitive (Teramura and Ziska, 1996). Many reports support the notion that UV-B has subtle effects on PS II redox components at both the donor and acceptor side (Vass *et al.*, 1996).

PS II is a multisubunit water/plastoquinone (PQ) oxido/reductase complex embedded in the thylakoid membrane of higher plants, algae and cyanobacteria. It uses light energy to catalyze a series of e^- transfer reaction resulting in the oxidative splitting of water into molecular oxygen, proton and electrons. The redox cofactors of PS II e^- transport are bound to or contained by the D1 and D2 proteins subunits, which form the reaction center of PS II (Nanba and Satoh, 1987). The oxidation of water is catalyzed by a cluster of four Mn ions, and electrons isolated during this process are transferred to the chlorophyll P 680 via a redox active tyrosine residue Tyr-z of D1 protein. Tyr-D on D2 proteins donates electron to P 680, but is not connected to water oxidizing complex. On the acceptor side of PS II, the e^- produced by the light induced charge separation event reduces a pheophytin molecule and then the first Q_A and second Q_B, PQ electron acceptor, whereas the protons yielded are the sources of protons gradient across the thylakoid membrane required for ATP formation. These reactions take place on an enormous scale being primarily responsible for the production of atmospheric oxygen and providing energy for the assimilation of carbon dioxide (Fig. 19.2).

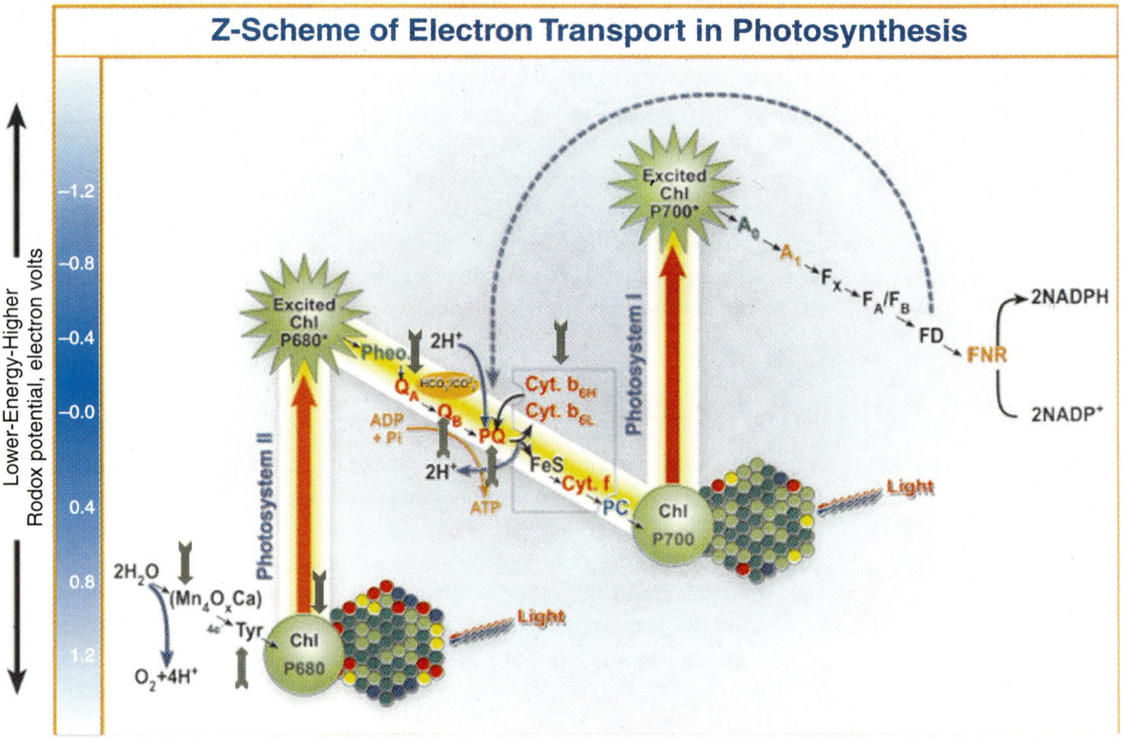

Fig. 19.2 A schematic representation of chloroplast electron transport. UV-B targets are shown by arrows. (Reference: from http:// www.life.illinois.edu/govindjee/page3.html, developed in collaboration with professor Govindjee of University of Illinois/ Urbana).

UV-B AND WATER OXIDATION COMPLEX

PS II components of thylakoid membrane of photosynthetic apparatus have been shown to be the most sensitive to UV-B irradiation by numerous investigations (Noorudeen and Kulandaivelu, 1982; Iwanzik *et al.*, 1983; Renger *et al.*, 1989; Kulandaivelu *et al.*, 1991; Melis *et al.*, 1992). PS II damage has thus been cited as the major limitation to photosynthesis under UV-B radiation. Damage to PS II is also evident in the photoinhibition of photosynthesis by excess photosynthetically active radiation (400-700 nm) (Baker and Bowyer, 1994). It is however suggested that the mechanism of UV-B induced damage may be different (Friso *et al.*, 1994; Jansen *et al.*, 1996).

PS II is a large pigment protein assembly embedded in the thylakoid membrane. It contains over 25 subunits and performs the exclusive function of light driven oxidation of water that serves as final electron source for all photosynthetic processes (Vass *et al.*, 2000). PS II is functionally described as a water/plastoquinone oxido/reductase embedded in the thylakoid membrane, which transfers the electrons liberated from light induced oxidation of water to membrane soluble plastoquinone molecules. In isolated spinach membrane particle, UV-B radiation results in rapid inhibition of oxygen evolution and the decline in variable fluorescence (Bischof *et al.*, 2002). Water oxidation is catalyzed by a cluster of four Mn ions which undergo light induced changes in their oxidation states called S states. The complex cycles through five states denoted as S_0 to S_4, and O_2 is released during the S_3- S_4- S_0 transition, in which S_4 is a short-lived intermediate (Messinger, 2004).

It is widely accepted that UV-B damages the acceptor and the donor side of PS II by inactivating the Mn cluster of water oxidation (Messinger, 2004). The loss of oxygen evolving capacity showed that the extent of UV-B damage is S-state dependent when PS II centers of isolated spinach thylakoids were synchronized to different distributions of the S_0, S_1, S_2 and S_3 states using packages of visible light flashes and then exposed to UV-B (Szilard *et al.,* 2002). UV-B sensitivity of PS II was significantly higher in the S_2 and S_3 states than in the S_1 and S_0 states. S-states transitions are accompanied by absorption changes in the UV-B range (Lavergne, 1991), which result in gradually increasing UV absorbing cross-section as the S-states cycle advances. This direct absorption of UV-B within the catalytic manganese cluster may contribute to inhibition of water oxidation (Szilard *et al.,* 2002).

Another mechanism of UV-B induced damage, as suggested by Szilard *et al.* (2002), could be the production of highly damaging hydroxyl radicals in the heart of the Mn cluster. The presence of bound hydrogen peroxide at the donor side of PS II has been reported (Klimov *et al.,* 1993). Hydroxyl radicals are the dominating reactive oxygen species induced by UV-B radiation in thylakoids (Hideg and Vass, 1996) and hydrogen peroxide is easily decomposed by UV light (<350 nm) producing hydroxyl radicals (Czapski, 1984). Some models of water oxidation assume the production of peroxidic intermediates in the higher S-states prior to the formation of S_4 (Hoganson and Babcock, 1997).

UV-B AND D1, D2 PROTEINS

A significant portion of D1 protein degradation in sunlight is due to UV-B, since exclusion of UV-B from the sunlight reduces the rate of degradation of D1 protein by 30% (Greenberg *et al.,* 1989). D1 and D2 proteins from the heart of the PS II reaction center contribute ligands to cofactor that mediates primary photochemistry (Anderson and Styring, 1991). Inhibition of PS II activity by UV-B radiation is accompanied by the degradation of D1 and D2 proteins subunits (Vass *et al.,* 2000). Intact photosynthetic cells respond to UV-B damage by repair of PS II via the *de novo* synthesis of the D1and D2 proteins (Sass *et al.,* 1997). In synechocystis 6803, out of the two psb A gene (psb A_2 and psb A_3) that encode identical D1 proteins, UV-B preferentially enhances the transcription of psb A_3 (Mate *et al.,* 1998). The rate of D1 production from psb A_3 increased with increasing UV-B intensities and was a transient phenomenon at low UV-B levels. Vass *et al.* (2000) have concluded that the main role of psb A_3 transcription activated by UV-B was to increase the size of the psb A_3 mRNA pool available for translation when a rapid repair of D1 protein is needed under UV-B stress.

PS II REDOX COMPONENTS AND UV-B

The donor side sensitivity to UV is well presented by tyrosine, as it absorbs at 280 nm in its reduced form and at 310 nm in its oxidized radical form (Vass *et al.,* 1996). Greenberg *et al.* (1989) showed that the degradation of D1 protein in UV is mainly sensitized by plastoquinone in one or more of its redox states. It appears likely that both the oxidizing and reducing sides of PS II complex are targeted by UV-B (Bornman, 1989; Vass, 1996, Vass *et al.*, 2007). On the acceptor side of PS II reaction center the quinone molecules (Q_A and Q_B), which have high absorption in the UV spectral region, are the likely candidates for UV damage (Greenberg *et al.,* 1989; Melis *et al.,* 1992; Jansen *et al.,* 1993)., UV-B predominantly affects the primary charge separation sequence from Z^- P680 QA to Z^- P680 Q_A^- (Noorudeen and Kulandaivelu, 1982, Renger *et al.,* 1986; Tevini *et al.,* 1988; Renger and Eckert, 1991; Grzymski *et al.,* 2001), causing modification or loss in the function of quinone (Q_A and Q_B) e^- acceptor (Renger, 1989; Melis *et al.,* 1992; Jansen *et al.,* 1996; Vass *et al.,* 1996, 1999).

Fluorescence induction curves indicate the inhibition of energy transfer within PS II reaction center (Noorudeen and Kulandaivelu, 1982; Iwanzik *et al.*, 1983). Polyphasic kinetics of the chlorophyll fluorescence rise from F_0 or 'O' via level 'J' and 'I' to the peak 'P' of the kautsky curve is highly affected by ambient UV-B light (Krause *et al.*, 1999) followed by strong reduction in the variable to maximal fluorescence values (Fv/Fm) of dark adapted leaves (Krause *et al.*, 1995; Unal *et al*, 2009) leading to decreased potential efficiency of PS II (Kolb *et al.*, 2001). A decrease in the potential efficiency of PS II infers that the PS II centers, which are able to reduce Q_A, are decreased due to the decreased absorption cross-section of the PS II associated light harvesting antenna or it may be due to the PQ pool that should be reduced before Q_A^- can accumulate.

High energy UV-B causes a conformational change in the structure of reaction center. This change results in either a small loss of efficiency of stable charge separation or an inability of the reaction center to dissipate exciton energy. Under no or low light conditions UV photons cause instability of Q_B^- which causes a reduction in the rate of e^- flow to PQ (Grzymski *et al.*, 2001). UV-B radiation results in an imbalance between the e^- flow leaving the reaction center to the acceptor side and the e^- flow coming to the reaction center from the donor side.

The rapid inhibition of oxygen evolution and the decline in variable fluorescence in isolated spinach PS II membrane particles by UV-B is accompanied by the loss of the multiline EPR signal that arises from the S_2 state of the water oxidizing complex and the induction of 'signal II fast' originating from stabilized Tyr-z^+ (Vass *et al.*, 1996). An EPR signal ($g = 2.94$) that arises from the dark oxidized form of cytochrome b-559 is enhanced, whereas the light inducible signal ($g = 3.04$) that arises from the photo-oxidizable population of cytochrome b-559 and the Q_AFe^{2+} is diminished by UV-B.

All the available evidences indicate that UV-B radiation affects photosystem II redox components at both the donor and acceptor side. According to Vass *et al.* (1996), the primary damage occurs at the water oxidizing complex. Modification and/or inactivation of tyrosine-D, cytochrome b-559 and the Q_A Fe^{2+} acceptor complex are subsequent events that coincide more closely with the UV-B induced damage to the proteins structure of the photosystem II reaction center.

UV-B AND INTEGRITY OF THYLAKOID

The integrity of the thylakoid membrane is much more sensitive to UV-B than the activities of the photosynthetic components bound to the membrane within the chloroplast. UV-B irradiation of leaves results in the loss of integrity of thylakoid membrane in various regions (Vu *et al.*, 1982) and dilation of thylakoid (Brandle *et al.*, 1977). An increase in ion permeability and a decrease in relaxation $t_{1/2}$ (dissipation rate) resulting in accelerated discharge of the thylakoid membranes as a consequence of UV-B exposure has been demonstrated (Chow *et al.*, 1992). ATP synthase was unaffected by UV-B (Iwanzik *et al.*, 1983), although UV-B accelerated the relaxation kinetics of the electrochromic shift and enhanced the permeability to ions. However, recovery towards a much lower ion-permeability state after UV-B is very slow, implying a partial uncoupling of ATP synthesis from e^-transport chain (Vass and Turcsanyi, 2000).

UV-B AND PHOTOSYNTHETIC GENES

UV-B radiation leads to decreased abundance of chloroplast m-RNA transcripts of some photosynthetic genes. The decline in mRNA transcripts by UV-B irradiance is more rapid for nuclear encoded genes compared to chloroplast encoded genes listed in Box 19.2

Box 19.2 Selective examples of photosynthetic genes affected by UV-B in plants.

Nuclear encoded	rbcs- small subunit of Rubisco
	atpc- γ subunit of ATP synthase
	cab- chl a/b binding protein of the light harvesting antenna of PS II
Chloroplast encoded	rbcL- Large subunit of Rubisco
	atpB and atpE- B and E subunit of ATP synthase
	pet B- cyt-b
	pet D- Subunit IV of cyt b/f complex
	psb A3- D1, D2 protein

The decreased efficiency of light reaction under UV-B stress is linked to cyt b/f complex and ATP synthase. Besides this D1 and D2 proteins of PS II complex UV-B influence the expression of both chloroplastic and nuclear genomes.

UV-B irradiation results in nonspecific DNA damage by the formation of cyclobutane pyrimidine dimer (CPD) (Hidema *et al.,* 1997).This damage is repaired by CPD photolyase enzyme activated under long wavelength (PAR 400-700 nm) light (Hidema and Kumagai, 2006). Substitution of one or two bases in CPD photolyase gene can alter the activity of the enzyme and confer changes in the associated resistance of plants to UV-B, as demonstrated in rice cultivars (Hidema and Kumagai, 2006). In the intact green spinach leaf cells, CPD is formed in the chloroplast as well as in the nucleus by UV-B, but the CPD in the chloroplast is not photorepaired while that in the nucleus is photorepaired (Hada *et al.*,1998). These results suggest that the UV-B induced DNA damage in the chloroplast may be fatal to maintaining the chloroplast integrity. The UV-B induced down-regulation of transcripts of photosynthetic genes, Lhcb and psb A in pea leaves was ameliorated by high light (PAR). The effect of UV-B was mimicked by free radical generators (Rose Bengal; Amino nitrazole) and prevented by antioxidants [pyrrolidine dithio carbamate PDTC; ascorbic acid] (Mackerness *et al.,* 1998). These results suggest that active oxygen species are involved in UV-B signal transduction, however, the sensitivity of photosynthetic genes under highlight (PAR) in these studies has been linked to higher photosynthetic rates, and hence higher antioxidant capacities in the cells together with the greater ability of these cells to further increase that capacity under UV-B stress. Since higher level of antioxidants leads to generate efficiency in the removal of AOS generated by UV-B exposure, AOS and photosynthesis capable of generating antioxidants may determine the extent of sensitivity of a plant to UV-B.

UV AND CALVIN CYCLE ENZYMES

UV-B inhibition of PS II photochemistry does not appear to be the only site of inhibition of photosynthesis in plants, since the inhibition of photosynthesis without an appreciable effect on PS II photochemistry has been observed in rice (Ziska and Teramura, 1992), soybean (Middleton and Teramura, 1993), pea (Nogues and Baker, 1995), *Brassica napus* (Allen *et al.,* 1997) and algae (Lesser, 1996). In pea (*Pisum sativum* L. cv. Meteor), exposure of mature leaves of glass house grown plants to high intensity (40 kJ $m^{-2}s^{-1}$) UV-B concurrent with a photosynthetically active photon flux density (PPFD > 450 μmol $m^{-2}s^{-1}$) light saturated CO_2 assimilation rates (Asat) decreased with no significant damage to PS II complexes (Nogues and Baker, 1995). Amongst the calvin cycle enzymes synthesis and degradation of Rubisco is the target of UV-B (Strid *et al.,* 1990; Rao *et al.,* 1995; Nogues and Baker, 1995; Allen *et al.,* 1997; Bassman *et al.* 2001; Keiller and Holmes, 2001).

In several higher plants and algae, UV-B induced reduction in both Rubisco activity and content (Allen *et al.,* 1998). A decline in Rubisco activity may be attributed to deactivation (i.e., a reduction in the specific activity) or loss of the enzyme (Quick *et al.,* 1991). In oil seed rape, decline in Rubisco activity in response to UV-B was due to a decline in the content of Rubisco, since no significant decline was observed in the activity of Rubisco carboxylation (Allen *et al.,* 1997). In oil seed rape, pea, tomato and tobacco a 66 KDa photomodified large subunit of Rubisco appears during UV-B exposure (Wilson *et al.*, 1995), since levels of m-RNA coding for both the large and small subunits of Rubisco have been reported to decline before any effect at the protein level was evident (Jordan *et al.,* 1992). Rubisco itself may not be the UV-B radiation photoreceptor. UV-B reduces sedoheptulose 1,7- bisphosphate content(but not chloroplastic fructose-1,6, bisphosphate or phosphoribulokinase) in the leaves of oil seed rape (Allen *et al.* 1998). A good correlation exists between regeneration of sedoheptulose 1,7- bisphosphatase and Rubisco activity in transgenic tobacco (Harrison *et al.,* 1998) (Fig. 19.3).

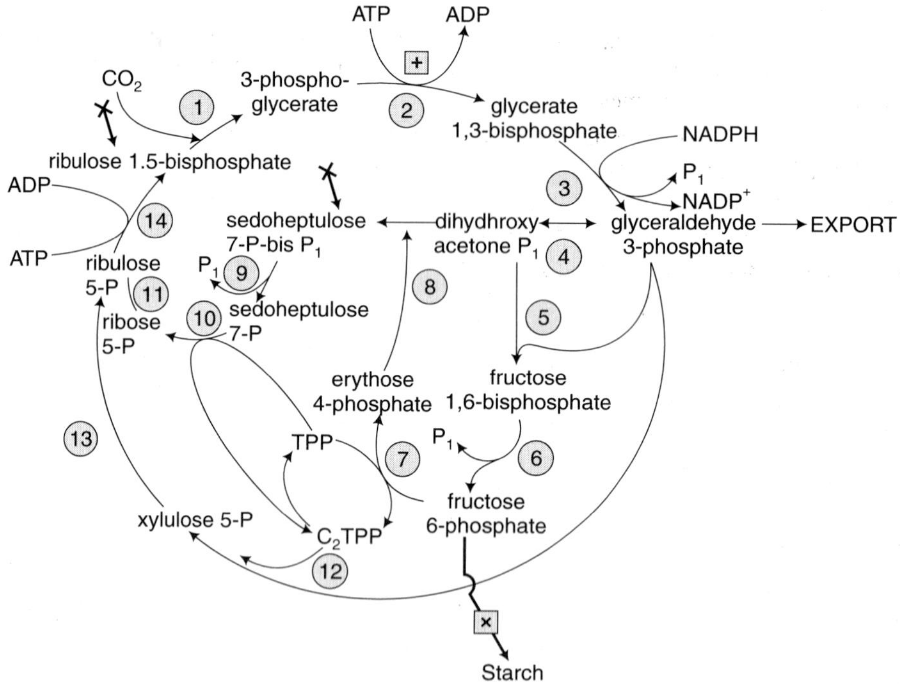

Fig. 19.3 Pathway of carbon metabolism in mesophyll cell of a C_3 plants. UV-B affected enzymes are Rubisco and Sedoheptulose 1,7-bisphosphate shown by dark arrow, source for calvin cycle "Photosynthesis" (third edition) David Lawlor- Bios scientific publishers Ltd., First Indian edition-2001.

The selective loss of specific Calvin cycle enzymes indicates that the loss is neither due to non-specific oxidation of leaf proteins resulting from UV-B absorption by these proteins nor due to UV-B induce production of reactive oxygen species. Alternatively, UV-B may induce proteases specific to the proteins that are degraded.

UV-B AND PLANT DEFENSE

UV-B is part of natural solar spectrum received at the earth's surface and higher plants have evolved strategies to reduce the adverse effect of UV-B on photosynthesis. These defense mechanisms include:

1. Scattering and reflection of UV-B radiation by epidermal and cuticular structures, other leaf optical properties such as waxy layer, leaf hairs and leaf bladder.

2. Absorption of UV-B radiation by pigments (flavonoids, anthocyanins) particularly in the subepidermal cells.
3. Photoreactivation of enzymes (photolyases), monomerization of dimers formed by DNA absorption of UV-B photons (photorepair)—a rapid process requiring sufficient PAR. A much faster process of phtorepair of CPD has been reported in Sorghum (Hada *et al.*, 1996).
4. Excision repair of DNA damage caused by UV-B irradiation – a slow process also occurring in dark.
5. Scavenging of radicals formed by absorption of UV-B photons—direct neutralization by ascorbic acid, glutathione, flavonoids (Jain *et al.,* 2004; Kataria *et al.,* 2007).
6. Production of polyamines, which stabilize injured membranes (Rozema *et al.,* 1997).

Although interest in UV-B was generated by the observation of ozone depletion and the concurrent rise in UV-B irradiance, which was perceived as a direct threat to photosynthetic productivity of crops and natural vegetation, several field studies on crops under natural solar radiation have revealed efficient defense mechanisms of plants to protect photosynthesis machinery. Leaves absorb about 90% of incident UV-B; leaf surface reflectance in this wavelength range is generally below 10% and there is negligible transmittance of UV-B through leaves (Robberecht *et al.,* 1980; Cen and Bornman, 1993; Gonzalez *et al.,* 1996). Epidermal UV-B penetration measured with a fiber optic microprobe has been found to be effectively zero in conifers needles, 3-12% in weedy dicotyledons and grasses and 18-41% in herbaceous dicotyledons (Day *et al.,* 1992). Epidermal UV-B penetration has been negatively correlated with UV-B absorbing pigments, concentration and epidermal thickness (Day, 1993). Reports of reduced growth and morphological changes in response to UV-B are more frequent than the reports of reduced photosynthetic competence. Plant productivity is a product of leaf area and the photosynthetic rate per unit area. Reductions in leaf area after exposure to UV-B have been observed in many plants species often in the absence of UV-B induced inhibition of photosynthesis. Reduced leaf area in response to UV-B is generally believed to be a protective modification as it reduces the amount of tissue exposed to UV-B irradiation (Bornman and Teramura, 1993). Ambient UV-B was found to significantly inhibit the growth of seedlings of *Datura ferox* at 34°S in Argentina (Ballare *et al.,* 1996). This was associated with a reduction in leaf area and stem elongation and not assimilation rate. Under field conditions, the inhibition of biomass accumulation in response to UV-B may be primarily through reduction in leaf area rather than direct damage to photosynthetic competence.

UV-B EXCLUSION STUDIES

Natural response of plants to UV-B can be measured only in field experiments with realistic UV-B, UV-A and PPFD irradiances. The components of solar radiation would vary with latitude and seasonal changes. Most of the experiments on UV-B damage to photosynthesis apparatus in the literature have used high intensities of UV-B. Much of the variability in plant responses to UV-B irradiation in these studies can be attributed to variations in the UV-B: UV-A: PPFD ratio in different experiments (Caldwell and Flint, 1994; Fiscus and Booker, 1995). Field experiment by excluding UV-B from solar spectrum with realistic UV-B, UV-A and PPFD irradiance would help to understand the impact of natural UV-B on photosynthesis and photomorphogenesis of plants. UV exclusion or below ambient solar UV-B is produced either by exclusion filters that selectively absorb UV-B or by special filters filled with ozone to reduce transmission of UV-B (Rozema *et al.,* 1997) along with filters that absorb both UV-A and UV-B have also been used to separate the impact of UV-A and UV-B (Amudha *et al.*, 2005; Varalakshmi *et al.,* 2003) (Fig. 19.4)

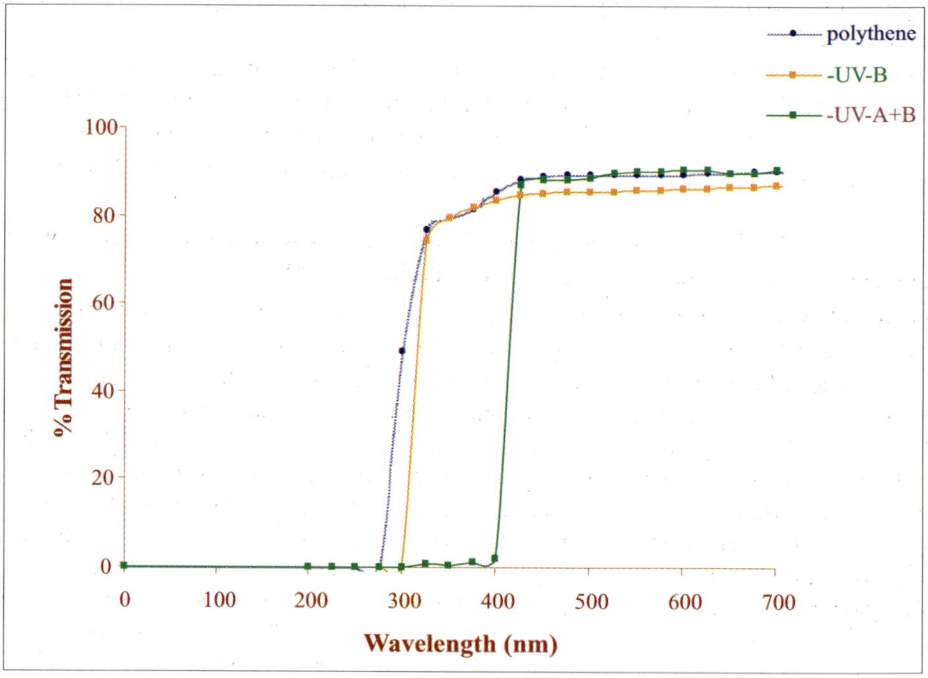

Fig. 19.4 Transmission spectra of the UV cut-off filters. The spectra between 200-700 nm by using an UV – visible spectrophotometer.

UV exclusion studies have indicated the prefound influence of the small component of UV-B in the solar spectrum on growth, biomass accumulation and carbon fixation. UV exclusion enhanced growth in plants like radish (Zavala and Botto, 2002), soybean (Varalakhmi *et al.,* 2003; Chouhan *et al.,* 2008), cucumber and cotton (Solanki *et al.,* 2006) and wheat (Pal *et al.,* 2006). UV exclusion increases photosynthetic pigments (Martz *et al.,* 2007), especially chlorophyll (Laposi *et al.,* 2002) but lowers carotenoids (Laposi *et al.,* 2002; Guruprasad *et al.,* 2007) and UV absorbing pigments (Tsormpatsides *et al.,* 2008). UV exclusion also lowers the synthesis of betacyanins in Amaranthus (Sharma and Guruprasad, 2009). Expansion of leaves is a common response of many dicotyledons (Solanki *et al.,* 2006) and monocot plants like maize and sorghum (unpublished data of Shine and Guruprasad, 2010; Sunita and Guruprasad, 2010) (Fig. 19.5). These observations support the earlier findings in field experiments that ambient UV-B reduces leaf area in several plants. Allen *et al.* (1998) have observed that in plants irradiated with UV-B throughout the development, a reduction in productivity is usually associated with a reduced ability to intercept light (smaller leaf area) and not an inhibition of photosynthesis competition. UV-B also affects stomatal conductance and causes stomatal closure in several plants. However, by supplementation of UV-B, the percentage decrease in light saturated photosynthesis that is attributable to stomatal conductance was not sufficient to account for the large depression in photosynthesis in oil seed rape (Allen *et al.,* 1998).

Under UV-B exclusion, there is an enhancement in the net rate of CO_2 in cotton (Guruprasad *et al.,* 2007, abstract), and in soybean there is an overexpression of Rubisco subunits (Guruprasad *et al.,* 2007) and increased PS II efficiency (Singh *et al.,* (2010) unpublished data). UV exclusion data available at present indicate that the UV-B and UV-A components of solar spectrum regulate photosynthesis by

Fig. 19.5 Photograph of mature soybean plants grown in normal and UV excluded sunlight. **(A)** Plant height in ambient and UV excluded condition **(B)** Leaf area in ambient and UV excluded condition, source: Guruprasad *et al.* (2007), photosynthetic research.

reducing the assimilation of CO_2. More data is needed on the exclusion of solar UV to understand the mechanism of regulation imposed by UV on photosynthesis besides reduction in the size of the leaves.

FUTURE PERSPECTIVE

Studies on UV-B emerged from the observations on ozone depletion, which has subsequently been controlled by international environmental laws. Ozone depletion is expected to be cured in a few decades provided the countries adhere to the regulatory methods about which a little skepticism prevails. In addition to the CFC, the main ozone destroyer, few other chemicals have been identified as potential destroyer like hexachloro butadiene, n-propyl bromide, 6-bromo-2-methoxyl-napthalene, halon-1202 identified by UN Environmental Program (2004). These observations imply that studies on plant defense methods against excess UV-B need to continue for finding methods to enhance the resistance to UV-B. The UV exclusion studies have opened up new avenues of research towards enhancing agricultural yields by shielding the plants from solar UV-B even in the present level. The enhancement in photosynthetic efficiency of plants in the absence of solar UV components also has the potential of enhanced rapid fixation of atmospheric carbon dioxide that can reduce global warming. Further investigations are needed to realize these possibilities.

REFERENCES

Allen DJ, Mckee IF, Farage PK, Baker NR (1997) Analysis of limitations to CO_2 assimilation on exposure of leaves of two *Brassica napus* cultivars to UV-B. Plant Cell Environ 20: 633-640

Allen DJ, Nogues S, Baker NR (1998) Ozone depletion and increased UV-B radiation: Is there a real threat to photosynthesis?. J Exp Bot 49: 1775-1788

Amudha P, Jayakumar M, Kulandaivelu G, (2005) Impact of ambient solar UV (280-400 nm) radiation on three tropical legumes. J of Plant Biol 48: 284-291

Anderson B, Styring S (1991) Photosystem II: molecular organization, function and acclimation. Curr Topics Bioenerg 16: 1-81

Baker NR, Bowyer JR (eds) (1994) Photoinhibition of Photosynthesis: From Molecular Mechanism to the Field. Oxford Bios Scintific Publishers

Ballare CL, Scopel AL, Stapleton AE, Yonovsky MJ (1996) Solar UV-B radiation affects seedling emergence, DNA integrity, plant morphology, growth rate, and attractiveness to herbivore insects in *Datura ferox*. Plant Physiology 112: 161-70

Bassman JH, Roberrecht R, Edwards GE (2001) Effects of enhanced UV-B radiation on growth and gas exchange in *Populus deltoides* Bartr. *Ex Marsh.* Int J Plant Sci 162: 103-110

Berkelaar EJ, Douglas PO, Hale BA (1996) The influence of photosynthetically active radiation on the effects of Ultraviolet-B radiation on *Arabidopsis thaliana.* Photochem and Photobiol 64(1): 110-116

Bischof K, Krabs G, Wiencke C, Hanelt D (2002) Solar UV radiation effects the activity of ribulose -1,5-bisphosphate carboxylase-oxygenase and the composition of photosynthetic and xanthophyll cycle pigments in the intertidal green alga *Ulva lactuca* L. Planta 215 (3): 502-509

Bornman JF (1989) Target sites of UV-B radiation in photosynthesis of higher plants. J Photochem Photobiol 4: 145-158

Bornman JF, Teramura AH (1993) Effects of ultraviolet-B radiation on terrestrial plants. In: Young AR, Bjorn LO, Moan J, Nultsch W, eds. Environmental UV photobiology, Plenum Press, New York, pp 427-72

Brandle JR, Campbell WF, Sissons WB, Caldwell MM (1977) Net photosynthesis, electron transport capacity and ultrastructure of *Pisum sativum* L. exposed to UV-B radtation. Plant Physiology 60: 165-169

Caldewell MM, Flint SD (1994) Stratospheric ozone reduction, solar UV-B radiation and tetrrestrial ecosystem. Climate Change 28: 375-94

Caldwell MM, Flint SD (1997) User of biological spectral weighting functions and the need of scaling for the ozone reduction problems. Plant Ecol 128: 66-76

Cen YP, Bornman JF (1993) The effect of exposure to enhanced UV-B radiation on the penetration of monochromatic and polychromatic UV-B radiation in leaves of *Brassica napus*. Physiologia Plantarum 87: 249-55

Chouhan S, Chauhan K, Kataria S, Guruprasad K N (2008) Enhancement in leghemoglobin content of root nodules by exclusion of solar UV-A and UV-B radiation on soybean. Journal of Plant Biology 51(12): 132-138

Chow WS, Strid A, Anderson JM (1992) Short term treatment of pea plants with supplementary UV-B radiation: Recovery time-course of some photosynthetic function and components. In: Murata N (eds) Research in Photosynthesis, Kluwer Academic Publishers, Dordrecht, Vol 4, pp 361-364,

Czapski G (1984) Reaction of OH, In: L. Pecker (ed), Methods in Enzymology. 105: 209-215

Day TA (1993) Relating UV-B radiation screening effectiveness of foliage to absorbing- compound concentration and anatomical characteristic in a diverse group of plants. Oecologia 95: 542-50

Day TA, Vogelmann TC, Delucia EH (1992) Are some plant life forms more effective than others in screening out ultraviolet radiation? Oecologia 92: 513-19

Edwards NT (1992) State-of-the-art field exposure techniques do not qualitatively simulate expected global increase in UV-B irradiance. Bull Ecol Soc America 73: 165

Farman JC, Gardiner BG, Shanklin JD (1985) Large losses of total ozone in Antarctica reveal seasonal ClOx/NOx interaction. Nature 328: 207-210

Figueroa FL, Gomez I (2001) Photosynthetic acclimation to solar UV radiation on marine red algae from warm-temperature coast of southern Spain: a review J Applied Phycol 13: 233-245

Fiscus EL, Booker FL (1995) Is increased UV-B a threat to crop photosynthesis and productivity? Photosynthetic Research 43: 81-92

Friso G, Spetea C, Giacometti GM, Vass I, Barbato R (1994) Degradation of photosystem II reaction center D1-protein induced by UV-B irradiation in isolated thylakoid. Identification and characterization of C and N-terminal breakdown product. Biochim Biophys Acta 1184: 78-84

Gonzalez R, Paul ND, Percy K, Ambrose M, McLaughlin CK, Barnes JD, Areses M, Wellburn AR (1996) Response to ultraviolet-B radiation (280-315 nm) of pea *(Pisum sativum)* lines differing in leaf surface wax. Physiologia Plantarum 98: 852-60

Greenberg BM, Gaba V, Canaani O, Malkin S, Mattoo AK, Edelman M (1989) Separate photo-sensitizers mediate degradation of the 23 kDa photosystem II reaction center protein in the visible and UV spectral region. Proc Natl Acad Sci USA 86: 6617-6620

Grzymski J, Orrico C, Schofield OM (2001) Monochromatic ultraviolet light induced damage to photosystem II efficiency and carbon fixation in the marine diatom *Thalassiosira pseudonana* (3H). Photosynthetic Research 68: 181-192

Guruprasad KN, Bhattacharjee S, Kataria S, Yadav S, Tiwari A, Baroniya S, Rajiv A, Mohanty P (2007a) Growth enhancement of soybean *(Glycine max)* upon exclusion of UV-B and UV-A components of solar radiation characterization of photosynthetic parameters in leaves. Photosynthetic Research 94: 299-306

Guruprasad KN, Kataria S, Jaiswal P, Dehariya P (2007b) Impact of ambient solar UV radiation (280-400 nm) in the growth and photosynthetic parameters in cotton (*Gossipium hirsutum*: cv- Khandwa-2 and Vikram) Abstract published in International symposium on light and life and XVIII biennial meeting of photobiology society, August 29-31, 2007 at University of Hyderabad, Hyderabad, India.

Hada M, Tsurumi S, Suzuki M, Wellmann E, Ashimoto T (1996) Involvement and non-involvement of pyrimidines dimmer formation in UV-B effects on Sorghum bicolour Moench seedlings. J Plant Physiol 148: 92-99

Hada M, Hashimoto T, Nikaido O, Shin M (1998) UV-B induced DNA damage and its photorepair in nuclei and chloroplasts of *Spinacia oleracea* L. Photochemistry and Photobiology, 68: 319-322

Hader DP, Porst M, Lebert M (2001) Photoinhibition in common Atlantic macroalgae measured in Gran Conaria. Helgol Mar Res 55: 67-76

Harison EP, Willingham NM, Lioyd JC, Raines CA (1998) Reduced sedoheptulose 1,7-bisphosphatase levels in transgenic tobacco lead to decreased photosynthetic capacity and altered carbohydrate accumulation. Planta 204: 27-36

Hideg E, Vass I (1996) UV-B induced free radical production in plant leaves and isolated thylakoid membranes. Plant Sci 115: 251-260

Hidema J, Kumagai T, Sutherland JC, Sutherland MB (1997) Ultraviolet B-sensitive rice cultivar deficient in cyclobutyl pyrimidine dimer repair. Plant Physiol 113: 39-44

Hidema J, Kumagai T (2006) Sensitivity of rice to ultraviolet- B radiation. Annals of Botany 97(6): 933-942

Hoganson CW, Babcock GT (1997) A metalloradical mechanism for the generation of oxygen from water in photosynthesis. Science 277: 1953-1956

Holm-Hansen O, Lubim D, Helbling EW (1993) Ultraviolet radiation and its effects on organisms in aquatic environments. In Environmental UV Photobiology. (Edited by Young AR, Bjorn LO, Moan J and Nultsch), Pienum Press, New York & London, pp 379-426

Iwanzik W, Tevini M, Dhont G, Voss M, Weiss W, Graber P, Renger G (1983) Action of UV-B on photosynthtic primary reactions in spinach chloroplasts. Physiol Plant 58: 401-407

Jain K, Kataria S, Guruprasad KN (2004) Oxyradicals under UV-B stress and their quenching by antioxidants. Indian J of Exp Biol 42: 884-892

Jansen MAK, Mattoo AK, Malkin S, Edelman M (1993) Direct demonstration of binding site competition *in vivo* between photosystem II inhibitors at the Q_B niche of the D1 protein. Pesticide Biochim Physiol 46: 78-83

Jansen MAK, Gaba V, Greenberg BM, Mattoo AK, Edelman M (1996) Low threshold level of UV-B in a background of photosynthetically active radiation trigger rapid degradation of D2 protein of photosystem II. Plant J 9(5): 693-699

Jordan BR, He J, Chow WS, Anderson JM (1992) Changes in the mRNA levels and polypeptide subunits of ribulose-1,5-bisphosphate carboxylase in response to UV-B radiation. Plant Cell Environ 15: 91-98

Kataria S, Jain K, Guruprasad KN (2007) UV-B induced changes in antioxidant enzymes and their isoform in cucumber (*Cucumis sativus* L.) cotyledons. Indian Journal of Biochemistry and Biophysics 44: 31-37

Keiller DR, Holmes MG (2001) Effects of long-term exposure to elevated UV-B radiation on the photosynthetic performance of five broad-leaved tree species. Photosythetic Reaserch 67: 229-240

Klimov VV, Ananyev GM, Zastryhnaya O, Wydrynski T, Renger G (1993) Photoproduction of hydrogen peroxide in photosystem II membrane fragments: a comparision of four signals. Photosynth. Res. 38: 409-416

Kolb CA, Kaser MA, Kopecky J, Zotz G, Riederer M, Pfundel EE (2001) Effects of natural intensities of visible and ultraviolet radiation on epidermal ultraviolet screening and photosynthesis in grapes leaves. Plant Physiol 127: 863-875

Krause GH, Virgo A, Winter K (1995) High susceptibility to photoinhibition of young leaves of tropical forest trees. Planta 197: 583-59

Krause GH, Schmude C, Hermann G, Koroleva O, Winter K (1999) Effects of solar ultraviolet radiation on the potential efficiency of photosystem II in leaves of tropical plants. Plant Physiology 121: 1349-1358

Kulandaivelu G, Nedunchezhian N, Annmalainathan K (1991) Ultraviolet-B (280-320 nm) radiation induced changes in photochemical activities and polypeptide components of C3 and C4 chloroplasts. Photosynthetica 25: 333-339

Laposi R, Veres SZ, Mile O, Meszaros I (2002) Photosynthesis-ecophysiological properties of beech (*Fagus sylvatica* L.) under the exclusion of ambient UV-B radiation. Acta Biologica Szegediensis 46(3-4): 243-245

Lavergne J (1991) Improved UV visible spectra of the S-transition in the photosynthetic oxygen-evolving system. Biochim Biophys Acta 1060: 175-188

Lesser MP (1996) Acclimation of phytoplankton to UV-B radiation: oxidative stress and photoinhibition of photosynthesis are not prevented by UV-absorbing compounds in the dinoflagellate *Prorocentrum micans*. Marine Ecology Progress Series 132: 289-97

Lingakumar K, Kulandaivelu G (1993) Changes induced by UV-B radiation in vegetative growth, foliar characteristics and photosynthetic activities in *Vigna unguiculata* L. Aust J Plant Physiol 20: 299-308

Mackerness AH, Surplus SL, Jordan BR, Thomas B (1998) Effect of supplementary ultraviolet-B on photosynthetic transcripts at different stages of leaf development and light levels in pea (*Pisum sativum* L.): Role of active oxygen species and antioxidant enzymes. Photochemistry and Photobiology 68(1): 88-96

Martz F, Sutinen M L, Derome K, Wingsle G, Julkunen-Tittoo R, Turunen M (2007) Effect of UV-exclusion on the seasonal content of photosynthetic and UV screening pigments of Scot pine needles. Global Change Biology 13(1): 252-265

Mate Z, Sass L, Szekeres M, Vass I, Nagy F (1998) UV-B induced differential transcription of psbA gene encoding the D1 protein of photosystem II in the cyanobacterium *Synechocystis* 6803. J Biol Chem 273: 17439-17444

Melis A, Nemson JA, Harrison MS (1992) Damage to functional components and partial degradation of photosystem II reaction center upon chloroplast exposure to UV-B radiation. Biochim Biophys Acta 1100: 312-320

Messinger J (2004) Evaluation of different mechanistic proposals for water oxidation in photosynthesis on the basis of Mn_4OxCa structures for the catalytic site and spectroscopic data. Phys Chem Chem Phys 6: 4764-4771

Middleton EM, Teramura AH (1993) The role of flavonol glycosides and carotenoids in protecting soybean from UV-B damage. Plant Physiology 103: 741-52

Mpoloka SW, Abratt VA, Mundree SG, Thomson JA, Musil CF (2007) Potential effects of prolonged ultraviolet radiation exposure in plants: Chloroplast DNA analysis. American-Eurasian J Agric Environ Sci 2(4): 437-441

Nanba O, Satoh K (1987) Isolation of photosystem II reaction center consisting of D1 and D2 polypeptides and cytochrome b-559. Proc Natl Acad Sci USA 84: 109-112

Nogues S, Baker NR (1995) Evaluation of the role of damage to photosystem II in the inhibition of CO_2 assimilation in pea leaves on exposure to UV-B radiation. Plant Cell Environ 18: 781-787

Noorudeen AM, Kulandaivelu G (1982) On the possible site of inhibition of photosynthetic electron transport by UV-B radiation. Physiol Plant 55: 161-166

Pal M, Zaidi PH, Voleti SR, Raj A (2006) Solar UV-B exclusion effects on growth and photosynthetic characteristics of wheat and pea. Journal of New Seeds 8 (1): 19-34

Quick WP, Schurr U, Scheibe R, Schulze ED, Rodermel SR, Bogorad L, Stitt M (1991) Decreased ribulose 1, 5-bisphosphate carboxylase-oxygenase in transgenic tobacco transformed with 'antisense' rbcS. Planta 183: 542-54

Rao MV, Paliyath G, Ormrod DP (1995) Differential response of photosynthetic pigments, Rubisco activity and Rubisco protein of *Arabidopsis thaliana* exposed to UV-B and ozone. Photochem Photobiol 62: 727-735

Renger G, Voss M, Graber P, Schulz A (1986) Effect of UV irradiation on different partial reactions of primary processes of photosynthesis. In Worrest RC and Caldwell M.M (eds) Stratospheric Ozone Reduction. Solar Ultraviolet Radiation and Plant Life, Springer-Verlag, Berlin, pp 171-184,

Renger G, Volker M, Eckert R, Fromme S, Hohm-Veit, Graber P (1989) On the mechanism of Photosystem II deterioration by UV-B irradiation. Photochem Photobiol 49: 97-105

Renger G, Eckert HJ (1991) Effects of UV-B and visible light on reaction mechanism of photosynthetic water cleavage. In: Abrol YP, Wattal PN, Gnanam A, Govindjee, Ort DR and Teramura AH (eds) Impact of Global climate changes on Photosynthesis and Plant Productivity, Oxford and IBH Publications, New Delhi, proceedings Indo-US workshop: pp 39-58

Robberecht R, Caldwell MM, Billings WD (1980) Leaf ultraviolet optical properties along a latitude gradient in the arctic alpine zone. Ecology 61: 612-19

Roh SK, Choi YB (2003) UV-B radiation affects chlorophyll and activation of Rubisco activase in *Canavalia ensiformis* L. leaves. J of Plant Biology 46(2): 117-121

Rozema J, Staaij JVD, Bjorn LO, Caldwell MM (1997) UV-B as an environmental factor in plant life: stress and regulation. Tree 12(1) 22-28

Sass L, Spetea C, Mate Z, Nagy F, Vass I (1997) Repair of UV-B induced damage of photosystem II via de novo synthesis of the D1 and S2 reaction centre subunits in *Synechocystis* sp. PCC 6803. Photosynth Res 54: 55-62

Schulze R (1977) UV-Strallenklima. In: Ultraviolette strahlen (Ed. Kiefer J), walter de grvyter, Berlin and New York

Sharma A, Guruprasad KN (2009) Similarity in the bochemical changes between solar UV exclusion and GA application in *Amaranthus caudatus*. Physiol and mol biol of plants. 15 (4): 367-370

Singh P, Pandey GP, Guruprasad KN (2010) Inhibitory effect of ambient solar UV on photosynthesis and growth of soybean var: JS-71-05. (Unpublished data)

Solanki R, Lakshmi N, Athale R, Baroniya S, Guruprasad K N (2006). Growth and chlorophyll contents as affected by UV-A and UV-B components in cucumber and cotton. Physiol and Mol Biol of Plants 12(4): 321-323

Strid A, Chow WS & Anderson JM (1990) Effect of supplementary ultraviolet-B radiation on photosynthesis in *Pisum sativum*. Biochim. Biophys. Acta 1020: 260-268

Szilard A, Sass L, Vass I (2002) Photoinactivation of photosystem II at low light intensity, mathematical model. Acta Biol Szeged 46: 167-169

Teramura AH, Ziska LH (1996) Ultraviolet –B radiation and photosynthesis. In: N.R. Baker, ed. Photiosynthesis and the Environment, Kluwer Academic Publishers, Dordrechi, The Netherlands, pp 435-450

Tevini M, Fieser G, Grusemann P (1988) Assessment of UV stress by chlorophyll fluorescence analysis. In: Lichtenthaler H.K. (ed.) Applications of Chlorophyll Fluroscence, Kluwer Academic, The Netherlands, pp 229-238

Tsormpatsidis E, Henbest RGC, Davis FJ, Battey NH, Hadley P, Wagstaffe A (2008) UV irradiance as a major influence on growth, development and secondary products of commercial importance in *Lollo Rosso* lettuce 'Revolution' grown under polyethylene films. Environmental and Experimental Botany 63: 232-239

Unal D, Tuney I, Dereboylu-Esiz A, Sukatar A (2009) The effect of UV-A (352 nm) stress on chlorophyll fluorescence, chl a content, thickness of upper cortex and determinate DNA damage in *Physcia semipinnata*. J of Photochem and Photobiol 94: 71-76

Varalakshmi D, Lakshmi N, Guruprasad KN (2003) Physiological changes in soybean CV. JS 71-05 after the exclusion of UV-A and UV-B from the solar radiation. Indian Journal of Plant Physiol (special issue): 602-606

Vass I (1996) Adverse effects of UV-B light on the structural and function of the photosynthetic apparatus. In: M. Passarakali (ed) Handbook of Photosynthesis, Marcel Dekker, Inc., New York, pp 931-950

Vass I, Sass L, Spatea C, Bakou A, Ghanotaki D, Petrouleas V (1996) UV-B induced inhibition of PS II electron transport studied by EPR and chlorophyll fluorescence, impairment of donor and acceptor side components. Biochemistry 35: 8964-8973

Vass I, Kirilovsky D, Etienne A (1999) UV-B radiation induced donor and acceptor side modification of PS II in the cyanobacterium *Synechocystis* sp. PCC 6803. Biochemistry 38: 12786-12794

Vass I, Kirilovsky D, Perewoska I, Mate Z, Nagy F, Etienne AL (2000) UV-B radiation induced exchange of the D1 reaction center subunits produced from the *psbA2* and *psbA3* genes in the cyanobacterium *Synechocystis* sp. PCC 6803. Eur J Biochim 267: 2640-2648

Vass I, Turcsanyi E (2000) Inhibition of photosynthetic electron transport by UV-A radiation targets the photosystem II complex. Photochemistry and Photobiology 72(4): 513-520

Vass I, Szilard A, Sass L, Deak Z (2007) The sensitivity of photosystem II to damage by UV-B radiation depends on the oxidation state of the water-splitting complex. Biochimica et Biophysica Acta 1767: 876-882

Vu CV, Allen LH, Garrad LA (1982) Effects of supplemental UV-B on primary photosynthetic carboxylation enzymes and soluble proteins in leaves of C_3 and C_4 crop plants. Physiologia Plantarum 55: 11-16

Wilson MI, Ghosh S, Gerhardt KE, Holland N, Babu TS, Edelman M, Dumbrof EB, Greenburg BM (1995) *In vivo* photomodification of ribulose 1,5-bisphosphate carboxylase oxygenase holoenzyme by ultraviolet-B radiation-formation of a 66-kilo Dalton variant of the large subunit. Plant Physiology 109: 221-9

Zavala JA, Botto JF (2002) Impact of solar UV-B radiation on seedling emergence, chlorophyll fluorescence, growth and yield of radish (*Raphanus sativus*). Funct Plant Biol 29: 797-804

Ziska LH, Teramura AH (1992) CO_2 enhancement of growth and photosynthesis in rice (*Oryza sativa*): modification by increased ultraviolet-B radiation. Plant Physiology 99: 473-81

Index